INNOVATIVE PACKAGING OF FRUITS AND VEGETABLES

Strategies for Safety and Quality Maintenance

INNOVATIVE PACKAGING OF FRUITS AND VEGETABLES

Strategies for Safety and Quality Maintenance

Edited by
Mohammed Wasim Siddiqui, PhD
Mohammad Shafiur Rahman, PhD
Ali Abas Wani, PhD

APPLE ACADEMIC PRESS

Apple Academic Press Inc.
3333 Mistwell Crescent
Oakville, ON L6L 0A2 Canada

Apple Academic Press Inc.
9 Spinnaker Way
Waretown, NJ 08758 USA

© 2018 by Apple Academic Press, Inc.
First issued in paperback 2021
No claim to original U.S. Government works
ISBN-13: 978-1-77463-138-6 (pbk)
ISBN-13: 978-1-77188-597-3 (hbk)

Library and Archives Canada Cataloguing in Publication

Innovative packaging of fruits and vegetables : strategies for safety and quality maintenance/edited by Mohammed Wasim Siddiqui, PhD, Mohammad Shafiur Rahman, PhD, Ali Abas Wani, PhD.
(Postharvest biology and technology book series)
Includes bibliographical references and index.
Issued in print and electronic formats.
ISBN 978-1-77188-597-3 (hardcover).--ISBN 978-1-315-14306-4 (PDF)
1. Fruit--Packaging. 2. Vegetables--Packaging. 3. Food--Preservation. I. Siddiqui, Mohammed Wasim, editor II. Rahman, Shafiur, editor III. Wani, Ali Abas, editor IV. Series: Postharvest biology and technology book series
TP440.I56 2018   664'.8   C2018-900473-8   C2018-900474-6

Library of Congress Cataloging-in-Publication Data

Names: Siddiqui, Mohammed Wasim, editor. | Rahman, Shafiur, editor. | Wani, Ali Abas, editor.
Title: Innovative packaging of fruits and vegetables : strategies for safety and quality maintenance / editors, Mohammed Wasim Siddiqui, PhD, Mohammad Shafiur Rahman, PhD, Ali Abas Wani, PhD.
Description: Toronto : Apple Academic Press, 2018. | Series: Postharvest biology and technology | Includes bibliographical references and index.
Identifiers: LCCN 2018001752 (print) | LCCN 2018006994 (ebook) | ISBN 9781315143064 (ebook) | ISBN 9781771885973 (hardcover : alk. paper)
Subjects: LCSH: Food--Packaging. | Food--Packaging--Quality control.
Classification: LCC TP374 (ebook) | LCC TP374 .I57 2018 (print) | DDC 664/.09--dc23
LC record available at https://lccn.loc.gov/2018001752

Apple Academic Press also publishes its books in a variety of electronic formats. Some content that appears in print may not be available in electronic format. For information about Apple Academic Press products, visit our website at **www.appleacademicpress.com** and the CRC Press website at **www.crcpress.com**

ABOUT THE EDITORS

Mohammed Wasim Siddiqui, PhD

Dr. Mohammed Wasim Siddiqui is an Assistant Professor and Scientist in the Department of Food Science and Post-Harvest Technology, Bihar Agricultural University, Sabour, India and author or co-author of more than 40 journal articles, more than; 40 book chapters, and several conference papers. He has 22 books to his credit published by Elsevier, CRC Press, Springer, and Apple Academic Press.

He is the Founder Editor-in-Chief of two book series, Postharvest Biology and Technology and Innovations in Horticultural Science, both being published by Apple Academic Press. He is also is a Senior Acquisitions Editor for Horticultural Science. Dr. Siddiqui established an international peer reviewed journal *Journal of Postharvest Technology*. He has been serving as an editorial board member and active reviewer of several international journals, including *Horticulture Research* (Nature Publishing Group), *Postharvest Biology and Technology* (Elsevier), *PLoS ONE,* (PLOS), *LWT- Food Science and Technology* (Elsevier), *Food Science and Nutrition* (Wiley), *Journal of Plant Growth Regulation* (Springer), *Acta Physiologiae Plantarum* (Springer), *Journal of Food Science and Technology* (Springer), and *Indian Journal of Agricultural Science* (ICAR).

Dr. Siddiqui has received numerous awards and fellowships in recognition of his research and teaching achievements. Recently, he was conferred with the Glory of India Award (2017), Best Researcher Award (2016), Best Citizens of India Award (2016), Bharat Jyoti Award (2016), Outstanding Researcher Award (2016), Best Young Researcher Award (2015), Young Scientist Award (2015), and the Young Achiever Award (2014) for outstanding contribution in research and teaching from several organizations of national and international repute. He was also awarded a Maulana Azad National Fellowship Award from the University Grants Commission, New Delhi, India. He is an honorary board member and life time author in the Society for Advancement of Human and Nature (SADHNA), Nauni, Himachal Pradesh, India. He has been an active member of the organizing committees of several national and international seminars/conferences/summits. He is one of key members in establishing the World Food Preservation Center (WFPC), LLC, USA. Presently, he is an active associate and supporter of WFPC, LLC, USA. Considering his outstanding contribution in science and technology, his biography has been published in *Asia Pacific Who's Who, Famous Nation: India's Who's Who, The Honored Best Citizens of India* and *Emerald Who's Who in Asia*.

Dr. Siddiqui acquired his BSc (Agriculture) degree from Jawaharlal Nehru Krishi Vishwa Vidyalaya, Jabalpur, India. He received his MSc (Horticulture) and PhD (Horticulture) degrees from Bidhan Chandra Krishi Viswavidyalaya, Mohanpur, Nadia, India, with specialization in postharvest biotechnology. He has received several grants from various funding agencies to carry out his research projects. He is dynamically involved in teaching (graduate and doctorate students) and research, and he has proved himself as an active scientist in the area of postharvest biotechnology.

Mohammad Shafiur Rahman, PhD

Mohammad Shafiur Rahman is a Professor at the Sultan Qaboos University, Sultanate of Oman and the author or co-author of over 300 technical articles, including 117 refereed journal papers, 107 conference papers, 66 book chapters, 34 reports, 13 popular articles, and 8 books. He is the author of the internationally acclaimed and award-wining *Food Properties Handbook*, published by CRC Press, which was one of the bestsellers from CRC Press in 2002. The second edition is now released under his editorship. He is also the editor of the popular book *Handbook of Food Preservation*, published by CRC Press. The first edition was one of the bestsellers from CRC Press in 2003, and the second edition is now an the market. The first edition was translated into Spanish. In addition, he is one of the editors of *Handbook of Food Process Design* (two volumes) published by Wiley-Blackwell in 2012. He was invited to serve as one of the associate editors for the *Handbook of Food Science, Engineering and Technology*, and one of the editors for the *Handbook of Food and Bioprocess Modeling Techniques* published by CRC Press.

Professor Rahman has initiated the *International Journal of Food Properties* (Marcel Dekker, Inc.) and has been serving as the founding editor for more than 15 years. In addition, he is serving on the editorial boards of eight international journals. He is a member in the Food Engineering Series Editorial Board of Springer Science, New York. He is serving as a Section Editor for the Sultan Qaboos University journal *Agricultural Sciences*. In 1998, he has been invited and continues to serve as a Food Science Adviser for the International Foundation for Science (IFS) in Sweden.

Professor Rahman is a professional member of the New Zealand Institute of Food Science and Technology and the Institute of Food Technologists, and a member of executive committee of the International Society of Food Engineering, (ISFE). He was involved in many professional activities, such as organizing international conferences, training workshops, and other extension activities. He has initiated and served as the Founding Chair of the International Conference on Food Properties (ICFP). He was invited as a key note/plenary speaker at eight international conferences.

He received BSc (Eng.) (Chemical) and MSc (Eng.) (Chemical) degrees from Bangladesh University of Engineering and Technology, Dhaka, an MSc degree in food engineering from Leeds University, England, and the a PhD degree in food engineering from the University of New South Wales, Sydney, Australia.

Professor Rahman has received numerous awards and fellowships in recognition of his research and teaching achievements, including the HortResearch Chairman's Award, the Bilateral Research Activities Program (BRAP) Award, CAMS Outstanding Researcher Award (2003), SQU Distinction in Research Award (2008), and the British Council Fellowship. In 2008 Professor Rahman ranked among the top five leading scientists and engineers of the 57 OIC member states in the agroscience discipline.

Professor Rahman is an eminent scientist and academic in the area of food processing. He is recognized for his significant contribution to the basic and applied knowledge of food properties related to food structure, food functionality, engineering properties, and food stability. His Google Scholar citations are more than 20,000 and h-index 63, respectively, which indicates the high impact of his research in the international scientific community.

Ali Abas Wani, PhD

Ali Abas Wani is a senior researcher at the Fraunhofer Institute for Process Engineering and Packaging, Freising, Germany. Born and educated in India, he has received MS (Food Technology) and PhD (Food Technology) degrees from leading Indian universities. Since 2006, he has been an assistant professor at the Islamic University of Science and Technology, Jammu and Kashmir, India, where he established the Department of Food Technology and also initiated several key food science programs in the region. Additionally, he is establishing a food testing center at the Islamic University of Science and Technology, Awantipora. Dr. Wani is also a co-chair for the European Hygienic Engineering Design Group (EHEDG) representing the regional section of India. He is the co-founder and editor-in-chief of *Food Packaging & Shelf Life*, published by Elsevier Science, UK.

Dr. Wani has published several international papers, book chapters, edited books, and numerous conference papers. His research focus is on the development of functional ingredients and measurement of food quality. In addition to close association with many scientific organizations in the area of food science and technology, he is an active reviewer for *Carbohydrate Polymers*, *Food Chemistry*, *LWT-Food Science and Technology*, *Journal of Agricultural Food Chemistry*, and many other scientific journals.

ABOUT THE BOOK SERIES: POSTHARVEST BIOLOGY AND TECHNOLOGY

As we know, preserving the quality of fresh produce has long been a challenging task. In the past, several approaches were in use for the postharvest management of fresh produce, but due to continuous advancement in technology, the increased health consciousness of consumers, and environmental concerns, these approaches have been modified and enhanced to address these issues and concerns.

The *Postharvest Biology and Technology* series presents edited books that address many important aspects related to postharvest technology of fresh produce. The series presents existing and novel management systems that are in use today or that have great potential to maintain the postharvest quality of fresh produce in terms of microbiological safety, nutrition, and sensory quality.

The books are aimed at professionals, postharvest scientists, academicians researching postharvest problems, and graduate-level students. This series is intended to be a comprehensive venture that provides up-to-date scientific and technical information focusing on postharvest management for fresh produce.

Books in the series will address the following themes:
- Nutritional composition and antioxidant properties of fresh produce
- Postharvest physiology and biochemistry
- Biotic and abiotic factors affecting maturity and quality
- Preharvest treatments affecting postharvest quality
- Maturity and harvesting issues
- Nondestructive quality assessment
- Physiological and biochemical changes during ripening
- Postharvest treatments and their effects on shelf life and quality
- Postharvest operations such as sorting, grading, ripening, de-greening, curing, etc.
- Storage and shelf-life studies
- Packaging, transportation, and marketing
- Vase life improvement of flowers and foliage

- Postharvest management of spice, medicinal, and plantation crops
- Fruit and vegetable processing waste/by products: management and utilization
- Postharvest diseases and physiological disorders
- Minimal processing of fruits and vegetables
- Quarantine and phytosanitory treatments for fresh produce
- Conventional and modern breeding approaches to improve the post-harvest quality
- Biotechnological approaches to improve postharvest quality of horticultural crops

We are seeking editors to edit volumes in different postharvest areas for the series. Interested editors may also propose other relevant subjects within their field of expertise, which may not be mentioned in the list above. We can only publish a limited number of volumes each year, so if you are interested, please email your proposal wasim@appleacademicpress.com at your earliest convenience.

We look forward to hearing from you soon.

Editor-in-Chief:

Mohammed Wasim Siddiqui, PhD
Scientist-cum-Assistant Professor | Bihar Agricultural University
Department of Food Science and Technology | Sabour | Bhagalpur | Bihar | India
AAP Acquisitions Editor, Horticultural Science
Founding/Managing Editor, *Journal of Postharvest Technology*

Email: wasim@appleacademicpress.com
        wasim_serene@yahoo.com

BOOKS IN THE POSTHARVEST BIOLOGY AND TECHNOLOGY SERIES:

Postharvest Biology and Technology of Horticultural Crops:
Principles and Practices for Quality Maintenance
Editor: Mohammed Wasim Siddiqui, PhD

Postharvest Management of Horticultural Crops:
Practices for Quality Preservation
Editor: Mohammed Wasim Siddiqui, PhD

Insect Pests of Stored Grain: Biology, Behavior, and Management Strategies
Ranjeet Kumar, PhD

Innovative Packaging of Fruits and Vegetables: Strategies for Safety and Quality Maintenance
Editors: Mohammed Wasim Siddiqui, PhD, Mohammad Shafiur Rahman, PhD, and Ali Abas Wani, PhD

Advances in Postharvest Technologies of Vegetable Crops
Editors: Bijendra Singh, PhD, Sudhir Singh, PhD, and Tanmay K. Koley, PhD

Plant Food By-Products: Industrial Relevance for Food Additives and Nutraceuticals
Editors: J. Fernando Ayala-Zavala, PhD, Gustavo González-Aguilar, PhD, and Mohammed Wasim Siddiqui, PhD

Emerging Technologies for Shelf-Life Enhancement of Fruits
Editors: Basharat Nabi Dar, PhD, and Shabir Ahmad Mir, PhD

Sensor-Based Quality Assessment Systems for Fruits and Vegetables
Editors: Bambang Kuswandi, PhD, and Mohammed Wasim Siddiqui, PhD

Emerging Postharvest Treatment of Fruits and Vegetables
Editors: Kalyan Barman, PhD, Swati Sharma, PhD, and Mohammed Wasim Siddiqui, PhD

CONTENTS

LIST OF CONTRIBUTORS

Oluwafemi J. Caleb
Department of Horticultural Engineering, Leibniz Institute for Agricultural Engineering (ATB), Max-Ethy-Allee 100, Potsdam 14469, Germany

O. P. Chauhan
Defence Food Research Laboratory, Siddarthanagar 570011, Mysore, India. E-mail: opchauhan@gmail.com

K. Chitravathi
Defence Food Research Laboratory, Siddarthanagar 570011, Mysore, India

Neeru Dubey
Amity International Centre for Post-Harvest Technology and Cold Chain Management, Amity University, Noida, Uttar Pradesh, India. E-mail: needub@gmail.com

Stefano Farris
Department of Food, Environmental and Nutritional Sciences, DeFENS, Packaging Division, University of Milan,Via Celoria 2 – 20133, Milano, Italy

Abhay Kumar Gaurav
Division of Floriculture and Landscaping, Indian Agricultural Research Institute, New Delhi, India

Martin Geyer
Department of Horticultural Engineering, Leibniz Institute for Agricultural Engineering (ATB), Max-Ethy-Allee 100, Potsdam 14469, Germany

Khalid Gul
Department of Processing and Food Engineering, Punjab Agricultural University, Ludhiana 141004, India

Amit Kumar
Department of Food Engineering and Technology, Sant Longowal Institute of Engineering and Technology, Longowal 148106, Punjab, India

Nirmal Kumar
Division of Food Science and Postharvest Technology, Indian Agricultural Research Institute, New Delhi, India

Simple Kumar
Amity International Centre for Post-Harvest Technology and Cold Chain Management, Amity University, Noida, Uttar Pradesh, India

Bambang Kuswandi
Chemo and Biosensors Group, Faculty of Pharmacy, University of Jember, Jl. Kalimantan 37, Jember 68121, Indonesia. E-mail: b_kuswandi@farmasi.unej.ac.id

Pramod V. Mahajan
Department of Horticultural Engineering, Leibniz Institute for Agricultural Engineering (ATB), Max-Ethy-Allee 100, Potsdam 14469, Germany

Vigya Mishra
Amity International Centre for Post-Harvest Technology and Cold Chain Management, Amity University, Noida, Uttar Pradesh, India. E-mail: vigyamishra.horticulture@gmail.com

Pallavi Neha
Division of Postharvest Technology, Indian Institute of Horticultural Research, Bengaluru, India

Astrid F. Pant
Department of Material Development, Fraunhofer Institute for Process Engineering and Packaging (IVV), Freising, Germany. E-mail: astrid.pant@ivv.fraunhofer.de

K. Prasad
Division of Food Science and Postharvest Technology, Indian Agricultural Research Institute, New Delhi, India

Kamlesh Prasad
Department of Food Engineering and Technology, Sant Longowal Institute of Engineering and Technology, Longowal 148106, Punjab, India. E-mail: profkprasad@gmail.com

Ovais Shafiq Qadri
Department of Post-Harvest Engineering and Technology, Faculty of Agricultural Sciences, Aligarh Muslim University, Aligarh, India
Department of Bioengineering, Integral University, Lucknow, India

D. V. Sudhakar Rao
Division of Post-Harvest Technology, ICAR-Indian Institute of Horticultural Research, Bengaluru 560089, India. E-mail: dvsrao@iihr.ernet.in; sudhadvrao@gmail.com

R. R. Sharma
Division of Food Science and Postharvest Technology, Indian Agricultural Research Institute, New Delhi, India

Shubhra Shekhar
Department of Food Engineering and Technology, Sant Longowal Institute of Engineering and Technology, Longowal 148106, Punjab, India

M. W. Siddiqui
Department of Food Science and Postharvest Technology, Bihar Agricultural University, Sabour, Bhagalpur 813210, Bihar, India

Neha Singh
Warner School of Food and Dairy Technology, SHIATS, Allahabad, Uttar Pradesh, India

Preeti Singh
Chair Food Packaging Technology, Technical University of Munich, Freising 85354, Weihenstephan, Germany

Abhaya Kumar Srivastava
Department of Post-Harvest Engineering and Technology, Faculty of Agricultural Sciences, Aligarh Muslim University, Aligarh, India

J. Thielmann
Department of Material Development, Fraunhofer Institute for Process Engineering and Packaging (IVV), Freising, Germany

Lakshmi E. Unni
Defence Food Research Laboratory, Siddarthanagar 570011, Mysore, India

Ali Abas Wani

Fraunhofer Institute for Process Engineering and Packaging IVV, Freising 85354, Germany
Department of Processing and Food Engineering, Punjab Agricultural University, Ludhiana 141004, India

Basharat Yousuf

Department of Post-Harvest Engineering and Technology, Faculty of Agricultural Sciences, Aligarh Muslim University, Aligarh, India. E-mail: yousufbasharat@gmail.com

LIST OF ABBREVIATIONS

ACC	aminocyclopropane-1-carboxylic acid
AG	Arabic gum
Ag	silver
AgNPs	silver nanoparticles
AITC	allyl isothiocyanate
ANN	artificial neural network
BS	bacterial cellulose
C_2H_4	plant hormone ethylene
CA	controlled atmosphere
CAS	controlled atmosphere storage
CD	cyclodextrines
CHW	chitin whiskers
CI	chilling injury
CIE	Commission International de L'eclairage
CMC	carboxymethyl chitosan
CNT	carbon nanotubes
CNW	cellulose nanowhiskers
COF	coefficient of friction
CVS	computer vision system
EMA	equilibrium modified atmosphere
Eos	essential oils
FCC	fluorescent chlorophyll catabolites
FFC	fresh-cut fruits and vegetables
FFV	fresh fruits and vegetables
FRF	frequency response function
GAC	granular-activated carbon
GAC-Pd	granular-activated carbon with or without a palladium catalyst
GRAS	generally recognized as safe
GTR	gas transmission rate
HACCP	hazard analysis and critical control points
ICC	image capturing camber
ICP-MS	inductively coupled plasma mass spectrometry
ILSS	inter laminar shear strength

LDH	layered double hydroxide
LDPE	low-density polyethylene
LLDPE	linear low-density polyethylene
LOX	lipid peroxidation
MAHP	modified atmosphere and humidity packaging
MAP	modified-atmosphere packaged
MB	methylene blue
MDA	malondialdehyde
MFC	microfibrillated cellulose
MMT	montmorillonite
Mo	molybdenum
MR	methyl red
MSE	mean squared error
OPP/PE	oriented polypropylene/polyethylene
OTR	oxygen transmission rate
PAL	phenylalanine ammonia-lyase
PBAT	aromatic co-polyesters
PBS	polybutylene succinate
PBSA	aliphatic co-polyesters
PBT	polybutyleneterephthalate
PC	personal computer
PCL	polycaprolactone
PE	polyethylene
PEA	polyesteramides
PET	polyethylene terephthalate
PHAs	polyhydroxyalkanoates
PHB	poly hydroxybutyrate
PHBA	polyhydroxybutyrate/valerate
PHBv	poly hydroxybutyrate co-hydroxyvalerate
PLA	polylactic acid
PLW	physiological loss in weight
PME	pectin methyl esterase
POSS	polyhedral oligomeric silsesquioxane
PP	polypropylene
PS	potassium sorbate
PSA	pressure swing adsorption
PTT	polytrimethyleneterephthalate
PURs	polyurethanes

PVC	polyvinyl chloride
RCO_2	respiration rate
RFID	radio frequency identification
RH	relative humidity
RMSE	root mean squared error
ROC	receiver operating characteristic
ROS	reactive oxygen species
RQ	respiratory quotient
SAM	s-adenosylmethionine
SB	sodium benzoate
SE	semperfresh coating
SMRBF	smote memetic radial basis function
SNC	starch nanocrystals
SPI	soy protein isolate
SSCs	soluble solids
STP	standard temperature and pressure
TA	titratable acidity
TBARMs	thiobarbituric acid
TBZ	thiabendazole
TR	transmission rate
TSS	total soluble solid
TTI	time temperature indicators
UV	ultraviolet
Vis	visible
WPI	whey protein isolate
WVTR	water vapor transmission rate

PREFACE

Consumers consider fruits and vegetables important and high priority in their food lists since they play an imperative role in a healthy lifestyle. Despite efforts in promoting health benefits of fresh fruits and vegetables, their short shelf-life remains an impediment to consumption. Owing to the ripening/ ageing process, these products undergo various biological reactions that continue mainly after harvest. These processes cause gradual reduction in the quality. Consumers prefer an excellent fresh product with improved environmental sustainability. It is common to use cold chains and preservatives to extend the shelf-life and quality. The cold chain involves precise control of temperature and energy consumption, while preservatives raise safety concerns. Innovative packaging technologies are the most important and challenging strategies to overcome these limitations. The new packaging technologies maintain the quality and freshness of the products, thus extending the shelf-life, reducing food losses, and facilitating commercialization and export. Extending the shelf-life by just 3–4 days using innovative packaging prevents large amounts of these perishables from spoiling during transportation and handling. When stored in an optimal atmosphere, their metabolism slows down, and thus they remain fresh and crunchy.

Innovative packaging solutions significantly increase the shelf-life of fresh fruits and vegetables. For instance, in active packaging, gas controllers can be incorporated in the packaging materials to delay senescence. Intelligent packaging, such as freshness or time-temperature indicators, monitors product quality and safety and enhances the marketing of the packaged fruits and vegetables.

These are examples of the innovative packaging that can contribute to the economy, consumer health, safety, and environment preservation. However, these innovative packaging must be fulfilled with a wide variety of different requirements. Producers, aside from focusing only on highest food standards and product protection, also expect maximum efficiency in their production process. Retailers are demanding maximum shelf-life, attractive optics to spur buying impulses by customers, and the best logistical characteristics such as stacking at point of purchase. All these new technologies require specific knowledge and appropriate training and understanding for rationally selecting the most suitable packaging for each product and intended use.

This book is composed of 13 chapters contributed by experts. Each chapter offers a thorough discussion along with recent updates on the topic. This book provides sustainable packaging solutions that deliver protection, branding, consumer attractiveness, and speed to market in a competitive retail environment.

The main features of the book are:
- an inclusive overview of fruit and vegetable requirements and available packaging materials and systems;
- an understanding of the fundamentals of the impact of packaging on the evolution of quality and safety of fruits and vegetables;
- coverage of the fundamental aspects of packaging requirements, including mathematical modelling and mechanical and engineering properties of packaging materials;
- an in-depth discussion of innovative packaging technologies, such as MA/CA packaging, active packaging, intelligent packaging, and eco-friendly materials, applied to fruit and vegetables;
- information aimed to build competence in packaging design for a better environmental and economic performance.

CHAPTER 1

BIO-BASED PACKAGING FOR FRESH FRUITS AND VEGETABLES

O. P. CHAUHAN*, K. CHITRAVATHI, and LAKSHMI E. UNNI

Defence Food Research Laboratory, Siddarthanagar 570011, Mysore, India

Corresponding author. E-mail: opchauhan@gmail.com

CONTENTS

ABSTRACT

Biodegradable packaging materials are gaining increased popularity over the years as synthetic plastic materials cause serious environmental concerns due to their non-biodegradation and depletion of natural resources. Biodegradable materials offer a possible alternative as they are abundant, renewable, inexpensive, and environment friendly. Biopolymers are produced from various natural resources through direct extraction, chemical synthesis, or fermentation by microorganisms, such as starch, cellulose, proteins, polylactide (PLA) and polyhydroxyalkanoates (PHAs), and these are considered as attractive alternatives for non-biodegradable petroleum-based plastics. Bio-based polymers applied as dispersion coatings on paper and paperboard could be used to provide sufficient barrier properties with respect to fats, but these provide moderate water vapor barrier properties. Other drawbacks of the bio-based packaging could be inferior mechanical properties, insufficient heat tolerance, and high moisture sensitivity as compared to petroleum-derived plastics. These problems could be overcome by physical, chemical or enzymatic treatments, and by blending with hydrophobic additives or making multilayer films with varied properties. In addition, fiber-based packaging has the advantage of lower weight, which is favorable for transportation, and the final products can generally be recycled. The bio-based packaging materials and edible coatings are also being applied to fruits and vegetables for maintaining their quality and extending shelf-life. Being bio-based in nature, most of these materials are also edible in nature and can be consumed along with the commodity.

1.1   INTRODUCTION

Augmented environmental concerns over the use of certain synthetic packaging and coatings in combination with consumer demands for both higher quality and longer shelf-life have led to increased interest in developing alternative packaging materials. The worldwide interest in bio-based polymers has accelerated in recent years due to the desire and need to find non-fossil fuel-based polymers. Bio-based polymers offer important contributions by reducing the dependence on fossil fuels and provide positive environmental impacts such as reduced carbon dioxide emissions. Nowadays, bio-based polymers are commonly found in many applications from commodity to hi-tech applications due to advancement in biotechnologies and public awareness. The term "biodegradable materials" is used to describe those materials which can be degraded by the enzymatic action of living organisms, such as bacteria, yeasts, and fungi. The ultimate end products of the degradation process are CO_2, H_2O, and biomass under aerobic conditions and hydrocarbons, methane, and biomass under anaerobic conditions (Babu et al., 2013). Figures 1.1 and 1.2 give a detailed picture of the end products produced after degradation and biodegradation process.

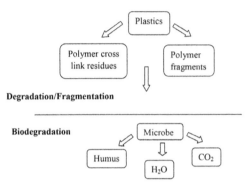

FIGURE 1.1 Degradation vs. biodegradation (*Source*: Narayan, 2005). (Reprinted with permission from Narayan, R. Biobased and Biodegradable Polymer Materials: Rationale, Drivers, and Technology Exemplars, Degradable Polymers and Materials. September 23, 2006, 282–306. DOI:10.1021/bk-2006-0939.ch018. © 2006. American Chemical Society.)

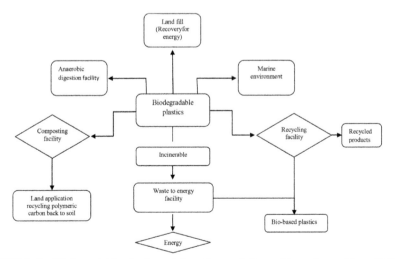

FIGURE 1.2 Biodegradable plastics—integration and disposable systems. (From © 2011, Renewable Resources and Renewable Energy: A Global Challenge, Second Edition, Paolo Fornasiero, Paolo Fornasiero, Mauro Graziani, Paolo Fornasiero, Mauro Graziani, Mauro Graziani. Reproduced by permission of Taylor and Francis Group, LLC, a division of Informa, PLC.)

Naturally, renewable biopolymers can be used as barrier coatings on paper packaging materials. These biopolymer coatings may retard unwanted moisture transfer in food products; and complement other types of packaging to minimize food quality deterioration and extending the shelf-life of foods (Debeaufort et al., 1998). These are good oxygen and oil barriers, biodegradable, and have potential to replace current synthetic paper and paperboard coatings (such as polyethylene, polyvinyl alcohol, rubber latex, and fluorocarbon) for food packaging applications (Chan & Krochta, 2001a, 2001b). Agriculturally derived alternatives to synthetic paper coatings provide an opportunity to strengthen the agricultural economy and reduce importation of petroleum and its derivatives (Khwaldia et al., 2010). Similarly, many bio-based polymers are biodegradable (e.g., starch and polyhydroxyalkanoates (PHAs)), but not all biodegradable polymers are bio-based (e.g., polycaprolactone). Biopolymer-based packaging materials originated from naturally renewable resources such as polysaccharides, proteins, and lipids or combinations of those components. These offer favorable environmental advantages of recyclability and reutilization compared to conventional petroleum-based synthetic polymers. Moreover, biopolymer-based films and coatings can act as efficient vehicles for incorporating various additives including antimicrobials, antioxidants, coloring agents, and nutrients (Han & Gennadios, 2005).

The association of biopolymers to paper provides interesting functionalities while maintaining environment-friendly characteristic of the material. Renewable biopolymers, such as caseinates (Khwaldia et al., 2005; Gastaldi et al., 2007; Khwaldia et al., 2009), whey protein isolate (WPI) (Han & Krochta 1999, 2001; Lin & Krochta 2003; Gallstedt et al., 2005), isolated soy protein (Park et al., 2000; Rhim et al., 2006), wheat gluten (Gallstedt et al., 2005), corn zein (Trezza & Vergano, 1994; Parris et al., 1998; Trezza et al., 1998), chitosan (Despond et al., 2005; Ham-Pichavant et al., 2005; Kjellgren & Engstrom, 2006), carrageenan (Rhim et al., 1998), alginate (Rhim et al., 2006), and starch (Matsui et al., 2004) have been investigated as paper-coating materials. Han and Krochta (1999, 2001) showed that whey-protein-coated paper improves paper-based packaging material performance by increasing oil resistance and reducing water-vapor permeability. Despond et al. (2005) and Kjellgren and Engstrom (2006) used paper coated with chitosan or chitosan/carnauba wax to obtain a packaging material with good barrier properties toward oxygen, nitrogen, carbon dioxide, and air. Rhim et al. (2006) indicated that water resistance of paper is improved by coating with soy protein isolate (SPI) or alginate.

The natural bio-based polymers can be used for packaging of fruits and vegetables, as bio-based and biodegradable plastics can form the basis for an environmentally preferable, sustainable alternative to current materials based exclusively on petroleum feed-stocks. Biodegradable materials mostly can be applied to foods requiring short-term chill storage such as fruits and vegetables, as bio-based materials present opportunities for producing films with variable CO_2 and O_2 selectivity and moisture permeability. Bio-based packaging of foods must be in compliance with the quality and safety requirements of the food product and meet legal standards. Additionally, the bio-based materials should preferably preserve the quality of the product better and longer to justify any extra material cost (Webber, 2000; Dilip et al., 2013). Various types of bio-based polymers are described in Figure 1.3.

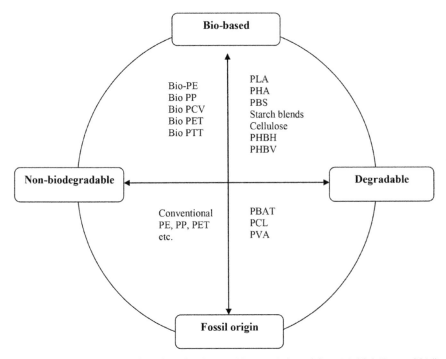

FIGURE 1.3   Categories of bio-based polymers (*Source*: Adapted from Muhl & Beyer, 2014).

1.2   CLASSIFICATION OF BIODEGRADABLE POLYMERS

Biodegradable polymers can be classified into four categories depending on the synthesis and on the sources which are shown in Figures 1.4 and 1.5.

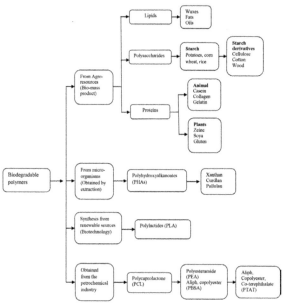

FIGURE 1.4 Classification of biodegradable polymers (*Source*: From C. J. Weber, V. Haugaard, R. Festersen, and G. Bertelsen, Production and applications of biobased packaging materials for the food industry. Food Additives & Contaminants Vol. 19, Iss. Sup1, 2002. Reprinted with permission from Taylor & Francis.)

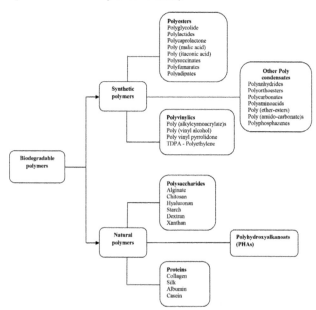

FIGURE 1.5 Classification of synthetic and natural biodegradable polymers (Adapted from Ghanbarzadeh & Almasi, 2013).

1.2.1 DIRECT EXTRACTION FROM BIOMASS

Direct extraction from biomass yields a series of natural polymer materials (cellulose, starch, and protein), fibers and vegetable oils, that can form the base on which polymer materials and products can be developed (Fornasiero & Graziani, 2006). They occur naturally in biomass/plants, then formulated or modified as necessary (e.g., plasticized) for use in plastic processes (e.g., plasticized starch-based plastics) (Jbilou et al., 2010).

1.2.1.1 CHITOSAN

Chitin is a beta-1, 4 linked linear polymer of 2-acetamido-2-deoxy-D-glucopyranosyl residue. It occurs as the major organic skeletal substance of invertebrates and as a cell wall constituent of fungi and green algae. Chitin and chitosan are the most abundant natural amino polysaccharide and valuable bio-based natural polymers. Currently, chitin and chitosan are produced commercially by chemical extraction process from crab, shrimp, and prawn wastes. The chemical extraction of chitin is quite an aggressive process based on demineralized by acid and deproteinated by the action of alkali followed by deacetylated into chitosan (Roberts, 1997). Chitin can also be produced by using enzyme hydrolysis or fermentation process, but these processes are not economically feasible on an industrial scale (Win & Stevens, 2001). Chitosan displays interesting characteristics including biodegradability, biocompatibility, chemical inertness, high mechanical strength, good film-forming properties, and low cost (Liu et al., 2012). Chitosan is being used in a vast array of widely varying products and applications ranging from pharmaceutical and cosmetic products to water treatment and plant protection. Chitosan is non-toxic and can form semi-permeable coating which can modify the internal atmosphere, thereby delaying ripening and decreasing transpiration rates in fruits and vegetables. It also has the ability to inhibit growth of fungi and phytopathogens (Ahvenainen, 2005).

1.2.1.2 CELLULOSE

Cellulose is the most abundant biopolymer in the world, serving as a renewable polymer. The most important raw material sources for the production of cellulosic plastics are cotton fibers and wood. Plant fiber is dissolved in alkali and carbon disulfide to create viscose, which is then reconverted to cellulose in cellophane form following a sulfuric acid and sodium sulfate bath. There are currently two processes used to separate cellulose from the other wood constituents (Yan et al., 2009). These methods, sulfite and pre-hydrolysis kraft pulping, use high pressure and chemicals to separate cellulose from lignin and hemicellulose attaining greater than 97% cellulose purity. The main derivatives of cellulose for industrial purposes are cellulose acetate, cellulose esters (molding, extrusion, and films), and regenerated cellulose for fibers. Cellulose is a hard polymer and has a high tensile strength of 62–500 MPa and elongation of 4% (Eichhorn et al., 2001).

In order to overcome the inherent processing problems of cellulose, it is necessary to modify, plasticize, and blend with other polymers. Eastman chemical is a major producer of cellulosic polymers. There are three main groups of cellulosic polymers that are produced by chemical modification of cellulose for various applications. Cellulose esters, namely cellulose nitrate and cellulose acetate, are mainly developed for film and fiber applications. Cellulose ethers, such as carboxymethyl cellulose and hydroxyethyl cellulose, are widely used in construction, food, personal care, pharmaceuticals, paint, and other pharmaceutical applications (Kamel et al., 2008). At present, regenerated cellulose is the largest bio-based polymer produced globally for fiber and film applications.

1.2.1.3 STARCH

Starch is a unique bio-based polymer because it occurs in nature as discrete granules. Starch forms as a granular energy reserve in plant cells. It is highly water soluble, readily biodegradable and consists of a mixture of linear amylose and highly branched amylopectin. Starch is present in maize, potatoes, wheat, and other plants as biopolymer. It is considered a very useful material in certain thermoplastic applications because of its biodegradability, availability, non-toxicity, high purity, and low cost. Starch promotes the biodegradability of a non-biodegradable plastic and it can also be used together with fully biodegradable synthetic polymers, producing biodegradable blends of low costs (Babu et al., 2013). The starch remains in granular form in the plastic matrix and so may act as filler. On mixing, starch enhances the biodegradability of the synthetic polymer, mainly because of the increase in polymer surface created after consumption of the starch by micro-organisms. The mechanical properties are dependent on such factors as filler volume, filler particle size and shape, and the degree of adhesion of the filler to the polymer matrix. The amount of starch in a blend plays an important role in its biodegradability, as indicated by the soil burial biodegradability test. Plasticized starch blends and composites and/or chemical modifications may overcome the issues like the phenomenon of retrogradation—a natural increase in crystallinity over time, leading to increased brittleness (Maurizio et al., 2005). Potential future applications could include foam loose-fill packaging and injection-molded products such as "take-away" food containers. Starch and modified starches have a broad range of applications both in the food and non-food sectors. In 2002, the total consumption of starch and starch derivatives was approximately 7.9 million tons in Europe and out of which 54% was used for food applications and 46% in non-food applications. Novamont is one of the leading companies in processing starch-based products (Li et al., 2009).

The edible films and coatings (xanthan, carrageenan, alginates, pectin, guar gum, gum acacia, etc.) act as a barrier toward respiratory gases and water vapors causing an internal modified atmosphere within the tissues, thereby improving the quality and shelf life of fresh fruits and vegetables. Shellac, an exudate of insect *Lacciferlacca*, has been widely used as surface coating for a variety of food applications. Shellac is non-toxic and physiologically harmless and therefore, listed as Generally Recognized as Safe (GRAS) substance by FDA (Okamoto & Ibanez, 1986). Certain efforts have been made to develop shellac based surface coatings for shelf-life extension of fresh green chillies (Chitravathi et al., 2014), tomatoes (Chauhan et al., 2013), apples (Chauhan et al., 2011), and citrus fruits (McGuire et al., 2001) (Tables 1.1 and 1.2).

TABLE 1.1 Advantages and Disadvantages of Using Cellulosic Fiber in Composites.

Advantages	Disadvantages
Renewable	Poor dimensional stability
Strong	Low biological resistance
Light weight	No thermal plasticity
Biodegradable	Low processing temperature
Inexpensive	Incompatible with hydrophobic thermoplastics

Source: Yam and Le (2012). Reprinted from S. Imam, G. Glenn and E. Chiellini, Chapter 21: Utilization of biobased polymers in food packaging: assessment of materials, production and commercialization in Emerging Food Packaging Technologies, eds. Kit Yam Dong Sun Lee. © 2012, with permission from Elsevier.

TABLE 1.2 Different Types of Bio-based/Biodegradable Polymers Used in Fruits and Vegetable Products Packaging.

Polymer name	Applications
Bio-PE	Plastics bags, milk and water bottles, food packaging films, and toys
PBS/PLA blend	Packaging films, dishware, fibers, and medical materials
PBS and co-polymers	Industrial applications
Chitosan	Coffee cups, food containers and trays, light weight foam pieces for cushioning, and coating of fruits and vegetables.
Cellulose	Films and sheets
Cellulose esters	Membranes for separation
Modified celluloses, cellulose whiskers, microfibrous cellulose	Barrier films, water preservation in food packing
PBS/starch	Barrier films
Polyhydroxyalcanoates (PHAs)	Food containers, pouches, tubs, bottles, films
Polylactic acid (PLA)	Bottles, films, labels, compostable thermoformed cups, eating utensils, and bowls
Polycaprolactone (PCL)	Adhesives, biodegradable bags, and containers
Bio-PE and blends	Agricultural mulch films
Modified starch	Food applications
Thermoplastic starch	Packaging, containers, mulch films, textile sizing agents, and adhesives

Source: Babu et al. (2013).

1.2.1.4 PROTEIN FILMS

Casein, whey protein, gelatin/collagen, fibrinogen, soy protein, wheat gluten, corn zein, and egg albumen have been processed into edible films. Protein-based films adhere well to hydrophilic surfaces, provide barriers for oxygen and carbon dioxide, but do not resist water diffusion. Proteins are desirable as films because of their nutritional value, excellent mechanical and barrier properties, solubility in water, and ability to act as emulsifiers. Considerable research has been made using protein-based films for packaging of fruits and vegetables and other dairy foods (Cutter, 2006). Gelatin is reported to have better oxygen barrier properties when combined with other types of films (Gennadios, 2002). Additional studies have demonstrated that gelatin can be used to carry antioxidants, to reduce oxidation, enhance color stability, to retain flavor, taste, and aroma of foods during refrigerated or frozen storage (Gennadios et al., 1997).

For improvements in edible film technologies, most of the researches have addressed film formulations using various combinations of edible materials. Two or more materials can be combined to improve gas barrier properties, adherence to coated products, or moisture vapor permeability properties. Composite films consisting of lipids and a mixture of proteins or polysaccharides take advantage of the individual component properties. These individual or combined films can be applied as emulsions or bilayer films (Cutter, 2006). Additionally, plasticizers can be used to modify film mechanical properties, thereby imparting desirable flexibility, permeability, or solubility to the resulting film. For example, adding glycerol, polyethylene glycol, or sorbitol to a film composition can reduce brittleness (Ben & Kurth, 1995). In another example of composite films, a combination of vegetable oils, glycerin, citric acid, and antioxidants prevented rancidity by acting as a moisture barrier, restricting oxygen transport, and serving as a carrier for antioxidants to various foods (Cutter & Sumner, 2002).

1.3 POLYMERS PRODUCED BY CLASSICAL CHEMICAL SYNTHESIS

Polymers are produced by classical chemical synthesis using renewable bio-based monomer (e.g., polylactic acid (PLA), a bio-polyester polymerized from lactic acid monomers) (Dilip et al., 2013).

1.3.1 LACTIC ACID

Lactic acid is used as an important molecule for the chemical and food industries for centuries. It is produced through anaerobic fermentation by many bacteria. Traditionally, its main applications are in the food industry where it is used as a natural acidifying agent. Recently, the scope of its applications was significantly enhanced by the synthesis of PLA as a new biodegradable and bio-based bioplastic (Berezina & Martelli, 2014).

1.3.2 POLYLACTIC ACID (PLA)

PLA is the most widely used biopolymer-based packaging material with the highest potential and it is synthesized from lactic acid, which is derived from renewable resources, such as corn or sugar beets (Lim et al., 2008). It can also be derived from fermentation process of any starch, which generates lactic acid. Even though PLA can be synthesized by direct polycondensation of lactic acid, polymers with higher molecular weights and lower polydispersity indexes are commonly obtained by ring opening polymerization of lactide (i.e., the cyclic dilactone of lactic acid). LL- and DD-lactide yield semi-crystalline polymers named P(L)LA and P(D)LA, while *rac*-lactide is used to synthesize amorphous polymers named P(LD)LA. Structurally, PLA is a linear aliphatic polyester, which is a thermoplastic with high-strength, high-modulus, good processability, being completely biodegradable, compostable, and biocompatible, and therefore, perfectly safe for the environment and for the food packaging application. PLA presents a medium water and oxygen permeability level comparable to polystyrene (Auras et al., 2004). In addition, PLA is safe and "Generally Recognized as Safe" (GRAS) for use in food packages (Conn et al., 1995). However, its high polarity, brittleness, stiffness and low deformation at break limit its use. Considerable efforts have been made to improve the properties of PLA so as to compete with low-cost and flexible commodity polymers. These attempts were made either by modifying PLA with biocompatible plasticizers or by blending PLA with other polymers (Luckachan & Pillai, 2011). One way to overcome such weakness of the biopolymer-based packaging materials is to associate them with a moisture resistant polymer with good mechanical properties, while maintaining the overall biodegradability of the product. For such purposes, multilayer films formed through coating or lamination methods have been widely used to combine the properties of two or more polymers into single multilayered structure (Rhim, 2013). PLA grades are available for injection moulding, extrusion, film or sheet and emulsion coating and it is used in wide variety of applications including packaging (e.g., water bottles). Modification of PLA is gaining popularity and the addition of groups such as maleic anhydride to the PLA chain enables better mixing with other polymers or fillers (Pilla, 2011).

1.4   PRODUCTION AND EXTRACTION OF BIO-BASED POLYMERS FROM MICRO-ORGANISMS/GENETICALLY MODIFIED BACTERIA

Production and extraction of bio-based polymers are prepared from micro-organisms/genetically modified bacteria, for example PHA, poly-hydroxy-butyrate (PHB), and poly hydroxybutyrate co-hydroxyvalerate (PHBv) (Pilla, 2011).

1.4.1 POLYHYDROXYALKANOATES (PHAS)

PHAs are a family of polyesters produced by microbial (bacterial) fermentation with the potential to replace conventional hydrocarbon-based polymers. The most common PHA is poly 3-hydroxybutyrate (PHB) which is produced as a carbon/energy store in a wide variety of bacteria (Pilla, 2011). PHB, the simplest PHA, was discovered in 1926 by Maurice Lemoigne as a constituent of the bacterium *Bacillus megaterium*. PHA can be produced by varieties of bacteria using several renewable waste feedstocks. A generic process to produce PHA by bacterial fermentation involves fermentation, isolation, and purification from fermentation broth. A large fermentation vessel is filled with mineral medium and inoculated with a seed culture that contains bacteria. The feedstocks include cellulosics, vegetable oils, organic waste, municipal solid waste, and fatty acids depending on the specific PHA required. The carbon source is fed into the vessel until it is consumed and cell growth and PHA accumulation are complete. In general, a minimum of 48 h is required for fermentation time. To isolate and purify PHA, cells are concentrated, dried, and extracted with solvents such as acetone or chloroform. The residual cell debris is removed from the solvent containing dissolved PHA by solid–liquid separation process. The PHA is then precipitated by the addition of an alcohol (e.g., methanol) and recovered by a precipitation process (Kathiraser et al., 2007). More than 150 PHA monomers have been identified as the constituents of PHAs (Steinbuchel & Valentin, 1995).

PHB possesses similar thermal and mechanical properties to those of polystyrene and polypropylene (Savenkova et al., 2000). However, due to its slow crystallization, narrow processing temperature range, and tendency to "creep," it is not attractive for many applications, and it requires further development in order to overcome these shortcomings (Reis et al., 2008). Several companies have developed PHA copolymers with typically 80–95% (R)-3-hydroxybutyric acid monomer and 5–20% of a second monomer in order to improve the properties of PHAs (Babu et al., 2013). It has become clear that PHA and its related technologies are forming an industrial value chain ranging from fermentation, materials, energy to medical fields (Chen, 2009).

1.4.2 BIO-POLYETHYLENE

Traditionally, ethylene is produced through steam cracking of naphtha or heavy oils or ethanol dehydration. With increase in oil prices, microbial PE or green PE is now being manufactured from dehydration of ethanol produced by microbial fermentation. Currently, bio-PE was produced on an industrial scale from bio-ethanol derived from sugarcane. Bio-ethanol is also derived from bio-renewable feedstocks, including sugar beet, starch crops such as maize, wood, wheat, corn, and other plant wastes through microbial and biological fermentation process. In a typical process, extracted sugarcane juice with high sucrose content is anaerobically fermented to produce ethanol. At the end of the fermentation process ethanol is distilled in order to remove water and to yield azeotropic mixture of hydrous ethanol. Ethanol is then dehydrated at high temperatures over a solid catalyst to produce ethylene and, subsequently, polyethylene (Luiz et al., 2010). Bio-based polyethylene has exactly the same chemical, physical, and mechanical properties as petrochemical polyethylene. Braskem (Brazil) is the largest producer of bio-PE with 52% market share, and this is the first certified bio-PE in the world. Similarly, Braskem is developing other bio-based polymers such as bio-polyvinyl chloride, bio-polypropylene, and their copolymers with similar industrial technologies. The current Braskem bio-based PE grades are mainly targeted toward food packing, cosmetics, personal care, automotive parts, and toys. Dow Chemical (USA) in cooperation with Crystalsev is the second largest producer of bio-PE with 12% market share. Solvay (Belgium), another producer of bio-PE, has 10% share in the current market. However, Solvay is a leader in the production of bio-PVC with similar industrial technologies. China Petrochemical Corporation also plans to set up production facilities in China to produce bio-PE from bio-ethanol (Haung et al., 2008). Bio-PE can replace all the applications of current fossil-based PE. It is widely used in engineering, agriculture, packaging, and many day-to-day commodity applications because of its low price and good performance (Table 1.3).

1.4.3 BACTERIAL CELLULOSE

Chemically pure cellulose can be produced from certain type of bacteria (*Acetobacter xylinum* and *Acetobacter pasteurianus*). Bacterial cellulose is processed under ambient conditions, in contrast to the harsh chemical and high-temperature treatment required for the processing of plant cellulose resulting in a higher degree of polymerization in the finished cellulose. Bacterial cellulose (bio-cellulose) is chemically pure and very strong; however, low yields and high costs are currently barriers to its large-scale production. Currently, applications for bacterial cellulose outside food and biomedical fields are rather limited because of its high price (Berezina & Martelli, 2014).

TABLE 1.3 Polysaccharides Produced from Micro-organisms.

Sl. No.	Type of micro-organism	Polysaccharide	Type	Main producing strain	Application
1.	Bacteria	Cellulose	Extracellular	*Acetobacter, Rhizobium, Rhizobacterium, Agrobacterium, Sarcina*	Paper, textile, food, cosmetics, medicine
		Curdlan	Extracellular	*Alcaligenes, Agrobacterium*	Food, pharmaceutical, agricultural, support for immobilization
		Dextran and derivatives	Extracellular	*Streptococcus, Leuconostoc*	Oil drilling, food, agriculture
		Hyaluronan and hyaluronic acid	Extracellular	*Streptococcus, Pasteurella*	Cosmetics, medicine
		Xanthan	Extracellular	*Xanthomonas*	Food, oil drilling
		Glycogen	storage (accumulated)	*E.coli, Clostridia, Bacillus, Streptomyces*	
		Succinoglycan	Extracellular	*Rhizobium, Agrobacterium, Alcaligenes, Pseudomonas*	Thickening, gel-forming, precipitation agent
		Alginate	Extracellular	*Pseudomonas, Azotobacter*	Food, pharmaceutical, biotechnology (immobilization)
		Glucuronan	Extracellular	*Rhizobium, Pseudomonas*	Cosmetics, agriculture, medicine
		Sphingan	Extracellular	*Sphingomonas*	Food, biotechnology (solid culture media & gel electrophoresis), construction (cement-based materials), oil drilling fluids
		Alternan	Extracellular	*Leuconostoc*	
		Levan	Extracellular	*Bacillus, Zymomonas, Aerobacter, Pseudomonas*	Cosmetics, food, pharmaceuticals

TABLE 1.3 *(Continued)*

Sl. No.	Type of micro-organism	Polysaccharide	Type	Main producing strain	Application
		Murein	Cytoplasm	Any	
		Teichoic and Teichuronic acids	Cell wall	Gram-positive bacteria	
2.	Fungi	Pullulan	Extracellular	*Aureobasidium, Pullularia, Dematium*	Food, pharmaceuticals, industry (adhesives)
		Chitin/chitosan	Cell wall	*Basidomycetes, Ascomycetes, Phycomycetes/Mucorales*	Absorption of coloring matters and metal, medicine
		Scleroglucan	Extracellular	*Sclerotium*	Food, medicine, oil drilling
		Schizophyllan (sizofilan, sizofiran)	Extracellular	*Schizophyllum*	Medicine (anti-tumor)
3.	Algae	Alginate	Structural	*Phaeophyceae* (brown algae)	Shear-thinning viscosifyer for textile, paper coating, can sealing, medicine, pharmacy, food
		Carrageenan	Cell wall	*Rhodophyceae* (red seaweeds)	Gelling, thickening, stabilizing agents
		Ulvan	Cell wall	*Ulva* sp.	Food, pharmaceuticals

Source: Berezina and Martelli (2014). http://pubs.rsc.org/en/content/chapterhtml/2014/bk9781849738989-00001?isbn=978-1-84973-898-9

1.5 POLYMERS CONVENTIONALLY OBTAINED FROM THE PETROCHEMICAL INDUSTRY BY CHEMICAL SYNTHESIS (BIO-DEGRADABLE POLYMERS FROM PETROLEUM SOURCES)

Polymers conventionally obtained from the petrochemical industry by chemical synthesis: for example, polycaprolactones (PCL), polyesteramides (PEA), aliphatic co-polyesters (e.g., PBSA), and aromatic co-polyesters (e.g., PBAT).

1.5.1 POLYCAPROLACTONE (PCL)

PCL is prepared by ring opening polymerization of ε-caprolactone using a catalyst such as stannous octoate. PCL is a biodegradable polyester with a low melting point of around 60°C and a glass transition temperature of about −60°C. The most common use of polycaprolactone is in the manufacture of specialty polyurethanes. Polycaprolactones impart good water, oil, solvent, and chlorine resistance to the polyurethane produced. This polymer is often used as an additive for resins to improve their processing characteristics and their end-use properties (e.g., impact resistance). Being compatible with a range of other materials, PCL can be mixed with starch to lower its cost and increase biodegradability or it can be added as a polymeric plasticizer to PVC (Labet & Thielemans, 2009).

1.5.2 POLYESTERAMIDES (PEA)

Poly(ester amide)s (PEAs) constitutes a promising family of biodegradable materials as, they have good degradability, high thermal stability, high modulus and high tensile strength of polyamides, afforded by hydrolyzable ester groups (—COO—) placed in the backbone. It possesses relatively good thermal and mechanical properties and gives the strong inter-molecular hydrogen bonding interactions that can be established between their amide groups (—NHCO—). Currently, a considerable variety of PEAs have been studied including the use of different monomers (e.g., α-amino acids, α,omega-amino alcohols, or carbohydrates) or different polymer microstructures (e.g., ordered blocky or random monomer distributions). The presence of hydrolytically cleavable ester bonds in the backbone and the lowering of the crystallinity also make the poly(ester amide)s as promising materials for their use in medical fields and packaging applications (Rodriguez-Galan et al., 2011).

1.5.3 ALIPHATIC CO-POLYESTERS (PBSA) AND AROMATIC CO-POLYESTERS (PBAT)

A large number of aliphatic copolyesters based on petroleum resources are biodegradable copolymers. They are obtained by the combination of diols such as 1,2-ethanediol, 1,3-propanediol, or 1,4-butanediol, and of dicarboxylic acids like adipic, sebacic, or succinic acid. The biodegradability of these products also depends on their structure. The addition of adipic acid, which decreases the crystallinity, tends to increase the compost biodegradation rate (Averous & Pollet, 2012). According to Ratto et al. (1999), the biodegradation results demonstrate that, while PBSA is inherently biodegradable, the addition of starch filler can significantly improve the rate of degradation. Compared to totally aliphatic co-polyesters; aromatic copolyesters are often based on terephthalic acid. However, Muller et al. (1998) reported that increase of terephthalic acid content tends to decrease the degradation rate (Tables 1.4–1.6).

TABLE 1.4 Overview of Currently Most Important Groups and Types of Bio-based Polymers.

Sl. No.	Bio-based polymer (group)	Type of polymer	Structure/production method
1.	Starch polymers	Polysaccharides	Modified natural polymer
2.	Polylactic acid (PLA)	Polyester	Bio-based monomer (lactic acid) by fermentation, followed by polymerization
3.	Other polyesters from bio-based intermediates	Polyester	
	a) Polytrimethyleneterephthalate (PTT)		Bio-based 1,3-propanediol by fermentation plus petrochemical terephthalic acid (or DMT)
	b) Polybutylene terephthalate (PBT)		Bio-based 1,4-butanediol by fermentation plus petrochemical terephthalic acid
	c) Polybutylene succinate (PBS)		Bio-based succinic acid by fermentation plus petrochemical terephthalic acid (or DMT)
4.	Polyhydroxyalkanoates (PHAs)	Polyester	Direct production of polymer by fermentation or in a crop (usually genetic engineering in both cases)
5.	Polyurethanes (PURs)	Polyurethanes	Bio-based polyol by fermentation or chemical purification plus petrochemical isocyanate

TABLE 1.4 *(Continued)*

Sl. No.	Bio-based polymer (group)	Type of polymer	Structure/production method
6.	Nylon	Polyamide	
a)	Nylon 6		Bio-based caprolactam by fermentation
b)	Nylon 66		Bio-based adipic acid by fermentation
c)	Nylon 69		Bio-based monomer obtained from a conventional chemical transformation from oleic acid via azelaic (di)acid.
7.	Cellulose polymers	Polysaccharides	a) Modified natural polymer
			b) Bacterial cellulose by fermentation

Source: Wolf (2005). Taken from a report available on public domain. http://ftp.jrc.es/EUR-doc/eur22103en.pdf

TABLE 1.5 Bio-based Packaging Applications for Fruits and Vegetables.

Sl. No.	Food	Packaging method	Packaging materials	Storage conditions
1.	Zucchini	Passive MAP Active MAP (%) $N_2/CO_2/O_2$:90/5/5; 75/10/15	1) Oriented PP bag 2) Bio-polymeric film (COEX)	At 5°C for 8–9 days
2.	Blackcurrant	Air	PP trays sealed with laminated PET/PP; inserted in PLA pouches (25 and 40 µm, holes area 7.9 cm^2) OPP (40 µm) + cardboard boxes placed in PLA film pouches (25 and 40 µm, holes area 7.9 cm^2)	At 4 ± 1°C for 24 days
3.	Grape	Air	Biodegradable monolayer films; NVT 100 (100 µm) NVT 50 (50 µm) Multilayer PET-based coextruded film, NVT 35 (35 µm) OPP (20 µm) PA layer with a polyolefin-based film (95 µm) Alu/PE (133 µm)	At 5°C for 35 days
4.	Lettuce	Air	Biodegradable PE-HD (51 µm); PP (61 µm); with or without anti-fog treatment PE/OPP (58 µm)	At 4.4°C 80% RH for 14 days

TABLE 1.5   *(Continued)*

Sl. No.	Food	Packaging method	Packaging materials	Storage conditions
5.	Sweet cherries	Air MAP:10% O_2/4%CO_2/86%N_2	OPP 20 (μm) Biodegradable co-extruded polyesters (COEX 35 μm)	At 0°C for 32–36 days
6.	Fresh cut 'Gold' pineapple	HiO_2:38–40% LoO_2:10–12% O_2/1%CO_2	Trays:PP, Wrap PP (64 μm) $P(O_2) = 110$ cm^3/m^2 day bar at 23°C, 0% RH	At 5°C for 20 days
		Air passive MAP Air	$P(CO_2) = 550$ cm^3/m^2 day bar at 23°C, 0% RH PP/PP PP/PP + coal: 1% w/v Alginate+glycerol+ sunflower oil+CaCl$_2$	

Note: Alu = Aluminum; OPP = oriented polypropylene ; P = permeability; PA = polyamide; PP = polypropylene; PE = polyethylene; PE-HD = high-density polyethylene; PET = polyethylene terephthalate; PLA = polylactide;   PS = polystyrene; PVC = polyvinyl chloride. (*Source*: From © 2013, Innovation in Healthy and Functional Foods, Dilip Ghosh, Shantanu Das, Debasis Bagchi, R.B. Smarta. Reproduced by permission of Taylor and Francis Group, LLC, a division of Informa, PLC.)

TABLE 1.6   Comparative Properties of Bio-based and Synthetic Polymers.

Polymer	Bio-based		
	Moisture permeability	Oxygen permeability	Mechanical properties
Cellulose acetate	Moderate	High	Moderate
Starch/polyvinyl alcohol	High	Low	Good
Proteins	High–medium	Low	Good
Cellulose/cellophane	High–medium	High	Good
Polyhydroxyalkanoates (PHAs) Polyhydroxybutyrate/valerate (PHBA)	Low	Low	Good
Polylactate	Moderate	High–moderate	Good
	Synthetic		
Low-density polyethylene	Low	High	Moderate–good
Polystyrene	High	High	Poor–moderate

Source: Pilla (2011).

1.6 BIO-BASED FIBERS

Current trends in the packaging industry are toward lighter weight materials for reduction of raw material use, transportation costs, and minimizing the amount of waste. Interest in sustainable materials combined with barrier property is tremendously increasing. Natural fibers can be defined as bio-based fibers or fibers from vegetable and animal origin. This includes all plant fibers (cotton, jute, hemp, kenaf, flax, coir, abaca, ramie, etc.) and protein-based fibers (wool and silk). Natural fibers are grouped into three types: seed hair, bast fibers, and leaf fibers depending upon the source. Some examples are cotton, coir (seed/fruit hairs), hemp, flax, jute, kenaf (bast fibers), sisal, and abaca (leaf fibers). Out of these fibers, jute, flax, hemp, and sisal are the most commonly used fibers for polymer composites (NabiSaheb & Jog, 1999).

Natural fibers in the form of wood flour have also been often used for preparation of natural fiber composites. Most cellulosic fibers are a renewable resource and the production requires little energy, and in the process, in a way CO_2 is used while O_2 is given back to environment. Additionally, thermal recycling is possible where glass and man-made fibers often cause problems in combustion processes. Compared to synthetic fibers, natural plant-based fibers have several specific advantages. These natural fibers have low cost, high specific mechanical properties, good thermal and acoustic insulation and bio-degradability. However, when the specific modulus of natural fibers (modulus/specific gravity) is considered, the natural fibers show values that are comparable to or better than those of glass fibers. These higher specific properties are one of the major advantages of natural fiber for making composites for applications wherein the desired properties also include weight reduction (Mitra, 2014). Natural fiber composites show superior performance at a given price. Natural fiber composites have significant potential in transportation and construction market. The chemical composition of natural fibers varies depending upon the type of fiber. Primarily, fiber contains cellulose, hemicellulose, pectin, and lignin. The properties of each constituent contribute to the overall properties of the fibers. Hemicellulose is responsible for the bio-degradation, moisture absorption and thermal degradation of the fiber as it shows the least resistance, whereas lignin is thermally stable but is responsible for the degradation on exposure to UV light (Mohanty et al., 2000).

1.7   FIBER-MATRIX INTERFACE

The mechanical properties of fiber reinforced polymer composites are of great importance in deciding their end applications. The inter-facial shear strength (IFSS) is a critical factor that controls the toughness, transverse mechanical properties, and inter-laminar shear strength (ILSS) of the composite materials. The interfacial adhesion can be controlled by (1) physicochemical interaction (2) mechanical interaction (interlocking), and (3) chemical interaction, or bonding. Good interfacial adhesion provides composites with structural integrity and good load transfer from broken fibers to intact fibers through the matrix, maintaining the integrity of the composites and resulting in strong composites (Netravali et al., 1989).

1.8   MODIFICATION OF NATURAL FIBERS

Natural fibers are incompatible with the hydrophobic polymer matrix and have a tendency to form aggregates. These hydrophilic fibers exhibit poor resistance to moisture. To eliminate the problems related to high moisture absorption, treatment of fibers with hydrophobic aliphatic and cyclic chemical structures are generally done. These structures contain reactive functional groups that are capable of bonding to the reactive groups in the matrix polymer. Chemical treatment of natural fibers can be achieved by de-waxing (de-fatting), de-lignifications, bleaching, acetylation, cyanoethylation, and chemical grafting and these treatments are used for modifying the surface properties of the fibers for enhancing their performance. Delignification (de-waxing) is generally carried out by extracting with alcohol or benzene and treatment with caustic soda, followed by drying at room temperature. Many oxidative bleaching agents such as alkaline calcium or sodium hypochlorite and hydrogen peroxide are commercially used. Bleaching generally results in loss of weight and tensile strength. These losses are mainly attributed to the action of the bleaching agent or alkali or alkaline reagent on the non-cellulosic constituents of fibers such as hemicellulose and lignin (Mitra, 2014). Natural fibers (lingo-cellulosics) are degraded by biological organisms since they can recognize carbohydrate polymers in the cell wall. Ligno-cellulosics exposed outdoors, undergo photochemical degradation caused by ultraviolet light. Resistance to biodegradation and UV radiation can be improved by bonding chemicals to the cell wall polymers or by adding polymer to the cell matrix. In addition to the surface treatment of fibers, use of compatibilizer or coupling agent for effective stress transfer across the interface are also explored (Mitra, 2014; Bledzki et al., 1996).

1.9 BIO NANO-COMPOSITES

Recently, polymer–clay nano-composites have received significant attention as an alternative to conventionally filled polymers. Because of their nanometer size dispersion, the polymer–clay nano-composites exhibit the large-scale improvement in the mechanical and physical properties compared with pure polymer or conventional composites. These include increased modulus and strength, decreased gas permeability, increased solvent and heat resistance and decreased flammability (Mitra, 2014; Helbert et al., 1996).

1.10 CONCLUSION

The use of biodegradable polymer reduces the environmental impact of non-biodegradable plastic. The approach on biopolymer blending can be a promising alternative to enhance the scope of their applications in bio-films and coatings. As single layer biopolymer films rarely have a competitive edge against synthetic films, therefore, the combination of individual layers to multilayer structures is necessary. By multilayer structures, the technical properties may be dramatically improved. Incorporation of nanoparticles is an excellent way to improve the performance of bio-based films. The rationale in implementing the use of biodegradable materials promises to expand the application of edible coatings and biodegradable films that reduce the packaging waste associated with processed foods that support the preservation of fresh foods by extending their shelf-life. Recent breakthroughs also concern the development of pure bio-based products identical to the petroleum-based materials such as PE, PP, PA-6, PA-6, polyisobutylene, etc. The economic outcomes of these initiatives are clearly dependent on the global energy supply. The recent enhancement of shale gas production, including the production of so-called "wet gases" such as butane, propane, and ethane will deeply influence the economic perspectives of bio-PE and bio-PP.

Biopolymer-coating on paper packaging materials are very promising systems for the future improvement of food packaging. The use of such bio-based packaging will open up potential economic benefits to farmers and agricultural processors. Biopolymer coatings on paper packaging materials are potential inclusion matrices of antimicrobial agents to develop biodegradable active packaging. Further research is needed to gain more knowledge regarding the interactions between the coating matrix, active compounds, and target microorganisms to evaluate the material's performance and to optimize the compositions of active coatings. Furthermore, the antimicrobial properties of coated papers are of great merit for future research. For food product applications, research is essential to evaluate the impact of active agents on organoleptic properties of the packaged food products. At present, bio-packaging materials are most suitable for foods with high respiration (fruit and vegetables) or for food with short shelf lives (bread, convenience food, etc.), because of poor water barrier properties of these packaging materials. The main concept for bio-based materials applications could be as organic food wrappings in bio-based food packaging materials that are environmental friendly. However, despite these advancements, there are still some drawbacks which prevent the wider commercialization of bio-based polymers in many applications. This is mainly due to performance and price when compared with their conventional counterparts and it still remains a significant challenge for bio-based polymers.

KEYWORDS

- **paperboard coatings**
- **fuel-based polymers**
- **biodegradation process**
- **modulus/specific gravity**
- **transportation costs**

REFERENCES

Ahvenainen, R. *Novel Food Packaging Techniques*; Wood Head Publishing Limited CRC Press: Cambridge, England, 2005; pp 80–81.

Auras, R.; Harte, B.; Selke, S. An Overview of Polylactides as Packaging Materials. *Macromol. Biosci.* **2004,** *4* (9), 835–864. ISSN 1616–5197.

Averous, L.; Pollet, E. Eds. *Environmental Silicate Nano-Biocomposites Green Energy and Technology;* Springer: London, 2012. DOI: 10.1007/978-1-4471-4108

Babu, Ramesh, P.; Connor, Kevin, O.; Seeram, R. Current Progress on Bio-based Polymers and Their Future Trends. *Prog. Biomater.* **2013,** *2*, 8. Doi: 10.1186/2194-0517-2-8

Ben, A.; Kurth, L. B. Edible Film Coatings for Meat Cuts and Primal. Meat. 95, The Austra-lian Meat Industry Research Conference. CSIRO, September 10–12, 1995.

Berezina, N.; Martelli, S. M. Bio-based Polymers and Materials (Chapter 1). In *Renewable Resources for Biorefineries;* Royal Society of Chemistry: London, 2014; pp 1–28. DOI: 10.1039/9781782620181-00001

Bledzki, A. K.; Reihmane, S.; Gassan, J. Properties and Modification Methods for Vegetable for Natural Fiber Composites. *J. Appl. Polym. Sci.* **1996,** *59* (8), 1329–1336.

Cutter, C. N. Opportunities for Bio-based Packaging Technologies to Improve the Quality and Safety of Fresh and Further Processed Muscle Foods. *Meat Sci.* **2006,** *74,* 131–142.

Chan, M. A.; Krochta, J. M. Grease and Oxygen Barrier Properties of Whey Protein – Isolate Coated Paper Board Solutions. *Tappi J.* **2001a,** *84* (10), 57.

Chan, M. A.; Krochta, J. M. Colour and Gloss of Whey-Protein Coated Paper Board Solu-tions. *Tappi J.* **2001b,** *84* (10), 58.

Chauhan, O. P.; Nanjappa, C.; Ashok, N.; Ravi, N.; Roopa, N.; Raju, P. S. Shellac and Aloe Vera Gel Based Surface Coating for Shelf Life Extension of Tomatoes. *J. Food Sci. Technol.* **2013,** *52* (2), 1200–1205. DOI 10.1007/s13197-013-1035-6

Chauhan, O. P.; Raju, P. S.; Singh, A.; Bava, A. S. Shellac and Aloe-Gel Based Surface Coat-ings for Maintaining Keeping Quality of Apple Slices. *Food Chem.* **2011,** *126,* 961–966.

Chen, G. Q. A Microbial Polyhydroxyalkanoates (PHA) Based Bio- and Materials Industry. *Chem. Soc. Rev.* **2009,** 38, 2434–2446. doi: 10.1039/b812677c

Chitravathi, K.; Chauhan, O. P.; Raju, P. S. Postharvest Shelf-Life Extension of Green Chillies (*Capsicum annuum* L.) Using Shellac-Based Edible Coatings. *Postharvest Biol. Technol.* **2014,** *92,* 146–148. http://dx.doi.org/10.1016/j.postharvbio. 2014.01.021

Conn, R. E.; Kolstad, J. J.; Borzelleca, J. F.; Dixler, D. S.; Filer, L. J.; La, Du, B. N.; Pariza, M. W. Safety Assessment of Polylactide (PLA) for Use As a Food-Contact Polymer. *Food Chem. Toxicol.* **1995,** *33* (4), 273–283.

Cutter, C. N.; Sumner, S. S. Application of Edible Coatings on Muscle Foods. In *Protein-based Films and Coatings;* Gennadios, A., Ed.; CRC Press: Boca Raton, FL, 2002; pp 467–484.

Debeaufort, F.; Quezada-Gallo, J. A.; Voilley, A. Edible Films and Coatings: Tomorrow's Packagings: A Review. *Crit. Rev. Food Sci. Nutr.* **1998,** *38* (4), 299–313.

Despond, S.; Espuche, E.; Cartier, N.; Domard, A. Barrier Properties of Paper–Chitosan and Paper–Chitosan–Carnauba Wax Films. *J. Appl. Polym. Sci.* **2005,** *98* (2), 704–710. DOI: 10.1002/app.21754

Dilip, G.; Shantanu, D.; Debasis B.; Smarta, R. B. *Innovation in Healthy and Functional Foods*; CRC Press, Taylor and Francis Group: Boca Raton, FL, 2013; p 227.

Eichhorn, S. J.; Baillie, C. A.; Zaferiropouls, N.; Mwaikambo, L. Y; Ansell, M. P; Dufresne, A.; Entwistle, K. M.; Herrera-Franco, P. J.; Escamilla, G. C.; Groom, L.;

Fornasiero, P.; Graziani, M. *Renewable Resources and Renewable Energy: A Global Chal-lenge;* 2nd ed.; CRC Press: Boca Raton, FL, 2006.

Gallstedt, M.; Brottman, A.; Hedenqvist, M. S. Packaging Related Properties of Protein and Chitosan Coated Paper. *Packag. Technol. Sci.* **2005,** *18* (4), 161–170. DOI: 10.1002/pts.685

Gastaldi, E.; Chalier, P.; Guillemin, A.; Gontard, N. Microstructure of Protein-Coated Paper as Affected by Physico-Chemical Properties of Coating Solutions. *Colloid Surf. A.* **2007,** *301,* 301–10.

Gennadios, A. Ed. Soft Gelatin Capsules. In *Protein-Based Films and Coatings;* CRC Press LLC: Boca Raton, FL, 2002; pp 393–443.

Gennadios, A.; Hanna, M. A.; Kurth, L. B. Application of Edible Coatings on Meat, Poultry and Sea Foods: A Review. *Lebensm. Wiss. Technol.* **1997,** *30,* 337–350.

Ghanbarzadeh, B; Almasi, H. Biodegradable Polymers. In. *Biodegradation-Life of Science;* Intec Open Access Publication: Croatia, 2013; pp 141–174. http://dx.doi.org/10.5772/56230

Ham-Pichavant, F., Sebe, G., Pardon, P., Coma, V. Fat Resistance Properties of Chitosan-based Paper Packaging for Food Applications. *Carbohyd. Polym.* **2005,** *61,* 259–265.

Han, J. H.; Gennadios, A. Edible Films and Coatings: A Review. In *Innovations in Food Packaging;* J. Han., Ed.; Academic Press, Elsevier: Cambridge, MA, 2005; pp 239–259.

Han, J. H; Krochta, J. M. Physical Properties and Oil Absorption of Whey-Protein Coated Paper. *J. Food Sci.* **2001,** *66* (2), 294–299.

Han, J. H.; Krochta, J. M. Wetting Properties and Water Vapor Permeability of Whey Protein-coated Paper. *Trans. ASAE.* **1999,** *42* (5), 1375–1382.

Haung, Y. M.; Li, H.; Huang, X. J.; Hu, Y. C.; Hu, Y. Advances of Bio-Ethylene. *Chin. J. Bioprocess Eng.* **2008,** *6,* 1–6.

Helbert, W.; Cavaille, J. Y.; Dufresne, A. Thermoplastic Nano-composites Filled with Wheat Straw Cellulose Whiskers. Part I: Processing and Mechanical Behavior. *Polym. Compos.* **1996,** *17* (4), 604–611.

Hughes, M; Hill C.; Rials, T. G.; Wild, P. M. Review: Current International Research into Cellulosic Fibres and Composites. *J. Mater. Sci.* **2001,** *36,* 2107–2131.

Jbilou, F.; Galland, S.; Ayadi, F. Biodegradation of Corn Flour-based Materials Assessed by Enzymatic Aerobic and Anaerobic Tests: Influence of Specific Surface Area. *Polym. Test.* **2010,** *30,* 131–139.

Kamel, S.; Adel, A. M.; El-Sakhawy, M.; Nagieb, Z. A. Mechanical Properties and Water Absorption of Low Density Polyethylene/Saw Dust Composites. *J. Appl. Polym. Sci.* **2008,** *107* (2), 1337–1342.

Kathiraser, Y.; Aroua, M. K.; Ramachandran, K. B.; Tan, I. K. P. Chemical Characterization of Medium-Chain-Length Polyhydroxyalkanoates (PHAs) Recovered by Enzymatic Treatment and Ultrafiltration. *J. Chem. Technol. Biotechnol.* **2007,** *82,* 847–855.

Khwaldia, K.; Elmira-Arab, T.; Desobry, S. Biopolymer Coatings on Paper Packaging Materials. *Compr. Rev. Food Sci. Food Saf.* **2010,** *9* (1), 82–91.

Khwaldia, K.; Linder, M.; Banon, S.; Desobry, S. Effect of Mica, Carnauba Wax, Glycerol and Sodium Caseinate Concentrations on Water Vapour Barrier and Mechanical Properties of Coated Paper. *J. Food Sci.* **2005,** *70* (3), 192–197.

Khwaldia, K.; Perez, C.; Banon, S.; Desobry, S.; Hardy, J. Functional Properties and Applications of Edible Films Made of Milk Proteins. Physical Properties of Polyol-Plasticized Edible Films Made from Sodium Caseinate and Soluble Starch Blends. *LWT Food Sci. Technol.* **2009,** *42,* 868–873.

Kjellgren, H.; Engstrom, G. Influence of Base Paper on the Barrier Properties of Chitosan-coated Papers. *Nord. Pulp. Pap. Res. J.* **2006,** *21* (5), 685–689.

Labet, M.; Thielemans, W. Synthesis of Polycaprolactone: A Review. *Chem. Soc. Rev.* **2009,** *38* (12), 3484–3504. Doi:10.1039/B820162P

Li, S.; Juliane, H.; Martin, K. P. Product Overview and Market Projection of Emerging Bio-based Products. *PRo-BIP.* **2009,** *1,* 1–24.

Lim, L.T.; Auras, R.; Rubino, M. Processing Technologies for Poly (Lactic Acid). *Prog. Polym. Sci.* **2008,** *33,* 820–852.

Lin, S.Y.; Krochta, J. M. Plasticizer Effect on Grease Barrier and Colour Properties of Whey-Protein Coatings on Paper Board. *J. Food Sci.* **2003,** *68,* 229–333.

Luckachan, G. E.; Pillai, C. K. S. Biodegradable Polymers—A Review on Recent Trends and Emerging Perspectives. *J. Polym. Environ.* **2011,** *19,* 637–676.

Liu, M.; Zhang, Y.; Wu, C.; Xiong, S.; Zhou, C. Chitosan/halloysite Nanotubes Bio-Nano-composites: Structure, Mechanical Properties and Biocompatibility. *Int. J. Biol. Macromol.* **2012,** 51, 566–575.

Luiz, A.; De-Castro, R.; Morschbacker. A Method for the Production of One or More Olefins, an Olefin, and a Polymer. US 2010/0069691A1, July 24, 2008.

Matsui, K. N.; Lorotonda, F. D. S.; Paes, S. S.; Luiz, D. B.; Pires, A. T. N.; Laurindo, J. B. Cassava Bagasse-Kraft Paper Composites: Analysis of Influence of Impregnation with Starch Acetate on Tensile Strength and Water Absorption Properties. *Carbohydr. Polym.* **2004,** 55, 237–243.

Maurizio, A.; Jan, J. D. V.; Maria, E. E.; Sabine, F.; Paolo, V.; Maria, G. V. Biodegradable Starch/Clay Nano-Composite Films for Food Packaging Applications. *Food Chem.* **2005,** 93 (3), 467–474.

McGuire, R. G.; Hagenmaier, R. D. Shellac Formulations to Reduce Epiphytic Survival of Coliform Bacteria on Citrus Fruit Postharvest. *J. Food Prot.* **2001,** 64 (11), 1756–1760.

Mitra, B. C. Environment Friendly Composite Materials: Biocomposites and Green Composites. *Def. Sci. J.* **2014,** 64 (3), 244–261. DOI: 10.14429/dsj.64.7323

Mohanty, A. K.; Misra, M.; Hinrichsen, G. Biofibres, Biodegradable Polymers and Biocomposites: An Overview. *Macromol. Mater. Eng.* **2000,** 276–277 (1), 1–24.

Muhl, S.; Beyer, B. Bio-Organic Electronics–Overview and Prospects for the Future. *Electronics.* **2014,** 3, 444–461. doi: 10.3390/electronics3030444

Muller, R. J.; Witt, U.; Rantze, E.; Deckwer, W. D. Architecture of Biodegradable Copolyesters Containing Aromatic Constituents. *Polym. Degrad. Stab.* **1998,** 59 (1–3), 203–208.

NabiSaheb, D.; Jog, J. P. Natural Fiber Polymer Composites: A Review. *Adv. Polymer Technol.* **1999,** 18 (4), 351–63.

Narayan, R. Bio-based and Biodegradable Polymer Materials: Rationale, Drivers, and Technology Exemplars Presented at the National American Chemical Society, Division of Polymer Chemistry Meeting, ACS Symposium Series Book Chapter: San Diego, CA, 2005.

Netravali, A. N.; Henstenburg, R. B.; Phoenix, S. L.; Schwartz, P. Interfacial Shear Strength Studies Using the Single –Filament –Composite Test I: Experiments on Graphite Fibers in an Epoxy. *Polym. Composite* **1989,** 10 (4), 226–241.

Okamoto, M. Y.; Ibanez, P. S. Final Report on the Safety Assessment of Shellac. *J. Am. Coll. Toxicol.* **1986,** 5, 309–327.

Park, H. J.; Kim, S. H.; Lim, S. T.; Shin, D. H.; Choi, S. Y.; Hwang, K. T. Grease Resistance and Mechanical Properties of Isolated Soy Protein-Coated Paper. *J. Am. Chem. Soc.* **2000,** 77, 269–273.

Parris, N.; Vergano, P. J.; Dickey, L. C.; Cooke, P. H.; Craig, J. C. Enzymatic Hydrolysis of Zein-Wax-Coated Paper. *J. Agric. Food Chem.* **1998,** 46, 4056–4059.

Pilla, S. *Handbook of Bioplastics and Biocomposites Engineering Applications*; John Wiley & Sons: Hoboken, NJ, 2011; pp 161–164, 121–140.

Ratto, J. A.; Stenhouse, P. J.; Auerbach, M.; Mitchell, J.; Farrell, R. Processing, Performance and Biodegradability of a Thermoplastic Aliphatic Polyester/Starch System. *Polymer* **1999,** 40 (24), 6777–6788.

Reis, K. C.; Pereira, J.; Smith, A. C.; Carvalho, C. W. P.; Wellner, N.; Yakimets, I. Characterization of Polyhydroxybutyrate-Hydroxyvalerate (PHB-HV)/Maize Starch Blend Films. *J. Food Eng.* **2008,** 89, 361–369.

Rhim, J. W. Effect of PLA Lamination on Performance Characteristics of Agar/κ-carrageenan/clay Bio-Nano-Composite Film. *Food Res. Int.* **2013,** 51, 714–722.

Rhim, J. W.; Hwang, K. T.; Park, H. J.; Kang, S. K.; Jung, S. T. Lipid Penetration Characteristics of Carrageenan-Based Edible Films. *Korean J. Food Sci. Technol.* **1998**, *30*, 379–384.

Rhim, J. W.; Lee, J. H.; Hong, S. I. Water Resistance and Mechanical Properties of Biopolymer (Alginate and Soy Protein) Coated Paperboards. *Lebensm. Wiss. Technol.* **2006**, *39*, 806–813.

Roberts, G. A. F. Chitosan Production Routes and Their Role in Determining the Structure and Properties of the Product. In *Advances in Chitin Science;* Domard, M., Roberts, A. F., Varum, K. M., Eds.; Jacques Andre: Lyon; National Taiwan Ocean University: Taiwan, 1997; Vol. 2, pp 22–31.

Rodriguez-Galan, A.; Franco, L.; Puiggali, J. Degradable Poly(ester amide)s for Biomedical Applications. *Polymers.* **2011**, *3*, 65–99. Doi:10.3390/polym3010065

Savenkova, L.; Gercberga, Z.; Nikolaeva, V.; Dzene, A.; Bibers, I.; Kalina, M. Mechanical Properties and Biodegradation Characteristics of PHB-Based Films. *Process Biochem.* **2000**, *35*, 537–579.

Steinbuchel, A.; Valentin, H. E. Diversity of Bacterial Polyhydroxyalkanoicacids. *FEMS Microbiol. Lett.* **1995**, *128*, 219–228.

Trezza, T. A.; Vergano, P. J. Grease Resistance of Corn Zein Coated Paper. *J. Food Sci.* **1994**, *59* (4), 912–915.

Trezza, T. A.; Wiles, J. L.; Vergano, P. J. Water-Vapor and Oxygen Barrier Properties of Corn Zein Coated Paper. *Tappi J.* 1998, *81* (8), 171–176.

Webber, C. J. *Food Stuffs Packaging Biopolymers. Bio-based Packaging Materials for the Food Industry: Status and Perspectives*; A European Concerted Action: Frederiksberg, Denmark, November 2000.

Weber, C. J.; Haugaard, V.; Festersen, R.; Bertelsen, G. Production and Applications of Bio-based Packaging Materials for the Food Industry. *Food Addit. Contam.* **2002**, *19*, 172–177.

Wolf, O. *Techno-Economic Feasibility of Large-scale Production of Bio-based Polymers in Europe;* European Science and Technology Observatory: Seville, Spain, 2005.

Win, N. N.; Stevens, W. F. Shrimp Chitin as Substrate for Fungal Chitin Deacetylase. *Appl. Microbiol. Biotechnol.* **2001**, *57*, 334–341.

Yam, K. L.; Le, D. S. *Emerging Food Packaging Technologies: Principles and Practice*; Woodhead Publishing: Abington, England, 2012; pp 450.

Yan, Y. F.; Krishnaiah, D.; Rajin, M.; Bono, A. Cellulose Extraction from Palmkernel Cake Using Liquid Phase Oxidation. *J. Eng. Sci. Technol.* **2009**, *4*, 57–68.

CHAPTER 2

MODIFIED ATMOSPHERE PACKAGING OF FRESH PRODUCE

ALI ABAS WANI[1,2*], KHALID GUL[2], and PREETI SINGH[3]

[1]*Fraunhofer Institute for Process Engineering and Packaging IVV, Freising 85354, Germany*

[2]*Department of Processing and Food Engineering, Punjab Agricultural University, Ludhiana 141004, India*

[3]*Chair Food Packaging Technology, Technical University of Munich, Freising 85354, Weihenstephan, Germany*

*Corresponding author. E-mail: waniabas@gmail.com

CONTENTS

ABSTRACT

Food packaging exists to make our lives easier. Modified atmosphere pack-aged (MAP) foods have become increasingly more common as food manu-facturers have attempted to meet consumer demands for fresh, refrigerated foods with extended shelf-life. MAP dramatically extends the shelf-life of packaged food products, and in some cases food does not require any further treatment or any special care during distribution. However, in most cases extending shelf-life and maintaining quality requires a multiple hurdle tech-nology system—for example, introducing temperature control as well as MAP is generally essential to maintain the quality of packaged foods. MAP technology offers the possibility to retard the respiration rate and extend the shelf-life of fresh produce and is increasingly used globally as value adding in the fresh and fresh-cut food industry. The attitude toward active and intel-ligent packaging is positive and there is still much potential for exciting innovation to come.

2.1   INTRODUCTION

Fruits and vegetables continuously respire in order to derive energy for driving cellular processes. While doing so carbon dioxide, water vapors, and heat are produced. The storage life of fruits and vegetables is inversely correlated to the respiration rate. The higher the respiration rate, the more perishable the product and vice versa. Prolonging the shelf-life of fresh fruits and vegetables after harvest is important for safe and nutritional diet. Thus, the optimization of quality and reduction of loss in the post-harvest chain of fresh fruits and vegetables are the main objectives of food scientists. The respiration of fresh fruits and vegetables can be reduced by many preservation techniques like low temperature, canning, dehydration, freeze-drying, controlled atmosphere and hypobaric, and modified atmosphere (Sandhya, 2010).

MAP is an indirect food preservation technique, where the gaseous composition of the package is modified such that microbial growth and chemical deterioration reactions are kept at minimal levels. The MA pack-aging of fresh produce relies on the modification of the atmosphere inside the package by a reduction in O_2 and an increase in CO_2 concentrations. This slows down the respiration rate of many fresh produce items and inhibits the plant hormone ethylene—the factors responsible for aging and ripening process (Fonseca et al., 2002; Kader et al., 1989). The technology of MAP seeks to create such an environment passively for shipping, storage, and marketing of produce. The MAP could be considered as one of the tech-nologies used to extend the shelf-life of fresh produce. However, MAP cannot solely protect the produce from pathogens and damage. The pack-aging system is not virtually without safety concerns and this alone is not sufficient to prevent the pathogen growth. This chapter will discuss various aspects of MAP and controlled atmosphere packaging for fresh fruits and vegetables.

2.2 INFLUENCE ON RESPIRATION RATE AND FRESH PRODUCE QUALITY

All unharvested plants respire, that is, different organic compounds, mainly sugar compounds, provide energy to other life processes in the cells. This process needs oxygen. Air contains 21% oxygen but the earth has a much lower concentration. When oxygen is available, the respiration is aerobic. Anaerobic respiration is an undesirable form of respiration which takes place without oxygen. Respiration is a complicated process which involves a series of enzymatic reactions. The entire aerobic process can be described in simplified form as:

Sugar + oxygen (O_2),

Carbon dioxide (CO_2) + energy + water.

The respiration rate is measured as generated mL CO_2/kg × hour or as used ml O_2/kg × hour

The primary goal of MAP for fresh produce is the extension of shelf-life. It should be stressed that this extension of produce shelf-life may allow for the growth of pathogenic bacteria to higher levels as compared with air stored samples. Since fruits and vegetables continue to respire after harvest, there are many other factors that affect the post-harvest shelf-life extension of fresh produce and the success of MAP (Day, 2001; Farber et al., 2003).

The physiological processes of produce (Mainly respiration and transpi-ration) play significant roles in the post-harvest quality of the MA packaged fresh fruits and vegetables. Respiration is a metabolic process that provides the energy needed for various plant biochemical reactions. The respira-tion rate of fruits and vegetables is inversely proportional to the achiev-able shelf-life; higher respiration rates are associated with shorter shelf-life (Day, 1993). Respiration can be measured by the production rate of CO_2 or by the consumption rate of O_2; it also results in the production of heat and water vapors too. While the respiration rate can be reduced by decreasing the O_2 concentration in the package, the process can induce a decrease in the activity of oxidizing enzymes, such as polyphenols oxidase, glycolic acid oxidase, and ascorbic acid oxidase (Kader, 1986; Caleb et al., 2013).

Decreasing respiration rate via modified atmosphere and lowering the temperature delays enzymatic degradation of complex substrates which reduces the sensitivity to ethylene synthesis (Saltveit, 2003; Tijskens et al., 2003), thereby extending the shelf-life and avoiding senescence of the produce. De Santana et al. (2011) evaluated the effect of MAP on respiration rate and ethylene synthesis during 6-day storage at 1 and 25°C. They reported that ethylene production was proportional to respiration rate for peaches during ripening at 25°C. However, lower ethylene synthesis and respiration rate were obtained at lower temperature in MAP treatments. This principle is a critical component to the successful application of MAP.

As respiration rates lower, postharvest shelf-life is increased. Using reduced O_2 and high CO_2 levels has been proved to effectively control enzymatic browning, firmness and decay of fresh-cut fruits, and vegetables. Besides, growth of aerobic spoilage microorganisms can be substantially delayed with reduced O_2 levels (Rojas-Graü et al., 2009). High CO_2 concentrations are also generally effective in controlling the growth of most aerobic microorganisms, specifically Gram-negative bacteria and molds, but fail to inhibit most yeasts (Al-Ati & Hotchkiss, 2002). However, removing O_2 from the package completely or keeping at very minimum levels is also not proper, since to prevent the growth of some anaerobic psychotropic pathogens, sufficient O_2 concentration in the package is required. Determining the optimum proportions of MAP gases itself is not enough for designing the best system. Selection of the packaging film has a dramatic effect on MAP because each polymeric film has unique O_2 and CO_2 permeabilities.

The exposure of fresh horticultural crops to low oxygen and/or elevated carbon dioxide atmospheres within the range tolerated by each commodity reduces their respiration and ethylene production rates; however, outside this range respiration and ethylene production rates can be stimulated indicating a stress response (Geeson, 1988). This stress can contribute to incidence of physiological disorders and increased susceptibility to decay. Elevated CO_2-induced stresses are additive to, and sometimes synergistic with, stresses caused by low O_2; physical or chemical injuries; and exposure to temperatures, RH, and/or C_2H_4 concentrations outside the optimum range for the commodity. Elevated CO_2 atmospheres inhibit activity of ACC synthase (key regulatory site of ethylene biosynthesis), while ACC oxidase activity is stimulated at low CO_2 and inhibited at high CO_2 concentrations and/or low O_2 levels (Kader, 2001). Ethylene action is inhibited by elevated

CO_2 atmospheres. Optimum atmospheric compositions retard chlorophyll loss (green color), biosynthesis of carotenoids (yellow and orange colors) and anthocyanins (red and blue colors), and biosynthesis and oxidation of phenolic compounds (brown color). Controlled atmospheres slow down the activity of cell wall degrading enzymes involved in softening and lignification leading to toughening of vegetables (Kader & Salveit, 2003). Low O_2 and/or high CO_2 atmospheres influence flavor by reducing loss of acidity, starch to sugar conversion, sugar interconversions, and biosynthesis of flavor volatiles. Storage of the produce at optimum atmosphere results in the retention of ascorbic acid and other vitamins leading to better nutritional quality. Severe stress CA conditions decrease cytoplasmic pH and ATP levels, and reduce pyruvate dehydrogenase activity while pyruvate decarboxylase, alcohol dehydrogenase, and lactate dehydrogenase are induced or activated. This causes accumulation of acetaldehyde, ethanol, ethyl acetate, and/or lactate, which may be detrimental to the commodities if they are exposed to stress CA conditions beyond their tolerance. Specific responses to CA depend upon cultivar, maturity and ripeness stage, storage temperature, and duration, and in some cases, ethylene concentrations.

The use of polymeric films in MAP serves as mechanical barrier to the movement of water vapor and this helps to maintain a high level of RH within the package, and reduce produce weight loss. However, an excessively high level of RH within the package can result in moisture condensation on produce, thereby creating a favorable condition for the growth of pathogenic and spoilage microorganisms (Zagory & Kader, 1988; Aharoni et al., 2008; Távora et al., 2004). Lowering the temperature and applying other technologies decrease the rate of physiological process which has a beneficial effect on preservation of fresh produce.

The key to successful MAP of fresh produce is to use a packaging film of correct intermediary permeability where a desirable equilibrium modified atmosphere (EMA) is established when the rate of oxygen and carbon dioxide transmission through the pack equals the produce respiration rate. Typically, optimum EMAs of 3–10% O_2 and 3–10% CO_2 can dramatically increase the shelf-life of fruits and vegetables. The EMA thus attained is influenced by numerous factors, such as the respiration rate, temperature, packaging film, pack volume, fill weight, and light. The respiration rate is affected by the variety, size, maturity, and intensity of produce preparation. Consequently, determining the optimum EMA of a particular item of produce is a complex problem that can only be solved through practical experimental tests.

2.3   TECHNOLOGY OF MODIFIED ATMOSPHERE PACKAGING

Food manufacturers are always looking for ways to extend shelf-life without altering the physical or chemical properties of foodstuffs or adding any unnatural ingredients. MAP is the ideal way to achieve this. It is a natural method that is rapidly growing in popularity worldwide. In many cases, it can also complement alternative preservation methods. The shelf-life or freshness of fruits and vegetables after harvest is important for ensuring a safe and nutritional diet at an affordable cost. Low-temperature storage is one of the major means to preserve the freshness of fresh produce. Membrane technology can be used to regulate the gaseous atmosphere surrounding produce during various phases of its distribution and is currently practiced in some instances. The storage of post-harvest produce under reduced O_2 and elevated CO_2 concentrations can provide an additional means of reducing the metabolic activity and increasing shelf-life.

There are different storage and packaging techniques that are based on the change in composition of the storage atmosphere. Conventional gas separation membrane modules can be used to create this gaseous atmosphere at a fixed storage site or on-board the transportation system used for shipping. The MAP is a passive means of creating an altered gas composition around the produce in packages during shipping and marketing, typically in small plastic packages. MAP is an active or passive dynamic process of altering gaseous composition with a package. This technology relies on the interaction between the respiration rate of the produce, and the transfer of gases through the packaging material, with no further control exerted over the initial gas composition (Farber et al., 2003).

MAPs slow down the biochemical changes inhibiting the processes of oxidation and the growth of microbes by decreasing the oxygen concentration and filling the packages with gases, such as N_2 and CO_2. A comparison of respiration rates as a measure of CO_2 in air and 3% O_2 is shown in Table 2.1. Controlled atmosphere storage (CAS) refers to exposing produce continuously to some desired, constant gas composition, typically during bulk storage, or shipping. CAS aims to control the optimum gas composition in a storage room within the specified tolerances. In CAS, a gas generator is usually used to create and control the modified atmosphere in a cold warehouse where the product is kept (Yam & Lee, 1995). However, in modified atmosphere storage initial gas composition is modified in an airtight storage room and the atmosphere changes with time due to respiratory activity and the growth of microorganisms. CAS is capital intensive and difficult to operate; thus, it is more appropriate for foods that are amenable to long-term

TABLE 2.1 Classification of Selected Fruit and Vegetables According to Their Respiration Rate and Degree of Perishability in Air and 3% O_2.

Commodity	In air			In 3% O_2			Relative respiration rate at 19°C
	0°C	10°C	20°C	0°C	10°C	20°C	
Onion (bedfordshire champion)	2	4	5	1	2	2	
Cabbage (decema)	2	4	11	1	3	6	
Beetroot (storing)	2	6	11	1	3	6	
Celery (white)	4	6	19	3	4	6	Low 3
Cucumber	8	7	8	3	4	6	<10
Tomato (Eurocross BB)	3	8	17	2	3	7	
Lettuce (kordaat)	5	9	21	4	6	14	
Peppers (green)	4	11	20	5	9	–	
Carrots (whole, peeled)	–	12	26	–	–	–	
Parsnip (Hollow Crown)	4	14	23	3	6	17	
Potatoes (whole, peeled)	–	14	33	–	–	–	Medium
Mango	–	15	61	–	–	–	10–20
Cabbage (primo)	6	16	23	4	8	17	
Lettuce (kloek)	8	17	42	8	13	25	
Cauliflower (April Glory)	10	24	71	7	24	34	
Brussels sprouts	9	27	51	7	19	40	
Strawberries (Cambridge Favorite)	8	28	72	6	24	49	High
Blackberries (Bedford Giant)	11	33	88	8	27	71	20–40
Asparagus	14	34	72	13	24	42	
Spinach (Prickly True)	25	43	85	26	46	77	

TABLE 2.1 *(Continued)*

Commodity	In air			In 3% O_2			Relative respiration rate at 19°C
	0°C	10°C	20°C	0°C	10°C	20°C	
Watercress	9	43	117	5	38	95	
Broad beans	18	46	82	20	29	45	Very high
Sweetcorn	16	48	119	14	32	68	40–60
Raspberries (Malling Jewel)	12	49	113	11	30	73	
Carrots (julienne-cut)	–	65	145	–	–	–	
Mushrooms (sliced)	–	67	191	–	–	–	Extremely high
Peas in pod (Kelvedon Wonder)	20	69	144	15	45	90	> 60
Broccoli (sprouting)a	39	91	240	33	61	121	

mg CO_2 converted to mL CO_2 using densities of CO_2 at 0°C = 1.98, 10°C = 1.87, 20°C = 1.77.
[a]Unless stated, produce is whole and unprepared.

changes, such as apple, kiwifruits, pears, and meat (Robertson, 2012). In MAP, the product is kept in a carefully designed permeable polymer package, and the modified atmosphere is created and maintained through the respiration of the product and the gas permeation of the package. MAP is a more affordable technology since a gas generator is not needed (Yam & Lee, 1995); however, it is also a more difficult technology to implement since the permeability of the packages to the gases should be considered for the best design. The combination of a high flux membrane (selective permeation of gases) patch with perforations (or holes that provide non-selective permeation of gases) in the package film offers a versatile route to create whatever O_2 and CO_2 environment may be needed for a given product. Detailed mathematical models have been developed for describing and designing modified atmosphere systems (Paul & Clarke, 2002).

2.4 MAP GASES

Food grade gases are defined as gases used as a processing aid and/or an additive in order to ensure that standards are complied with safety and legislation. Food-grade industrial gases are an effective and natural way of meeting rising consumer demands for quality, variety and freshness in the food, and beverages industry. Increasingly, consumers are looking for low or zero-additive alternatives to conventional preservation techniques. In particular, gases are proving to be indispensable in the growing market for convenience, home-inspired foods.

The gases predominantly used in MAP storage are CO_2, nitrogen (N_2), and O_2. The gas properties and the interaction of gases with the food ingredients, for example, solubility in the foodstuff, should be taken into account when choosing the gas or gas composition. Food grade gases CO_2, N_2, O_2, and other gases authorized for foodstuffs premixed, either as individual gases in cylinders under high pressure or as liquids in insulated tanks for subsequent mixing in the packaging machine. Food grade gases conform to "food grade" regulations, for example, the EC directive 96/77/EC on food additives within the EU countries and the FDA guidelines in the United States. N_2 and O_2 are separated from the atmospheric air. CO_2 is taken from natural wells or as a byproduct, for instance, fermentation processes (wine, beer), or ammonia production. Sometimes it may be more effective and practical to produce nitrogen on site using pressure swing adsorption (PSA) or a permeable membrane plant. If a PSA/membrane system is used, a back up gas supply system is recommended.

These gases are used either alone or in mixtures. By reducing the O_2-level and increasing the CO_2-level, ripening of fruits and vegetables can be delayed, respiration and ethylene production rates can be reduced, softening can be retarded, and various compositional changes associated with ripening can be slowed down. Oxygen is essential when packaging fresh fruits and vegetables as they continue to respire after harvesting. The absence of O_2 can lead to anaerobic respiration in the package which accelerates senescence and spoilage. Too high levels of O_2 do not retard respiration significantly and it is around 12% of O_2 where the respiration rate starts to decrease. So oxygen is used in low levels (3–5%) for positive effect. When packaging meat and fish, the high CO_2-levels are effective bacterial and fungal growth inhibitors. In the case of vegetables and fruits, CO_2 is not a major factor since CO_2-levels above 10% are needed to suppress fungal growth significantly. Unfortunately, higher levels than 10% of CO_2 are working phytotoxic for fresh produce. N_2 is used as filler gas since it neither encourages nor discourages bacterial growth. Used singly or in combination, these gases are commonly used to balance safe shelf-life extension with optimal organoleptic properties of the food. Noble or "inert" gases, such as argon are commercially used for products such as coffee and snack; however, the literature on their application and benefits is limited. Experimental use of carbon monoxide (CO) and sulfur dioxide (SO_2) has also been reported. Gas mixtures are a wide and diverse group of products developed for use in specific industries. Food gas mixtures are used to delay the deterioration of packaged food by substituting the air in the package with a protective gas mixture. Gas mixtures are supplied as food grade. CO_2, N_2, and O_2 along with other gases are authorized for foodstuffs as individual gases in cylinders under high pressure as well as liquids in insulated tanks for subsequent mixing at the packaging machine. Gas mixtures are either mixed continuously on site from pure gases or they can be supplied as premixed products in a range of package sizes. Table 2.2 shows some recommended gas mixtures for extending the shelf-life of variety of products.

2.5 ACTIVE, PASSIVE, AND SMART MAP

Modified atmospheres can passively evolve within a hermetically sealed package as a consequence of a commodity's respiration, that is, O_2 consumption and CO_2 evolution. If a commodity's respiration characteristics are properly matched to film permeability values, then a beneficial modified atmosphere can be passively created within a package. If a film of correct

TABLE 2.2 Recommended Gaseous Mixtures for MA Packaging.

Product	O_2 (%)	CO_2 (%)	N_2 (%)
Fruits			
Apple	1–2	1–3	95–98
Apricot	2–3	2–3	94–96
Avocado	2–5	3–10	85–95
Banana	2–5	2–5	90–96
Grape	2–5	1–3	92–97
Grapefruit	3–10	5–10	80–92
Kiwifruit	1–2	3–5	93–96
Lemon	5–10	0–10	80–95
Mango	3–7	5–8	85–92
Orange	5–10	0–5	85–95
Papaya	2–5	5–8	87–93
Peach	1–2	3–5	93–96
Pear	2–3	0–1	96–98
Pineapple	2–5	5–10	85–93
Strawberry	5–10	15–20	70–80
Vegetables			
Artichoke	2–3	2–3	94–96
Beans, snap	2–3	5–10	87–93
Broccoli	1–2	5–10	88–94
Brussels sprouts	1–2	5–7	91–94
Cabbage	2–3	3–6	81–95
Carrot	5	3–4	91–95
Cauliflower	2–5	2–5	90–96
Chili peppers	3	5	92
Corn, sweet	2–4	10–20	76–88
Cucumber	3–5	0	95–97
Lettuce (leaf)	1–3	0	97–99
Mushrooms	3–21	5–15	65–92
Spinach air	Air	10–20	–
Tomatoes	3–5	0	95–97
Onion	1–2	0	98–99

Source: Sandhya (2010). Reprinted from Sandhya. Modified Atmosphere Packaging of Fresh Produce: Current Status and Future Needs. *LWT – Food Sci. Technol.* **2010,** *43* (3), 381–392. © 2009 Elsevier Ltd. Used with permission.

intermediary permeability is chosen, then a desirable equilibrium modified atmosphere is established when the rates of O_2 and CO_2 transmission through the package equals a product's respiration rate.

Packaging material of the correct permeability must be chosen for the successful MAP of fresh fruits and vegetables. If the products are sealed in an insufficiently permeable film, undesirable anaerobic conditions (< 1% O_2 and > 20% CO_2) will develop with subsequent deterioration in quality. Conversely, if fruits and vegetables are sealed in a film of excessive permeability, little or no modified atmosphere will result and moisture loss will also lead to accelerated deterioration in quality. Examples of materials that can be used for MAP of fresh produce (fruits and vegetables) are microporous film or LDPE/OPP.

By pulling a slight vacuum and replacing the package atmosphere with a desired mixture of CO_2, O_2, and N_2, a beneficial equilibrium atmosphere may be established more quickly than a passively generated equilibrium atmosphere. This is known as active packaging method. Another active packaging technique is the use of CO_2, O_2, or ethylene scavengers/emitters. Such scavengers/emitters are capable of establishing a rapid equilibrium atmosphere within hermetically sealed produce packages. The passive MAP can be generated inside a package by relying on the natural processes of produce respiration and film permeability while active MAP is a rapid process of gas replacement or displacement or the use of gas scavengers, or absorbers to establish a desired gas mixture within a package (Caleb et al., 2013; Charles et al., 2003).

The introduction of "indicators" and "sensors" in packages, which enables the end user to trace the current state and past conditions of a product, is regarded as smart packaging. The recorded information provides everyone involved in the farm to fork chain with necessary information to make an informed decision about the product quality. Under this type of packaging, different types of time-temperature indicators (TTIs), gas sensors, microbial growth biosensors, breakage indicators, ripeness indicators, and radio frequency identification (RFID) tags can be included.

2.6   MICROBIOLOGICAL SAFETY OF MAP SYSTEMS

Although much information exists in the general area of MAP technology, research on the microbiological safety of these foods is still lacking. The great vulnerability of MAP foods from a safety standpoint is that with many modified atmospheres containing moderate to high levels of carbon dioxide, the aerobic spoilage organisms which usually warn consumers of spoilage are inhibited, while the growth of pathogens may be allowed or even stimulated. Microbial quality assurance for MA-packaged fresh and fresh-cut fruit and vegetables is invaluable. Considering the critical points for contamination from farm to fork, this includes postharvest handling, contaminated processing equipment or transportation vehicles, and cross-contamination.

In the past, the major concerns have been the anaerobic pathogens, especially the psychrotrophic, non-proteolytic Clostridia. However, because of the emergence of psychrotrophic pathogens such as *Listeria monocytogenes*, *Aeromonasky drophila*, and *Yersinia enterocolitica*, new safety issues have been raised. This stems mainly from the fact that the extended shelf-life of many MAP products may allow extra time for these pathogens to reach dangerously high levels in the food. Concerns have been expressed on the influence of MAP on the survival and growth of pathogenic microorganisms which may render fresh and fresh-cut produce unsafe while still being edible (Jay, 1992; Philips, 1996; Farber et al., 2003). Additionally, MAP may significantly inhibit spoilage organisms or eradicate desirable produce microflora due to the non-selective antimicrobial effect of CO_2 (Farber et al., 2003). As the interaction between natural microflora and food pathogens may play a significant role in product safety. Studies have shown that optimal gas composition, as well as the presence and competitive effect of background microflora have inhibitory effect on foodborne pathogens.

Antimicrobial packaging systems have been developed and well researched.

Antimicrobial packaging is a system that can kill or inhibit the growth of microorganisms and thus extend the shelf-life of perishable products and enhance the safety of packaged products. Antimicrobial packaging can kill or inhibit target microorganisms (Han, 2003). Among many applications, such as O_2-scavenging packaging and moisture-control packaging, antimicrobial packaging is one of the most promising innovations of active packaging technologies. It can be constructed by using antimicrobial packaging materials and/or antimicrobial agents inside the package space or inside foods. Most food packaging systems consist of food products, headspace atmosphere, and packaging materials. Anyone of these three components of food packaging systems could possess an antimicrobial element to increase antimicrobial efficiency. Various antimicrobial agents could be incorporated into conventional food packaging systems and materials to create new antimicrobial packaging systems.

Antimicrobial additives have been used successfully for many years. The direct incorporation of antimicrobial additives in packaging films is a convenient methodology by which antimicrobial activity can be achieved. Literature provides evidence that some of these additives may be effective as indirect food additives incorporated into food packaging materials. Several agents have been proposed and tested for antimicrobial packaging using this method. However, the use of such packaging materials is not meant to be a substitute for good sanitation practices, but it should enhance the safety of food as an additional hurdle for the growth of pathogenic microorganisms. MAP has been successfully used to maintain the quality of fresh and fresh-cut fruits and vegetables (Yahia, 2006; Mangaraj et al., 2009; Sandhya, 2010). However, the effect of MAP on microorganisms can vary depending on the type of produce packaged (Farber et al., 2003). For instance, increased CO_2 and decreased O_2 concentrations used in MAP generally favor the growth of lactic acid bacteria. This can accelerate the spoilage of produce sensitive to lactic acid bacteria such as carrots, chicory leaves, and lettuce (Nguyen-the & Carlin, 1994). Furthermore, oxygen concentrations below 1–2% can create a potential risk for the growth of pathogens such as *C. botulinum* (Charles et al., 2003; Farber et al., 2003). Therefore, it is necessary to highlight some foodborne pathogens that can be potential health risks due to the vulnerability of MA-packaged produce.

2.7   INTELLIGENT MA PACKAGING

Intelligent packaging can be defined as the packaging that contains an external or internal indicator to provide information about aspects of the history of the package and/or the quality of the food (Robertson, 2006). Intelligent packaging is an extension of the communication function of traditional packaging and communicates information to the consumer based on its ability to sense, detect, or record external or internal changes in the product's environment. This system is an emerging technology that uses the communication function of the package to facilitate decision making to achieve the benefits of enhanced food quality and safety. The intelligent MA packaging makes use of smart package devices. These devices are small, inexpensive labels and tags which can be attached on to primary packaging or more often onto secondary packaging to facilitate communication throughout the supply chain so that appropriate actions may be taken to ensure food quality and safety enhancement (Yam et al., 2005).

Intelligent packaging also makes use of barcodes, radio frequency identification tags (RFID), time, temperature indicators, gas indicators, and biosensors for real-time monitoring of the traceability, tracking, and record keeping of product flow throughout the production process, and supply chain. A major issue is that most active and intelligent packaging systems require that food be in direct contact with a sensor of some kind, and substances from the sensor may migrate into the food. Whether these migrations are intentional or unintentional, the substance, amount of the substance, and possible health effects of the substance must be determined in order for the substances to be allowed and regulated. In addition, the cost of active and intelligent packaging limits its commercial use. Most active or intelligent systems add cost to the package, so innovations in packaging must have a final beneficial outcome that outweighs the extra expenses of adding the technology. The attitude toward active and intelligent packaging is positive and there is still much potential for exciting innovation to come.

2.8   TESTING OF MAP

2.8.1   HEADSPACE ANALYSIS

The major testing performed for the MAP food is to analyze the gas headspace for the mixtures used. Headspace analysis is mainly done by gas analyzers which indicate the results for O_2 or CO_2. Headspace analysis ensures that the residual oxygen within the headspace gas does not exceed certain defined limits: excessive O_2 within the headspace can result in the growth and proliferation of microorganisms, such as bacteria or molds, which causes spoilage and results in a lower shelf-life of the product. Headspace analysis usually focuses primarily on oxygen measurements, but in many cases CO_2 measurement is also important.

A thin needle is inserted into the package and a pump draws a small sample of precise volume of headspace gas into the analyzer equipment. The headspace gas comes into contact with a sensor that can measure the concentration of residual O_2 or CO_2 in the headspace gas sample. The headspace analysis equipment—the headspace gas analyzer—provides a reading of the concentration of the gas that is being measured. In cases where products are coming off a line very rapidly, the speed at which headspace analysis can be carried out can be an important factor. In these situations it is much more efficient for an operator to be able to carry out a headspace gas analysis on a sample in, say, 5 s than in 30. Several companies are selling the gas analyzers in the market and Dansensor is one of the companies who still hold the major market share. The headspace or gas analyzers are accurate and easy to use, Dansensor offers CheckPoint, the portable CheckPoint II, or, for an even higher level of convenience and traceability, the advanced CheckMate three. More details are available on the company website.

2.8.2   SEAL INTEGRITY

The seal integrity is a vital aspect of quality control in MAP Sealing integrity is possibly the single most important element of quality control in MAP; if a package's seal integrity is compromised, a leak will cause loss of the protective atmosphere and the shelf-life of the food will be reduced, resulting in costly returns. Therefore, a package leak detector is a vital part of MAP quality control. In recognition of this, Dansensor has developed state-of-the-art package leak detection equipment for testing the integrity of the seal. Seal integrity can be compromised through a number of different routes. For example, particles of the product being packaged may come between the tray and the seal. Another potential source of a breach in seal integrity is that the heating element that creates the thermal seal may be defective or misaligned, or there may be issues with the pressure, heat and duration of the sealing process. All these can result in seal integrity being compromised. It is vital for quality control and quality assurance that leak detection equipment works as quickly and as accurately as possible. If seal integrity problems go undetected, issues are likely to arise with the product resulting in unwanted returns and potential damage to the relationship with a customer. Therefore, a reliable leak detector is vital to ensure seal integrity. Dansensor's leak detection systems offer both online and offline leak detection of MAPs and whole shipping crates for seal integrity assurance. Furthermore, Dansensor's innovative leak detection equipment allows package leak detection non-destructively.

2.8.3   PACKAGE BARRIER PROPERTIES

The barrier properties of packaging materials have significant influence on the shelf-life, safety, and quality of MAP food products. Gas, light, and water vapor barrier properties of packaging materials are main considerations while designing a packaging material for a specific end use.

Water vapor barrier: Food products tend to dry out or gain moisture if they are not packed appropriately. In fresh produce due to poor barrier properties, condensation may occur leading to increased microbial spoilage. Therefore, fresh produce requires packaging materials with certain amount of mass transfer to keep an optimal level of oxygen required by the produce to respire and to prevent excessive condensation within the package. Water vapor transmission rate (WVTR), a standard practice to measure water permeability is the packageability to allow water vapor to pass through it.

Oxygen barrier: Oxygen may be kept away by appropriate packaging to avoid many undesirable changes in foods. Oxidation causes discoloration of fresh and processed meat products, off-flavor development or rancidity in products rich in oil, mold growth in cheese and bread, and accelerates spoilage in several food products. However, the oxygen is used for respiration and excess levels of oxygen lead to increased respiration rates. Therefore, oxygen sensitive food products require packaging with adequate oxygen barrier properties. Oxygen transmission rate (OTR) is measured as milliliter of oxygen passed through meter square in 24 h at 1 atmospheric pressure. Packaging materials with values of 10–100, 1–10, and <1 are considered as good, very good, and extremely good, respectively. Oxygen may also find its way inside the package through inappropriate seals, folds, or through a damaged package. Factors like temperature and humidity influence the oxygen barrier properties and must be considered during the selection of a specific packaging material.

2.9 SUMMARY

The MAP technology plays a major role in the preservation of the quality of a wide range of fresh produce. MAP technology provides great opportunities as well as challenges to scientists, engineers and food manufacturers. Markets will continue to grow for high quality, fresh or near-fresh, and convenient foods. As society continues to advance, the expectations of the consumer will continue to advance. The use of active and intelligent modified packaging will likely become more popular as more technologies make their way to the market, innovative packaging in active and intelligent systems will become more common place. The growth area with respect to MAP products is in minimally processed produce, especially vegetables and fruits. The major challenge is how to produce the products so that we can ensure quality, safety, and reliability at competitive prices. Strict application of quality assurance programs is crucial to the success of MAP operations. MAP products are not sterilized; therefore, their quality and safety during the extended storage life are dependent on how strictly the operations adhere to quality assurance programs.

MAP still has not reached its full potential in research and development and commercialization. Limited information is available on the effects of MAP on various spoilage and pathogenic microorganisms. Also, the interactions of various microorganisms present on MAP products and how they affect product shelf-life and safety is limited. Hurdle technology is important for MAP applications, since the modified atmosphere provides an unnatural gas environment that can create serious microbial problems such as the growth of anaerobic bacteria and the production of microbial toxins. Therefore, an included temperature control system is very important for quality preservation and microbial control.

KEYWORDS

- energy
- carbon dioxide
- respiration
- fruits
- vegetables
- quality

REFERENCES

Al-Ati, T.; Hotchkiss, J. H. Applications of Packaging and Modified Atmosphere to Fresh-cut Fruits. In *Fresh-Cut Fruits and Vegetables: Science, Technology, and Market;* Lami-kanra, O., Ed.; CRC Press: Boca Raton, FL, 2002; pp 305–338.

Aharoni, N.; Rodov, V.; Fallik, E.; Porat, R.; Pesis, E.; Lurie, S. In *Controlling Humidity Improves Efficacy of Modified Atmosphere Packaging of Fruit and Vegetables,* Proceedings of EURASIA Symposium on Quality Management in Postharvest Systems, Acta Horticulturae, Kanlayanarat. S, Wangs-Aree, C., Eds.; 2008; Vol. 804, pp 189–196.

Caleb, O. J.; Mahajan, P. V.; Al-Said, F. A.; Opara, U. L. Modified Atmosphere Packaging Technology of Fresh and Fresh-cut Produce and the Microbial Consequences—A Review. *Food Bioprocess Tech.* **2013,** *6,* 303–329.

Charles, F.; Sanchez, J.; Gontard, N. Active Modified Atmosphere Packaging of Fresh Fruits and Vegetables: Modeling with Tomatoes and Oxygen Absorber. *J. Food Sci.* **2003,** *68* (5), 1736–1742.

Day, B. P. F. Fruit and Vegetables. In *Principles and Applications of MAP of Foods;* Parry, R. T., Ed.; Blackie Academic and Professional: New York, NY 1993; pp 114–133.

Day, B. P. F. *Fresh Prepared Produce: GMP for High Oxygen MAP and Nonsulphite Dipping*, Guideline No 31, CCFRA: Chipping Campden, Gloucestershire, UK, 2001.

De Santana, L. R. R.; Benedetti, B. C.; Sigrist, J. M. M.; Sato, H. H. Effect of Modified Atmosphere Packaging on Ripening of 'Douradão' Peach Related to Pectolytic Enzymes Activities and Chilling Injury Symptoms. *Rev. Bras. Fruticultura*. **2011**, *33* (4), 1084–1094.

Farber, J. N.; Harris, L. J.; Parish, M. E.; Beuchat, L. R.; Suslow, T. V.; Gorney, J. R.; Garrett, E. H.; Busta, F. F. Microbiological Safety of Controlled and Modified Atmosphere Packaging of Fresh and Fresh-cut Produce. *Compr. Rev. Food Sci. Food Saf.* **2003**, *2,* 142–160.

Fonseca, S. C.; Oliveira, F. A. R.; Brecht, J. F. Modelling Respiration Rate of Fresh Fruits and Vegetables for Modified Atmosphere Packages: A Review. *J. Food Eng.* **2002**, *52,* 99–119.

Geeson, J. D. Modified Atmosphere Packaging of Fruits and Vegetables. In *International Symposium on Post Harvest Handling of Fruits and Vegetables*, Leuven, Belgium, 1988; pp 143–147.

Han J. H. Antimicrobial Food Packaging. In *Novel Food Packaging Technique;* Ahvenainen, R., Ed.; Woodhead Publishing: Cambridge, 2003; pp 50–70.

Jay, J. M. Microbiological Food Safety. *Crit. Rev. Food Sci. Nutr.* **1992**, *31,* 177–190.

Kader, A. A. Biochemical and Physiological Basis for Effects of Controlled and Modified Atmospheres on Fruit and Vegetables. *Food Technol.* **1986**, *40,* 99–104.

Kader, A. A.; Zagory, D.; Kerbel, E. L. Modified Atmosphere Packaging of Fruits and Vegetables. *Crit. Rev. Food Sci. Nutr.* **1989**, *28,* 1–30.

Kader, A. A. Physiology of CA Treated Produce. In *VIII International Controlled Atmosphere Research Conference,* 2001; Vol. 600, pp 349–354.

Kader, A. A.; Saltveit, M. E. Atmosphere Modification. In *Postharvest Physiology and Pathology of Vegetables;* Marcel Dekker: New York, 2003; pp 229–246.

Mangaraj, S.; Goswami, T. K.; Mahajan, P. V. Application of Plastic Films for Modified Atmosphere Packaging of Fruits and Vegetables: A Review. *Food Eng. Rev.* **2009**, *1,* 133–158.

Nguyen-the, C.; Carlin, F. The Microbiology of Minimally Processed Fresh Fruit and Vegetables. *Crit. Rev. Food Sci. Nutr.* **1994**, *34* (4), 371–401.

Paul, D. R.; Clarke, R. Modeling of Modified Atmosphere Packaging Based on Designs with a Membrane and Perforations. *J. Memb. Sci.* **2002**, *208,* 269–283.

Philips, C. A. Review: Modified Atmosphere Packaging and Its Effects on the Microbiological Quality and Safety of Produce. *Int. J. Food Sci. Technol.* **1996**, *31,* 463–479.

Robertson, G. Modified Atmosphere Packaging. In *Food Packaging: Principles and Practice;* 3rd Ed.; CRC Press: Boca Raton, FL, 2012; pp 313–340.

Robertson, G. L. Active and Intelligent Packaging. In *Food Packaging: Principles and Practice;* 2nd Ed; CRC Press: Boca Raton, Florida, 2006.

Rojas-Graü, M. A.; Oms-Oliu, G.; Soliva-Fortuny, R.; Martín-Belloso, O. The Use of Packaging Techniques to Maintain Freshness in Fresh-cut Fruits and Vegetables: A Review. *Int. J. Food Sci. Technol.* **2009**, *44* (5), 875–889.

Saltveit, M. E. Is It Possible to Find an Optimal Controlled Atmosphere? *Postharvest Biol. Technol.* **2003**, *27,* 3–13.

Sandhya. Modified Atmosphere Packaging of Fresh Produce: Current Status and Future Needs. *LWT-Food Sci. Technol.* **2010**, *43,* 381–392.

Távora, L. N.; Raghavan, G. S. V.; Orsat, V. Storage of Cranberries in Plastic Packaging. *J. Food Technol.* **2004**, *2* (1), 28–34.

Tijskens, L. M. M.; Konopacki, P.; Simcic, M. Biological Variance, Burden or Benefit? *Postharvest Biol. Technol.* **2003**, *27,* 15–25.

Yam, K. L.; Lee, D. S. Design of Modified Atmosphere Packaging for Fresh Produce. In *Active Food Packaging;* Rooney, M. L., Ed.; Chapman & Hall: London, 1995.

Yam, K. L.; Takhistov, P. T.; Miltz, J. Intelligent Packaging: Concepts and Applications. *J. Food Sci.* **2005,** *70,* 1–10.

Yahia, E. M. Modified and Controlled Atmospheres for Tropical Fruits. *Stewart Postharvest Rev.* **2006,** *5* (6), 1–10.

Zagory, D.; Kader, A. A. 1988 Modified Atmosphere Packaging of Fresh Produce. *Food Technol.* **1988,** *42* (9), 70–74, 76–77.

CHAPTER 3

ACTIVE PACKAGING OF FRESH AND FRESH-CUT FRUIT AND VEGETABLES

ASTRID F. PANT* and J. THIELMANN

Department of Material Development, Fraunhofer Institute for Process Engineering and Packaging (IVV), Freising, Germany

Corresponding author. E-mail: astrid.pant@ivv.fraunhofer.de

CONTENTS

ABSTRACT

Extensive research has been put into novel techniques, such as "modified atmosphere packaging" (MAP) and "active packaging" in order to fulfill the specific needs of metabolically active products. These efforts were dedicated to developing new food packaging systems on the one hand, and to prove their effectiveness in maintaining product quality on the other hand. This chapter presents an overview of active packaging systems for fresh fruit and vegetables, with a special focus on their effect on product quality and shelf-life. Scientific studies on the storage of fresh produce in active packages are collected and reviewed.

3.1   INTRODUCTION

Fresh and fresh-cut fruit and vegetables are sensitive products with a limited shelf-life. This is mainly due to their susceptibility to microbial spoilage and to their active post-harvest metabolism. Therefore, fresh and fresh-cut fruit and vegetables pose unique challenges to packaging technology. During the last decades, packaging technology for fresh produce has evolved considerably. Extensive research has been put into novel techniques, such as "modified atmosphere packaging" (MAP) and "active packaging" in order to fulfill the specific needs of metabolically active products. These efforts were dedicated to developing new food packaging systems on the one hand, and to prove their effectiveness in maintaining product quality on the other hand. This chapter presents an overview of active packaging systems for fresh fruit and vegetables, with a special focus on their effect on product quality and shelf-life. Scientific studies on the storage of fresh produce in active packages are collected and reviewed.

3.2   PACKAGING REQUIREMENTS OF FRESH FRUIT AND VEGETABLES

Packaging is an integral part of the food supply chain, especially in view of a globalized market which includes long-distance distribution systems. Traditionally, food packaging fulfills various functions (Ahvenainen, 2003; Robertson, 2012):

- protecting the packed food from environmental influences such as light, oxygen, moisture, dust, microorganisms, and mechanical stress;

- facilitating food handling and identification;
- providing product information;
- providing convenience for the customer.

Furthermore, within the recent global "Save Food" debate, protecting food quality and minimizing food losses have been highlighted as important tasks for food packaging (Gustavsson et al., 2011). The term "fruit and vegetables" refers to a large and inhomogeneous group of food products that originate from different parts of plants, for example, stems, leaves, and fruits. Therefore, the requirements for storage (temperature and relative humidity) and packaging (gas permeability) differ considerably, depending on the respective commodity (Kader et al., 1989). The common feature of all fresh fruit and vegetables is their active post-harvest metabolism resulting in respiration, ripening, and senescence processes. Each fruit or vegetable is characterized by an optimum ripening stage with optimum levels of quality attributes such as firmness, flavor, color, and nutritional value. Hence, packaging solutions should be designed to maintain this optimum stage throughout distribution and commercialization.

The shelf-life of fresh fruit and vegetables is inherently limited by their post-harvest metabolism: respiration, ripening, and senescence can only be decelerated but never be stopped. Product respiration is influenced by product-specific parameters such as tissue type and species on the one hand and extrinsic conditions such as temperature, gas atmosphere, relative humidity, and mechanical stress on the other hand. Fresh-cut products show considerably higher respiration rates due to stress from mechanical processing (Kader & Saltveit, 2003b). While most of these parameters can be seen as pre-determined from a packer's point of view, the potential of packaging design lies in creating a gas atmosphere that fits the requirements of the respective product. The effect of storage gas composition on the quality of fresh produce has been described thoroughly in literature; recommendations for optimum oxygen (O_2) and carbon dioxide (CO_2) concentrations are available for most fruit or vegetables (Gorny, 2003; Kader & Saltveit, 2003a; Gross et al., 2004). Depending on the respective commodity, O_2 levels of 0.5–5% have to be present during storage in order to maintain aerobic respiration. Lower O_2 concentrations result in anaerobic metabolism, which, in turn, leads to the development of off-flavors and to rapid product deterioration. However, compared to air (20.95% v/v O_2), a reduced O_2 concentration can have a positive effect on the shelf-life of fresh produce by slowing down product ripening (Kader et al., 1989). Elevated concentrations of CO_2 are generally considered beneficial in terms of a decreased respiration rate, even though fruit and vegetables differ considerably in their CO_2 tolerance levels. While, for example, CO_2 levels of 2% must not be exceeded for tomatoes and grapes, melons and berries tolerate CO_2 levels of up to 15% (Kader et al., 1989; Cameron et al., 1995). Antimicrobial effects related to elevated CO_2 concentrations, preferably above 20% (McMillin, 2008), can therefore not be established in packages for fruit and vegetables without adverse effects on product quality.

The storage of fresh produce in altered gas atmospheres finds broad application in commercial warehouse storage, but has also been introduced into packaging technology. MAP relies on establishing a headspace gas atmosphere different from air. For non-respiring food products, the package headspace is flushed with the desired gas mixture in order to create beneficial storage conditions for the respective product. In sealed plastic packages with fresh produce, the in-package atmosphere is continuously modified by the product itself, due to its natural respiration. This implies a decreased O_2 concentration and an increased CO_2 concentration compared to air. However, to avoid eventually arising anoxic conditions in the package, the permeability of the packaging films has to be matched carefully with the respiration rate of the packed product. "Equilibrium MAP" is an engineering design that aims at creating an optimum equilibrium gas atmosphere inside the package. The equilibrium gas atmosphere is a function of gas transfer through the packaging material and respiration rate, which are both temperature dependent. Since this is a very complex interplay of packaging and product characteristics, simulation tools have been developed to facilitate the development of tailored MAP for fresh and fresh-cut fruit and vegetables, for example, PredOxyPack®, Tailorpack®, and PackInMAP® (Mahajan et al., 2007; Cagnon et al., 2013; Vermeulen et al., 2013). While these software tools resulted from research projects, industry has also recognized the need for tailored packaging solutions and provides application-oriented systems such as PerfoTec® (PerfoTec B.V., The Netherlands) and StarMAP® (Laser Micro Rofin-BaaselLasertech, Germany). These systems focus on microperforation of packaging films in order to ensure appropriate gas permeability. Beside these software tools calculating the needed number and size of perforations, these companies also provide the laser equipment for inline perforation of the films.

Another important factor for product shelf-life is the microbial load on fruit and vegetables which is unequally dependent on a wide variety of factors (Olaimat & Holley, 2012). Type, morphology, and metabolic properties of the plant product are intrinsic factors determining the microbial pattern which is further affected by external conditions as geographical and climate situation, agronomic practice, point of harvest, transport, and handling. The number and variety of microorganisms found on fruit and vegetables are, therefore, highly variable (Burnett & Beuchat, 2000; Ponce et al., 2002; Tournas, 2005). Although this natural microbiota is usually non-pathogenic to humans and may be ingested without concerns, it highly contributes to a rapid decay of fresh produce especially when inadequate storage conditions are applied (Ahvenainen, 1996). But contaminations, especially from soil or manure, can also lead to high loads of pathogenic bacteria and fungi (Ahvenainen, 1996; Beuchat, 2002). Therefore, decontamination steps prior to packing contribute to ensure shelf-life and food safety (Ramos et al., 2013). It has to be outlined that contamination routes along the supply chain remain incompletely elucidated for fresh produce (Berger et al., 2010). Regarding the packing of fresh produce, it has to be ensured that the packaging material itself does not contaminate the product and sufficiently protects it from external microbial cross contaminations. In addition, the application of active packaging technologies, such as antimicrobial packaging and MAP, aims to delay the spread of microorganisms and to prolong microbial shelf-life and safety of the products.

3.3   ACTIVE PACKAGING TECHNOLOGY

The term "active packaging" was first introduced by Labuza and Breene (1989) more than two decades ago and defined as "packaging which performs some desired function other than merely providing a barrier to the external environment" (Labuza & Breene, 1989; Rooney, 1995). This rather broad definition includes technologies such as microwave susceptors and self-heating or -cooling functions (Rooney, 2005). According to EU legislation, active packaging refers to materials that "are intended to extend the shelf-life or to maintain or improve the condition of packaged food; they are designed to deliberately incorporate components that would release or absorb substances into or from the packaged food or the environment surrounding the food" (European Commision, 2009).

Release and absorber systems can be further divided based on the respective substances that are being released (emitted) or absorbed (scavenged). The most important release systems include antimicrobial systems and gas release systems for CO_2 as well as ethylene. Absorber systems are used to remove O_2, CO_2, ethylene, volatiles, water vapor, and liquid water from the package. Ahvenainen (2003) provided a comprehensive overview of active systems for a wide range of food products.

In contrast to traditional passive packaging materials, active packaging systems can actively contribute to the retention of product quality due to their releasing or absorbing properties. Thereby, packaging can be designed to target specific food deterioration mechanisms. Usually, this will be the product property that is most critical for product shelf-life (Rooney, 2005). Table 3.1 gives an overview of typical factors for quality loss of fresh and fresh-cut fruit and vegetables and how they can be addressed by packaging technology. The main characteristics of fresh produce are their post-harvest metabolic activity and their susceptibility to microbial spoilage. Thus, active packaging for fresh produce has to deal with both microbial growth and respiration products, that is, CO_2 and water. In addition, physiological mechanisms such as ethylene susceptibility and sprouting can be addressed by active packaging.

TABLE 3.1 Quality Concerns of Packed Fruit and Vegetables and Related Packaging Functions.

Quality concern	Packaging function
Microbial spoilage	Antimicrobial release
	Atmosphere modification (CO_2 release)
Condensation	Humidity buffering
Premature ripening	Ethylene scavenging
	Atmosphere modification (CO_2 release)
Sprouting	Ethylene release

Active packaging should not be confused with intelligent packaging, which refers to systems that are able to monitor the conditions of the packed food, providing information about food quality (Ahvenainen, 2003). Intelligent systems include indicators (such as time–temperature indicators, gas indicators, and freshness indicators) and data carriers (such as barcode labels and radio frequency identification (RFID) tags) (Lee et al., 2008). A detailed description of intelligent systems and their applications has been presented by Han et al. (2005) and more recently by Realini and Marcos (2014).

Regarding the application of active packaging solutions on fresh fruit and vegetables, a large number of ideas have been presented in scientific literature. The following sections aim to review the state of the art of active packaging technologies for fresh produce. Focusing on application studies, the practicability and effectiveness of these technologies will be assessed.

3.3.1 OXYGEN SCAVENGERS

O_2 is well known to have a detrimental effect on the quality of most food products, for example, sausages, nuts, fruit juices, beer, etc. O_2-induced deterioration reactions include oxidative rancidity, discoloration, and nutrient degradation. Moreover, O_2 supports the growth of aerobic spoilage microorganisms (Souza et al., 2012). Therefore, research and development activities have focused on the lock-out and removal of O_2 from food packages. As a result, O_2 scavenging systems are the most developed among active packaging technologies and are available on the market in different forms (Brody et al., 2001). The systems can be divided into insert types (sachets, labels) and polymer structure types (monolayer and multilayer materials) (Lee et al., 2008). For a recent overview of O_2 scavenging systems, see Realini and Marcos (2014).

Fresh and fresh-cut fruit and vegetables are a field with limited application potential for O_2 scavengers since the complete depletion of O_2 in the package headspace is not desired. Depending on the respective commodity, 1–5% headspace O_2 must be available in the package to avoid anaerobic respiration (Kader et al., 1989). However, the application of O_2 scavengers as an extension of MAP technology has been discussed in literature (Gontard & Guillaume, 2009). As described above, the equilibrium O_2 and CO_2 concentrations are a function of packaging permeability and product respiration and will establish after a transient period. The duration of this transient period depends on several factors such as package volume, product mass, and temperature. One method to reduce the time until the equilibrium gas composition is reached is to flush the package with the desired gas composition during the packing process. This might be a costly step as gas flushing equipment is required (Scully & Horsham, 2007).

Recently, some studies reported O_2 scavengers to be used to shorten this transient period, thereby reducing the time the product is exposed to suboptimal gas conditions. For MAP of tomatoes, Charles et al. (2003) found O_2 scavengers to substantially reduce the transient period compared to MAP without O_2 scavengers. Tomatoes were packed with iron-based ATCO® LH-100 O_2 scavengers in LDPE pouches and stored at 20°C. Time until the equilibrium gas atmosphere was established was less than 50 h in comparison to about 100 h for the control packages. The equilibrium gas concentration was not affected. In addition to that, there was no respiration-induced CO_2 peak during the transient period, which on the contrary was observed in MAP without O_2 scavenger (Charles et al., 2003). The absence of the CO_2 peak was explained by a lowered respiration rate due to the fast depletion of

O_2 by the scavenger. This effect might be important for MAP of CO_2-sensitive products. A similar experiment was conducted for fresh endive. This study also showed a 50% reduced transient period compared to passive MAP (Charles et al., 2005). Greening and browning processes that are limiting for quality of fresh endive were significantly delayed in packages with scavengers. This retention of product quality was attributed to the shorter transient period (Charles et al., 2008).

Other groups followed this approach and tested O_2 scavengers for packaging of fresh strawberries. Different types of packaging were used in combination with ATCO®-100 or ATCO®-210 scavengers (Kartal et al., 2012; Aday & Caner, 2013). Unfortunately, the headspace gas concentrations did not reach equilibrium in these studies. These results can therefore not be discussed concerning a reduction of the transient period and the resulting effects on product quality. Moreover, anaerobic conditions were reached in some absorber containing packages (Aday & Caner, 2013). This highlights the importance of carefully matching absorber kinetics with product respiration rate and packaging permeability in order to design packages that provide a beneficial equilibrium gas composition.

To date, no commercial O_2 scavenger applications for fresh produce packaging have been reported. Studies investigating the beneficial effects of reduced transient periods on product quality remain scarce. This effect will have to be proven before the technology can gain commercial interest.

3.3.2 ETHYLENE CONTROL

The plant hormone ethylene (C_2H_4) has various effects on the growth, development, and senescence of fruit and vegetables. Even at lowest concentrations (ppm to ppb) it is still effective (Vermeiren et al., 2003). Especially for climacteric fruit, accumulation of ethylene in packages should be avoided. At the end of their growth, climacteric fruit undergo a transient increase in respiration during ripening, which is marked by both, an increase in production and sensitivity to ethylene (Zagory & Kader, 1988). In general, vegetables, that is, plant organs like stems, roots, and leaves are less sensitive to ethylene exposure than fruit or fruit like vegetables such as tomatoes, broccoli, and cucumbers (Vermeiren et al., 2003).

Techniques to minimize the effect of ethylene during storage of horticultural products comprise the storage at low temperatures, reduction of respiration rates (and thereby ethylene susceptibility) by controlled atmospheres and the use of filters and scrubbers to remove ethylene from the storage environment. In addition to that, 1-MCP (1-Methylcyclopropene) can be used in storage rooms. This substance blocks the ethylene receptors in the plant tissue, thereby preventing the activity of ethylene (Scully & Horsham, 2007).

For packaging applications, different substances have been described as possible ethylene scavengers. Vermeiren et al. (2003) provide a list of commercially available systems. These scavengers are either available as sachet or film applications with potassium permanganate, activated carbon, minerals or zeolites being the most important active substances. For a detailed overview of substances that were tested as ethylene scavengers see Zagory (1995).

There are few studies that focus on the effect of ethylene scavengers on the quality and shelf-life of packed products. Abe and Watada (1991) tested paper sachets containing charcoal with palladium chloride as ethylene absorber for the storage of fresh-cut kiwifruit, bananas, broccoli, and spinach leaves. The different products were stored in model packages (metal trays with glass cover) at 20°C for two days, with or without 10 g absorber sachets. The authors reported that the absorbers prevented the accumulation of ethylene, thereby reducing the rate of softening in kiwifruit and bananas. The rate of chlorophyll loss was reduced in spinach leaves but not in broccoli. Unfortunately, the size of the trays and the headspace O_2 concentrations were not specified. Therefore, it is not possible to sufficiently evaluate the effect of the headspace gas atmosphere on product quality and to compare the results with those from other packaging tests. Bailén et al. (2006) used granular-activated carbon with or without a palladium catalyst (GAC or GAC-Pd) as ethylene scavenger for the storage of tomato fruit. Oriented PP bags ($d = 20$ μm; $A = 30 \times 20$ cm) were filled with four tomatoes each and stored in chambers at 8°C and 90% relative humidity (RH). Storage without absorbers was compared to storage involving either 5 g sachets of GAC or 5 g sachets of GAC-Pd. During the first 2 weeks of storage, headspace ethylene concentrations were significantly lower in packages with absorbers than in control packages. Ethylene concentration in control packages reached a maximum of almost 50 μL/L after 1 day and then decreased continuously, eventually reaching levels of packages with GAC absorbers. This decrease was explained by the high CO_2 concentrations which decelerated ethylene production. The initial increase in ethylene content was less pronounced in packages with absorbers (ca. 20 μL/L at day one, for both types) and remained in the range of 25–18 μL/L for GAC and 20–10 μL/L for GAC-Pd with a slightly downward tendency. Samples with absorbers also showed altered equilibrium gas atmospheres with higher O_2 and lower CO_2 levels compared to control packages. This might be attributed to lower respiration rates due to lower ethylene concentrations. However, there were no significant differences between the two kinds of absorbers in spite of the differences in ethylene concentration. Product quality evaluation revealed delayed ripening in terms of color, softening, and weight. Decay was measured as visible fungal growth. In packages with absorbers, fungal growth was significantly delayed compared to control packages. Earlier studies with fruit associated fungi revealed that spore germination can be stimulated by increasing ethylene concentrations (El-Kazzaz et al., 1983). Furthermore, the absorbers prevented the development of ethylene-related off-flavor, which could be detected in control packages from the first day. Illeperuma and Nikapitiya (2002) assessed the storage of "Pollock" avocado fruit in LDPE bags ($d = 50$ μm) with added absorbers, namely granular charcoal and potassium permanganate. The authors described charcoal as "carbon dioxide absorber" and potassium as "ethylene absorber." However, both absorbers are well known to lack absorption selectivity so that a combined evaluation seems more reasonable. Packages with and without absorbers were stored at 12°C and 94% RH. Product quality was analyzed in terms of weight loss and firmness; both parameters were reported to be significantly improved for avocados stored with absorbers. Based on sensory evaluation, a shelf-life of 29 days was indicated for avocados stored with absorbers compared to 17 days without absorbers. In comparison to control packages without absorber, higher O_2 and lower CO_2 concentrations were reported for avocado packages with absorbers. The ethylene level was significantly lower in packages with absorbers (1.6–1.8 μL/L).

Esturk et al. (2014)_extruded ethylene scavenging films (LDPE + 4% w/w zeolites, $d = 60$ μm) and tested them for the storage of fresh broccoli florets in comparison to neat LDPE films ($d = 60$ μm) and open storage. The film with zeolites showed higher permeability for O_2 and CO_2 than neat LDPE films and led to a headspace ethylene concentration below 1 ppm in the packages. In contrast, the ethylene concentration in LDPE packages increased to around 60 ppm after 20 days of storage. While the O_2 concentration in LDPE packages reached anaerobic conditions very quickly, an equilibrium gas concentration of 2% O_2 was established in packages with ethylene scavengers. CO_2 concentrations of both packaging systems were reported to remain in the range of 9–11% during storage. Control packages showed slightly lower CO_2 contents than scavenger packages. The shelf-life of the packed broccoli florets was determined to be 20 days for bags with ethylene scavenger compared to 5 days for LDPE bags. The short shelf-life of broccoli in LDPE bags was attributed to the anaerobic conditions that evolved in these packages and led to the formation of off-flavors. It remains an open question, if the beneficial effect of the scavenging bags was due to the ethylene scavenging activity or to the higher gas permeability. This highlights the importance of carefully designed storage tests, where single effects can be analyzed individually.

In these studies, different ethylene scavengers are used with a variety of products. All studies report reduced levels of ethylene and positive effects on quality or a delayed ripening, which can be seen as a first "proof of concept" of ethylene absorption technology in packaging applications. However, further purposefully designed storage tests are necessary to describe the ethylene relations in inactive packages quantitatively. For example, absorption rate and capacity of the absorbers, ethylene permeability of the packaging film as well as the ethylene tolerance of the product should be taken into consideration. The active packaging system can only be fully understood and evaluated if all these parameters are known.

The applications described above focus on the removal of ethylene. In other cases, fresh produce is deliberately exposed to ethylene. For example, ethylene is used during storage of fruits such as bananas and tomatoes to induce their rapid ripening at a desired point of time. These fruits are harvested in a premature state and exposed to high ethylene concentrations of 100 ppm right before marketing (Zagory, 1995). Ethylene can also be used to prevent sprouting for commodities such as potatoes and onions. Briddon (2006) reviewed the use of ethylene for potato sprout control with a focus on the British potato industry. In Great Britain, ethylene is approved as sprouting suppressant for potatoes and onions; different systems are available for ethylene generation in commercial controlled atmosphere storage (Briddon, 2006). Oshida et al. (2013) patented a method for the control of potato sprouting that includes ethylene-desorbing sachets manufactured by Mitsubishi Gas Chemical Company, Inc. The sachets contain zeolites with absorbed ethylene. To the authors' knowledge, no further application studies are available for these sachets and there is no commercial application so far. Nonetheless, ethylene-release provides an additional method for quality retention of sprouting tubers.

3.3.3   CARBON DIOXIDE CONTROL

CO_2 plays a major role in packaging of fresh produce as it is both a product of respiration and an important parameter that influences the respiration rate and thereby the product shelf-life. Consequently, CO_2 transport should always be considered in MAP design. In fresh produce packages, it is usually desired to reach CO_2 levels of 5–10% to decrease the respiration rate. Higher levels of CO_2 are, in contrast, detrimental for product quality as they can lead to off-flavors and decay (Zagory & Kader, 1988; Kader & Saltveit, 2003b). The gas atmosphere in packages with fresh fruit or vegetables is a function of product respiration and film permeability. CO_2 absorbers or emitters can potentially be used to adjust the CO_2 concentration in the package.

Commonly, CO_2 absorbers are used to prevent inflation of packages due to high amounts of CO_2, formed after the packaging process. Examples are roasted coffee (non-enzymatic browning reactions) or fermented products such as kimchi, pickled products, and some dairy products (Lee et al., 2001; Han, 2005). Several CO_2 absorbers are commercially available; their main principle is a carbonatation reaction (Charles et al., 2006; Aday et al., 2011).

Charles et al. (2005) tested ATCO® CO-450 CO_2 scavenger sachets (paper coated with perforated polypropylene) containing sodium hydroxide for the storage of fresh endive in LDPE bags. During storage at 5°C, the scavengers only affected the headspace gas composition at a very late stage of the storage test when the maximum shelf-life had already been reached. At 20°C the CO_2 scavenger leads to reduced CO_2 levels after 30 h of storage. Therefore, it was concluded that CO_2 scavengers have limited interest for MA packaging at chilled temperatures. At higher temperatures, products which are very sensitive toward CO_2 (e.g. mushrooms) might benefit from reduced CO_2 levels. Aday et al. (2011) used two CO_2 absorbers from EMCO packaging systems differing in their concentrations of sodium carbonate peroxyhydrate and sodium carbonate for packaging of fresh strawberries. 200 g of strawberries were sealed in PLA trays with absorber sachets (amount of absorber or capacity not stated) and stored for four weeks at 4°C. CO_2 levels were significantly reduced in packages with absorbers compared to control packages. CO_2 was completely removed from the headspace in the first week; after two weeks a slow increase up to values of 14–16 kPa was noticed. In control packages CO_2 continuously increased from 0 to 46 kPa. Product quality was determined in terms of overall appearance and percentage of decay, but unfortunately not correlated precisely to the effect of the absorbers. Therefore, this study does not provide a clear insight into the effect of CO_2 scavengers on the quality retention of fresh strawberries.

CO_2 emitters can be used to compensate for CO_2 losses from the package headspace due to dissolution in the product or to permeation through the packaging material. The CO_2 emission is usually generated by a reaction of sodium bicarbonate and citric acid which is started by water, for example, drip loss from the packed product. The active substance can be integrated in pads that absorb moisture. This technology is mainly used for meat and poultry products which require high levels of CO_2 in order to reduce surface microbial growth and prolong product shelf-life (Realini & Marcos, 2014). Regarding fruit and vegetables, the application of CO_2 emitters seems limited as they generate CO_2 themselves and are sensitive toward high CO_2 concentrations.

3.3.4 HUMIDITY AND MOISTURE CONTROL

Relative humidity is an important parameter for the storage of fresh fruit and vegetables, which are characterized by high water activities (a_w) of 0.85–0.95. Storage conditions of 85–95% RH are, therefore, recommended for fresh produce. Low-humidity storage leads to quality defects such as wilting and shriveling. In saturated atmospheres condensation is likely to occur; water stress at product surfaces and microbial growth can be the consequences (Kader et al., 1989; Ben-Yehoshua & Rodov, 2003).

Common plastic packaging provides good protection against water loss, but usually leads to water-saturated atmospheres inside the package. This can be attributed to the low water vapor transmission rates of most plastic films. These cannot compensate for the amount of water vapor produced by product transpiration and respiration. Even with micro-perforations, the water vapor transmission rates of polymer films cannot be increased significantly (Ben-Yehoshua et al., 1998; Mahajan et al., 2008).

"Humidity buffering" refers to reducing the in-package RH and thereby the product surface moisture content to a product-specific level. This can possibly be achieved by humidity absorbers (Vermeiren et al., 1999; Han, 2005). Several studies have addressed this topic during the past decades and a variety of hygroscopic substances has been tested as humidity absorbers, either applied as sachets or integrated into the packaging material. These include polyols such as sorbitol and xylitol, deliquescent salts such as sodium chloride (NaCl), potassium chloride (KCl), or calcium chloride ($CaCl_2$) as well as bentonite (Shirazi & Cameron, 1992; Mahajan et al., 2008). A special feature of deliquescent salts is the rapid absorption of high amounts of water vapor when a specific threshold of RH is exceeded (e.g. 75% RH for NaCl). This absorption process is reversible, that is, the water vapor is released again if RH falls below this threshold. Hence, deliquescent salts have been proposed as substances that not only reduce but actively regulate the in-package RH, thereby buffering humidity fluctuations (Singh et al., 2010; Sängerlaub et al., 2013).

Although a lot of absorbers and packaging designs for humidity buffering have been introduced in scientific studies, only a few studies actually describe storage tests with fresh fruit or vegetables in order to assess the effect of humidity absorbers on product quality and shelf-life.

Fresh mushrooms (*Agaricusbisporus*) have been the subject of several application studies with humidity absorbers. Tyvek® sachets with different amounts of sorbitol or NaCl were used for storage of mushrooms, either in trays packed into polyolefin pouches or in trays wrapped with perforated PVC films (Roy et al., 1995, 1996). The RH in the packages was significantly lowered by both absorbers; as could be expected, higher amounts of the respective absorber led to lower RH in the package. For same amounts of absorber, NaCl resulted in a higher reduction in RH than sorbitol and thereby also to lower surface water contents of the mushrooms. Higher amounts of absorbers led to higher weight loss. For mushrooms stored with 25 g sorbitol, a weight loss of 21% was reported compared to 7% for control packages without absorber. Regarding product quality, it was reported that the absorbers had no significant influence on microbial growth and on product maturity. Storage conditions with 15 g sorbitol were reported to be most suitable as mushrooms showed the best color (measured as L and ΔE value) at these conditions and the surface moisture content was considered to be optimal. Although not quantified in days, a prolonged shelf-life of mushrooms due to storage with sorbitol was claimed (Roy et al., 1995, 1996). Mahajan et al. (2008) used a mixture of 20% $CaCl_2$, 55% bentonite, and 25% sorbitol for the storage of *Agaricus* mushrooms. A total of 250 g were stored at 10°C for five days in trays covered with perforated PVC film. Sachets with 0, 5, 10, and 15 g of the absorber were applied. As for the studies described above, product weight loss increased with increasing amount of desiccant and less condensation was reported for packages with absorbers than for control packages. Sensory evaluation of the overall appearance after five days showed the best quality for mushrooms packed with 5 g of absorber. This result was confirmed by color measurements (browning index). The authors described the quality of these mushrooms as still marketable after 5 days. Unfortunately, the experiment was not continued to determine the day when the mushrooms showed signs of decay. Singh et al. (2010) described the storage of 100 g *Agaricus* mushrooms in PP trays containing different amounts of NaCl (6–18% w/w) in the so-called active layer. The storage conditions were 5°C for 8 days. Weight loss was higher for trays with NaCl than for control trays, where higher amounts of NaCl led to higher weight loss. Microbial growth was reported to be significantly lower in packages with 12 or 18% NaCl compared to the trays with lower concentrations or the control trays. Color measurements showed higher lightness values for trays with NaCl than for control trays. Although data was not provided, sensory evaluation was reported to reveal the best overall appearance after eight days for mushrooms stored in trays with 18% NaCl. Rodov et al. (1995) stored red bell pepper (*Capsicum annuum* L.) in LDPE bags containing Tyvek®sachets with different amounts of NaCl. After storage at 8°C and 85% RH for two to three weeks, the packages were exposed to 17°C and 85% RH for four to six days to simulate marketing conditions. Again, the in-package RH was lowered by the addition of the absorber; the adjusted RH was in the range of 87–98%, depending on the amount of absorber while saturated conditions prevailed in control packages. Condensation was reduced or completely prevented, depending on the amount of absorber. Weight loss increased with increasing amounts of NaCl. The best quality after three weeks of storage at 8°C and three days at 17°C was reported for peppers stored with10 g NaCl. For this treatment, less than 20% of the fruit exhibited visible decay compared to almost 50% in control packages without absorber. Shirazi and Cameron (1992) reported an extended shelf-life for red-ripe tomatoes stored with NaCl absorbers. Tyvek® sachets with 20 g NaCl were used in perforated LDPE film packages (surface area of 600 cm²) containing 450 g of red-ripe tomatoes stored at 20°C. The extended shelf-life of 15–17 days compared to five days for control packages was attributed to retarded growth of surface molds. Unfortunately, the RH that developed in these packages was not reported.

The positive effect of humidity absorbers on product quality, if applied in suitable amounts, can be seen as a shared result of all analyzed studies. However, not in all cases the claimed prolongation of shelf-life could be quantified as a timespan and be correlated to a specific quality criterion. One major conclusion from these empirical studies is the importance of defining suitable markers for quality evaluation of the stored products. These markers have to be assessed repeatedly throughout the storage period. Moreover, a limit has to be defined for each marker, as this is the only way to quantify shelf-life. Regarding humidity absorbers, more research, that is, properly designed storage tests, is necessary to show the effect of humidity absorbers on fresh produce quality.

As described above, high relative humidity can lead to condensation in plastic packages with fresh fruit or vegetables resulting in water droplets on product and packaging surfaces. Furthermore, liquids can leak from damaged fruit.

To prevent the formation of condensation droplets on the inner surface of the package, so-called antifog films have been developed. The principle of these films relies on lowering the interfacial tension between the packaging surface and the condensation droplets. As a consequence, droplets coalesce and form an invisible film. This creates transparency and enables the customer to clearly see the product. This technology improves the appearance of the package but does not affect the humidity conditions in the package. Therefore, other approaches have been developed where the relative humidity in the package is reduced and condensation is prevented (Rooney, 1995; Brody et al., 2001).

Drip loss from damaged fruit can be absorbed by drip absorbing pads which find broad applications in packaging of fresh meat (Realini & Marcos, 2014). Basically, these pads consist of super absorbing polymers encapsulated between two layers of a microporous or non-woven polymer. Polyacrylates are the most popular absorbers for such systems (Rooney, 1995). In the field of fresh produce, such kind of absorbers can be found in berry packages. Berries are highly susceptible to mechanical damage and the pads absorb drip loss from destroyed fruit thereby ameliorating the package appearance.

3.3.5 ANTIMICROBIAL PACKAGING

Antimicrobial packaging is a form of active packaging designed to interact with the food product or the package headspace, to positively influence microbial product quality, shelf-life, and safety (Appendini & Hotchkiss, 2002). Although antimicrobial packaging systems can also be based on immobilized antimicrobial agents, only materials truly releasing preservative agents toward the food are considered active food packaging materials (Restuccia et al., 2010). In general, antimicrobial packaging can be divided into contact or non-contact materials. Non-volatile agents can only cross the interface between carrier material and food product surface, wherefore both have to be in constant contact. Volatile agents, on the contrary, can easily cross the headspace between active packaging material and product. Hence the carrier material does not need to be in direct contact with the product. Antimicrobial food packaging systems can be designed as follows:

- Sachets containing volatile antimicrobial agents;
- Packaging films containing volatile or non-volatile antimicrobial agents;
- Coatings containing volatile or non-volatile antimicrobial agents.

Antimicrobial contact materials containing non-volatile agents may be favorably applied as skin packaging for vacuum-packed products, for example, single-packed steaks (Hauser & Wunderlich, 2011; Ferrocino et al., 2013). Volatile antimicrobial systems are principally suitable for any packaging geometry and are applicable for tray-, bag- or tube-packed food products, such as portioned meat, cheese or fresh-cut produce (Han, 2003).

In terms of sophisticated preservation concepts, antimicrobial packaging is recognized as a forceful hurdle. It is designed to persistently delay the outgrowth of microbial contaminations on the product. This can lead to prolonged shelf-life periods and increased microbial food safety. For the sake of completeness, it has to be outlined that antimicrobial packaging concepts are most efficient when preceding decontamination, treatments are applied as fruit and vegetables are naturally contaminated products (Ahvenainen, 1996). It is generally necessary to ensure high hygienic standards at all processing and distribution stages to avoid cross contaminations. As fresh produce is sensitive to physical stress, gentle washing procedures appear to be the method of choice to reduce initial microbial loads. This process can be supported with antimicrobial additives such as hydrogen peroxide (H_2O_2), chlorine (Cl_2), chlorine dioxide (ClO_2), or ozone (O_3), but legal restrictions still have to be considered (Goodburn & Wallace, 2013). Furthermore, novel mild physical treatments such as gamma irradiation, pulsed-light irradiation or cold-plasma processing appear to be on the rise (Gomes et al., 2011; Ramos et al., 2013; Allende & Gil, 2014).

Due to extensive scientific efforts in the last two decades, a wide variety of different antimicrobial packaging designs has been presented. Multiple antimicrobial agents such as organic acids, bacteriocins, enzymes, crude plant extracts, single secondary plant metabolites, polysaccharides, and bacteriophages have been incorporated into several food contact materials and were tested for their antimicrobial activity in vitro or on food. Furthermore, there are studies focusing on gaseous or volatile antimicrobials such as ethanol, essential oils (EOs), sulfur dioxide, CO_2, and ClO_2. Table 3.2 presents a selection of reviews on antimicrobial packaging systems for different applications.

Regarding the applicability of antimicrobial packaging systems, literature reveals fewer studies on fresh fruit and vegetables than on meat or fish. Findings are predominantly based on in vitro results. Food application trials remain scarce. More general approaches, such as the direct application of antimicrobial agents on food products prevail. Nonetheless, there are promising approaches and developments regarding the shelf-life elongation and food safety improvement of fresh and fresh-cut fruit and vegetables by antimicrobial packaging solutions. These developments are mostly based on novel antimicrobial agents such as crude plant extracts, selected plant metabolites or metal ions, but the incorporation of traditional preservative agents as for example sorbic or benzoic acid is mostly rejected. This may be due to the claim to ensure the overall green and natural character of fresh produce. Most studies presented in the following focus on volatile antimicrobial agents, as these are able to diffuse across the headspace between packaging material and product. Studies focusing on immobilized agents were excluded from this review as they are not considered as active packaging materials (European Commision, 2009, 2011).

TABLE 3.2   Reviews on Antimicrobial Packaging.

Year	Author	Focus
2015	Malhotra et al.	General review, emphasis on potential and pitfalls
2014	Realini & Marcos	Antimicrobial packaging for meat products
2014	Tawakkal et al.	Polylactic acid-based antimicrobial films
2014	Jideani & Vogt	Antimicrobial packaging for bread
2013	Sung et al.	Antimicrobial agents
2013	Cruz-Romero et al.	Chitosan, organic acids and nano-sized solubilizes
2011	Suppakul	Natural extracts in antimicrobial packaging
2011	Bastarrachea et al.	Incorporation mechanisms and release physics
2011	Baldevraj & Jagadish	Antimicrobial agents and incorporation techniques
2008	Coma	Antimicrobial packaging for meat products
2006	Kerry et al.	Antimicrobial packaging for meat products
2004	Cha & Chinnan	Biopolymer-based antimicrobial packaging
2003	Suppakul et al.	General review of antimicrobial packaging systems
2003	Han	General review of antimicrobial packaging systems
2002	Quintavalla and Vicini	Antimicrobial packaging for meat products
2002	Appendini & Hotchkiss	General review of antimicrobial packaging designs

3.3.5.1   ANTIMICROBIAL PACKAGING FOR LEAFY GREENS

Seo et al. (2012) developed an antimicrobial sachet containing allyl isothiocyanate (AITC), a volatile antimicrobial secondary plant metabolite from mustard essential oil, encapsulated in alginate beads, and investigated its inhibitory effect against *Escherichia coli* on spinach leaves. At 25°C, the AITC vapor was able to inactivate 5.7 log of an initial *E. coli* population within 5 days. At 4°C only 2.6 log was inactivated within the same time. The reduced antimicrobial activity at chilling temperatures may be attributed to a reduced AITC-release from the sachets (Seo et al., 2012). Muriel-Galet et al. (2012) developed an active EVOH coating, containing either citral, a volatile secondary plant metabolite from lemongrass (*Cymbopogonci-tratus*), or EO of oregano (*Origanum vulgare*). The coatings were applied on PP films from which bags were produced to pack fresh-cut salad. The antimicrobial activity of the films in combination with MAP (12% CO_2, 4% O_2) was assessed against the pathogenic microorganisms *Escherichia coli*, *Salmonella enterica,* and *Listeria monocytogenes* as well as the originating microbiota. Results proved increasing antimicrobial activity with increasing concentrations of the antimicrobial agents in the film, which was also accompanied by increasing sensory off-flavors (Muriel-Galet et al., 2012). However, sensory studies revealed citral on fresh-cut salad to be accept-able for consumers (Muriel-Galet et al., 2013). Citral-based films were slightly more effective than the materials containing oregano EO. The total aerobic count was reduced initially about 1.23 log with citral and 1.08 log with oregano oil. Gunduz et al. (2012) had preliminarily shown that direct application of 500 ppm oregano EO on inoculated iceberg lettuce leaves was able to inactivate 2.28 log of *Salmonella* spp. within 10 min. But they also discovered a rapid browning and softening of the leaves due to the high oregano load. The incorporation of essential oils or single secondary plant metabolites as active agents for antimicrobial packaging appears to be one of the most addressed approaches (Suppakul, 2011; Patrignani et al., 2015). But application studies with fresh-cut salad or other fresh produce remain scarce, nonetheless. Lu et al. (2015) investigated the applicability of antimicrobial packaging systems to improve the microbial quality of fresh-cut lettuce. The authors used different ClO_2-, CO_2- and AITC-generators in comparison and in combination with O_2 scavengers. Artificially contaminated lettuce was stored for 3 weeks at 4, 10, and 22°C. Results showed the highest effec-tivity for ClO_2 against *E. coli*, especially at 22°C. The authors furthermore highlighted the importance of sufficiently high CO_2 concentrations in the package headspace. The applied iron-based O_2 scavenger also absorbed CO_2 and consequently limited the growth inhibitory potential of CO_2 (Lu et al., 2015). A study focusing on polylactic acid (PLA) films, containing up to 1.0% of silver ions, which were applied in contact to lettuce leaves and stored at 4°C revealed a 4 log-reduction of an artificial *Salmonella* contami-nation within six days of storage (Martínez-Abad et al., 2013). The disad-vantage of this approach clearly is the necessity of direct contact between film and product as silver ions are unable to diffuse within the headspace.

The presented studies reveal the general suitability of volatile antimicrobial agents to decrease the microbial load on fresh and fresh-cut leafy greens. Especially the natural terpenoidcitral may be considered as an interesting candidate for future research because of its beneficial sensorial impact when used in suitable concentrations. ClO_2 might be an effective antimicrobial substance, but appears to be less effective at low temperatures. This finding opposes the application at refrigeration temperatures which is necessary for storage of fresh and fresh-cut lettuce or other leafy greens.

3.3.5.2　ANTIMICROBIAL PACKAGING FOR WHOLE AND FRESH-CUT FRUIT

Antimicrobial packaging of fresh and fresh-cut fruit such as strawberries, apples, or melons is also gaining scientific interest. Due to the fruit's shape and the large headspace volumes of their packaging, volatile antimicrobial agents appear to be the most suitable. Rodríguez et al. (2007) assessed the antimicrobial activity of different natural essential oils released from paraffin-based active coatings for paper food packaging materials. Coatings containing cinnamon (*Cinnamomumzeylanicum*) EO proved to be highly active against a variety of fungi in vitro. Hence, their efficiency was tested with two varieties of strawberries. Whereas the control groups were completely colonialized by molds within seven days at 4°C, the active paper packaging proved to completely inhibit the development of surface-associated fungi for up to seven days. Focusing on bacterial contaminations in fresh strawberry puree, Jin et al. (2010) produced PLA films (24 cm^2) containing a combination of the traditional preservatives potassium sorbate (PS) (0.94 mg/cm^2) and sodium benzoate (SB) (1.56 mg/cm^2) as well as films containing nisin (5.2 mg/cm^2) or EDTA (5.2 mg/cm^2) and a film combining all agents. Direct addition of PS (0.45 mg/mL) and SB (0.75 mg/mL) were used as reference method to preserve the strawberry puree samples (50 mL). All films proved to be antimicrobial, but the combinatory treatment was most effective, inactivating 3.5 log cfu/mL of *E. coli* within 14 days at 10°C and within one day at 22°C, respectively. Although the effectivity of the films was proven, their applicability for whole or cut strawberries is questionable as the films need direct contact to the product. For whole strawberries preservation Cagnon et al. (2013) combined a specifically designed MAP system with the volatile, antimicrobial compound 2-nonanone which was incorporated into a wheat-gluten coating. The coating was applied on paper sheets, which were used to seal PET trays containing strawberries. Samples were stored at 20°C and their quality was continuously monitored. The system was able to delay the occurrence of visible microbial spoilage for about three days, in comparison to untreated control samples (Cagnon et al., 2013). Although the number of studies regarding the antimicrobial packaging of strawberries is quite small, they show the potential of this application. Strawberries are widely consumed, highly perishable products, undergoing long routes of transport. Shelf-life elongations of only a few days can help to counter product loss. Still, there is the necessity of further research exploring the antimicrobial effectivity of other antimicrobial packaging systems on strawberries and also the impact on their quality.

Tomatoes, whole and fresh-cut, have also been topic of application studies with antimicrobial packaging systems. Sachets containing garlic EO, encapsulated in β-cyclodextrines (CD), were added to trays with fresh tomato slices. The CD capsules were loaded with 202.4 mg garlic oil per gram. At 100% relative humidity and 20°C the capsules released 30% of their load within three weeks. Sachets with 1 g of the capsules were used to treat 100 g tomato. These were able to elongate the lag phases of bacteria and fungi, extending product shelf-life from 16 to 21 days at 5°C. Furthermore, the garlic flavor positively influenced the panel's sensorial perception throughout storage. In comparison, directly applied garlic oil (200 µg/slice) had higher antimicrobial potential, but its sensorial impact was mostly negative (Ayala-Zavala & Gonzalez-Aguilar, 2010). Also focusing on natural essential oils as antimicrobial agents Rodriguez-Lafuente et al. (2010) developed paraffin coatings for paper and board containing different EOs. The cinnamon coating (6%) proved to be most efficient to inhibit the fungus *Altemariaaltemate* in vitro. Moreover, the coating achieved complete inhibition of the spoilage fungus on artificially inoculated cherry tomatoes. Trans-cinnamaldehyde was detectable in the tomatoes, reaching maximum concentration (432.0 µg/kg) within two days. Concentration then decreased continuously throughout storage. The coating positively influenced pH, weight loss, water activity, and color in comparison to untreated controls. Sensorial evaluation revealed the panel to be able to detect the cinnamon treatment within the first three days, but afterwards cinnamon recognition decreased again (Rodriguez-Lafuente et al., 2010). Although only few studies investigated the applicability of antimicrobial packaging on tomatoes it may be prudently concluded that EO-based antimicrobial packaging solutions are suitable for the application on tomato fruit. As tomatoes have an intensive, aromatic taste the addition of sensorial active compounds, as they are found in essential oils, may not be as disadvantageous in terms of smell and taste as for other less aromatic products. Another approach was chosen by Ray et al. (2013). They produced ClO_2-releasing PLA films by incorporating sodium chlorite and citric acid. Activated by moisture from the packed tomatoes, the generated ClO_2 was able to reduce artificial *Salmonella spp.* and *Escherichia coli* contaminations about 3 log. The tomatoes, stored at 10 or 21°C, were not affected in terms of color or texture changes. The trials revealed total ClO_2 release and the release rate to be dependent on the concentration of reactants in the film and the film thickness. Higher reactant concentrations resulted in higher release rates as well as a greater total release, whereas increasing film thickness led to reduced release rates (Ray et al., 2013). The authors demonstrated the general feasibility of ClO_2-based antimicrobial packaging systems to reduce the microbial load on tomatoes, but the sensorial impact on the product was not tested.

Further studies presented additional approaches to store various fruit products in antimicrobial packaging systems. Fernández et al. (2010) loaded absorbent pads with a silver nitrate solution (1%) to reduce the microbial load on fresh-cut melon pieces. After 10 days at 4°C, the total colony counts remained 3 log lower for the treated samples than for the untreated controls. Lower microbial counts resulted in overall increasing quality parameters such as sugar content and juiciness (Fernández et al., 2010). Investigating the applicability of antimicrobial packaging on peach fruit, Montero-Prado et al. (2011) developed coatings and labels containing cinnamon EO for PET and PP trays. The fruit were packed in these trays and stored at room temperature for 12 days. Cinnamon EO was found to be highly beneficial as only 13% of the treated fruit showed mold outgrowth in contrast to untreated reference samples (86%). Further quality parameters as enzymatic activity and lipid oxidation were also affected positively. Other studies focused on antimicrobial packaging for fruit applications based on nanoparticles. Coatings on polyvinyl chloride films containing zinc-oxide-nanoparticles, were shown to be able to inhibit *E. coli* in vitro and on artificially inoculated apple slices (Li et al., 2011). Titanium-dioxide nanoparticles which were included in LDPE films were tested in vivo on fresh-cut pears. After photocatalytic activation the films continuously decreased the mesophilic bacterial count over 17 days at 5°C (Bodaghi et al., 2013). Although these ion-based antimicrobial packaging solutions proved to be partially effective their major disadvantage is their contact-based design. Candir et al. (2012) investigated the effect of ethanol vapor and sulfur dioxide emitters on table grapes, artificially inoculated with *Botrytis cinerea*. The grapes were stored in sealed bags with different modified atmospheres at 0°C for four months. The sachets, containing either 3, 6, and 8 g of ethanol powder or 7 g sodium metabisulfite were highly effective. Untreated grapes had a shelf-life of three months before mold spoilage became evident. The SO_2 sachet reduced fungal decay by 99% after four months, despite high O_2 (19 kPa) and low CO_2 (1 kPa) concentrations in the headspace. The ethanol sachets were less effective because ethanol headspace concentrations rapidly decreased after two months. Nonetheless their inhibitory effect on *B. cinerea* was significant in comparison to untreated controls. These results are comparable to those of Chervin et al. (2005) who also assessed the effect of SO_2 pads and self-made paper-based ethanol sachets on the shelf-life and quality of table grapes.

3.3.5.3 ANTIMICROBIAL PACKAGING FOR WHOLE AND FRESH-CUT VEGETABLES

Besides the presented studies focusing on fruit, there are further works investigating the applicability of antimicrobial packaging on vegetables. Gamage et al. (2009)_developed soy protein coatings incorporating either AITC, trans-cinnamaldehyde, garlic oil or rosemary oil (0.6–1.2% v/v). These were spread on oriented polypropylene/polyethylene (OPP/PE) packaging materials. Storage trials with alfalfa, broccoli, and radish (5 days, 10°C) revealed the produced packaging bags to inhibit microbial development, depending on the product, the antimicrobial compound used and its concentration in the film. AITC proved to be most effective, inhibiting microbial outgrowth on alfalfa and radish at all concentrations. Microbiota on broccoli was inhibited to the same extent only at the highest concentration (1.2% v/v) (Gamage et al., 2009). Also focusing on broccoli, Takala et al. (2013) successfully applied a methylcellulose coating containing undefined mixtures of antimicrobial essential oils and organic acids. The films were kept loose inside the packages containing artificially inoculated broccoli. The packages were stored at 4°C for 12 days. Although the inhibition of *S. thyphimurium* and *E. coli* was significant, *L. monocytogenes* as well as the mixed microbiota were unaffected (Takala et al., 2013). Qin et al. (2015) used cinnamaldehyde (0, 3, and 9% w/w) which was incorporated into a bio-based poly lactic-acid/poly ε-caprolactone (PLA/PCL) blend film to treat button mushrooms while stored at 4°C for 16 days. By assessing several quality parameters they showed the highest cinnamaldehyde concentration to be effective regarding microbial count reduction and color preservation. The observed increase in water loss (3.08%) was attributed to the high water vapor permeability of the blend films, which became porous by EO addition (Qin et al., 2015).

Regarding the general applicability of antimicrobial packaging on fresh and fresh-cut fruit or vegetable products, scientific basis remains fragmentary. A great number of studies rather follow an empirical trial and error approach. Especially for fresh produce, it is difficult to conduct elaborate shelf-life studies as the number of factors affecting product quality is often too large to be considered. However, studies on the applicability of antimicrobial packaging systems on fresh produce should at least consider the most important factors affecting product quality during storage. These factors, for example, temperature and gas composition should be chosen in regard of realistic storage conditions. Furthermore, sensorial properties should always be investigated, since antimicrobial agents, especially plant-derived substances, possess high sensorial activities.

3.4   CONCLUSION AND FUTURE ASPECTS

Many different active packaging systems have been proposed for fresh and fresh-cut fruit and vegetables and an increasing number of research activities can continuously be noticed. Future designs might be able to combine the unique beneficial effects of different active packaging technologies. However, previous to a successful introduction into the market, the effectiveness of active systems in maintaining food quality and prolonging shelf-life has to be demonstrated. As reviewed in this chapter, the functionality of active systems has been assessed for a variety of fruit or vegetables.

O_2 scavengers have been shown to reduce time until equilibrium gas concentrations are reached in packages of fresh produce. This approach relies on the assumption that the product benefits from a shortened transient period, wherein the product is subjected to non-optimum storage conditions. However, there are only a few studies showing a distinct relation between shortened transient periods and elongated shelf-life. In terms of MAP design, which is already challenging due to high variations in product respiration rates, O_2 scavengers may additionally increase the risk of anoxic conditions.

The application of CO_2 scavengers in fresh produce packages is based on the idea of reducing CO_2 concentrations in order to prevent physiological disorders leading to quality loss. Studies on CO_2 scavengers and their effect on product quality remain scarce. Especially their applicability at low temperatures, which are required for the storage of most fresh produce, has to be further investigated.

CO_2 releasing systems are available for storage of meat products where high levels of CO_2 are favored. Their applicability on fresh produce is limited as high levels of CO_2 should rather be avoided.

Ethylene scavengers have been demonstrated to successfully reduce ethylene levels in packages containing fresh fruit and vegetables. Furthermore, ethylene removal is a well-established technology in commercial warehouse storage of fresh produce. However, information on the beneficial effect of ethylene scavengers in consumer packages remains fragmentary. Further insight is necessary on the effect on product quality in short-term storage.

For ethylene release systems, no scientific studies evaluating their application are available. Regarding humidity regulation, different absorber systems have been described and proven to be effective in terms of lowering the in-package relative humidity. There are studies that provide a "proof of concept," but research is still necessary to better understand humidity-regulating packages as a whole. Especially absorption kinetics has to be fully investigated to enable tailor-made packaging solutions preventing both, condensation and dehydration of the product.

As reviewed, the applicability of antimicrobial packaging developments on fruit and vegetables is also only partially investigated. Among the introduced systems, materials containing essential oils prevail. These natural plant extracts fit the "green" character of fresh and fresh-cut fruit and vegetable products. Most studies focus on the general proof of concept in terms of the antimicrobial activity. But, further quality parameters might also be affected by antimicrobial packaging systems and should, therefore, be evaluated simultaneously. Concerning fresh produce, studies on antimicrobial packaging should always be performed in regard of the product-specific optimal packaging solution in terms of headspace composition, polymer permeability, perforation and other parameters which affect product quality as well as microbiology. Although antimicrobial packaging is generally able to improve shelf-life and to increase food safety, it can only act as an additional hurdle in a sophisticated preservation concept, but can never replace good manufacturing practices and high process hygiene standards.

Active packaging technology comprises some useful approaches to optimize fruit and vegetable packaging. Nonetheless, market implementations can only be achieved when effort into application studies is extended. To evolve in understanding active systems and to pursue their implementation, greater importance must be attached to the storage test design beforehand. This includes:

- Quantitative description of the release or absorption behavior of the active system, where the active system not only consists of the active substance, but also includes the way of its integration in the package, for example, a surrounding polymer matrix.

- Comprehensive knowledge on the product's response to the atmospheric changes that are induced by the active system.
- A realistic set-up for testing that reflects typical package dimensions and storage conditions. This also includes a comprehensive characterization of packaging properties (especially gas permeability) to allow for comparison among packaging designs.

However, active packages containing fresh fruit or vegetables are complex systems where the consequences of changing one or more properties (e.g., the amount of absorber) are difficult to capture without time-consuming, costly storage tests. To overcome these difficulties, mathematical models have been proposed to describe all processes relevant in packages. These processes include:

- gas transfer through packaging films;
- respiration and transpiration of the packed product;
- growth and respiration of microorganisms;
- absorption or release of active components.

KEYWORDS

- **fresh fruit and vegetables**
- **fresh-cut fruit and vegetables**
- **packaging**
- **packaging technology**
- **active packaging systems**

REFERENCES

Abe, K.; Watada, A. E. Ethylene Absorbent to Maintain Quality of Lightly Processed Fruits and Vegetables. *J. Food Sci.* **1991,** *56,* 1589–1592.

Aday, M. S.; Caner, C. The Shelf Life Extension of Fresh Strawberries Using an Oxygen Absorber in the Biobased Package. *LWT-Food Sci. Technol.* **2013,** *52,* 102–109.

Aday, M. S.; Caner, C.; Rahvalı, F. Effect of Oxygen and Carbon Dioxide Absorbers on Strawberry Quality. *Postharvest Biol. Technol.* **2011,** *62,* 179–187.

Ahvenainen, R. New Approaches in Improving the Shelf Life of Minimally Processed Fruit and Vegetables. *Trends Food Sci. Technol.* **1996,** *7,* 179–187.

Ahvenainen, R. Active and Intelligent Packaging: An Introduction. In *Novel Food Packaging Techniques;* Ahvenainen, R., Ed.; Woodhead Publishing Ltd: Cambridge, UK, 2003; pp 5–21.

Allende, A.; Gil, M. I. Suitability of Physical Methods to Assure Produce Safety. *Stewart Postharvest Rev.* **2014,** *10,* 1–5.

Appendini, P.; Hotchkiss, J. H. Review of Antimicrobial Food Packaging. *Innov. Food Sci. Emerg. Technol.* **2002,** *3,* 113–126.

Ayala-Zavala, J. F.; Gonzalez-Aguilar, G. A. Optimizing the Use of Garlic Oil as Antimicrobial Agent on Fresh-Cut Tomato through a Controlled Release System. *J. Food Sci.* **2010,** *75,* 398–405.

Bailén, G.; Guillén, F.; Castillo, S.; Serrano, M.; Valero, D.; Martínez-Romero, D. Use of Activated Carbon inside Modified Atmosphere Packages to Maintain Tomato Fruit Quality During Cold Storage. *J. Agric. Food Chem.* **2006,** *54,* 2229–2235.

Baldevraj, R. S. M.; Jagadish, R. S. Incorporation of Chemical Antimicrobial Agents into Polymeric Films for Food Packaging. In *Multifunctional and Nanoreinforced Polymers for Food Packaging;* Lagarón, J. M., Ed.; Woodhead Publishing: Cambridge, UK, 2011; pp 368–420.

Bastarrachea, L.; Dhawan, S.; Sablani, S. S. Engineering Properties of Polymeric-Based Antimicrobial Films for Food Packaging: A Review. *Food Eng. Rev.* **2011,** *3,* 79–93.

Ben-Yehoshua, S.; Rodov, V. Transpiration and Water Stress. In *Postharvest Physiology and Pathology of Vegetables;* Bartz, J. A., Brecht, J. K., Eds.; Marcel Dekker Inc.: New York, NY, 2003; pp 119–173.

Ben-Yehoshua, S.; Rodov, V.; Fishman, S.; Peretz, J. Modified-Atmosphere Packaging of Fruits and Vegetables: Reducing Condensation of Water in Bell Peppers and Mangoes. In *Postharvest '96 - Proceedings of the International Postharvest Science Conference;* Bieleski, R., Laing, W. A., Clark, C. J., Eds.; 1998; pp 387–392.

Berger, C. N.; Sodha, S. V.; Shaw, R. K.; Griffin, P. M.; Pink, D.; Hand, P.; Frankel, G. Fresh Fruit and Vegetables as Vehicles for the Transmission of Human Pathogens. *Environ. Microbiol.* **2010,** *12,* 2385–2397.

Beuchat, L. R. Ecological Factors Influencing Survival and Growth of Human Pathogens on Raw Fruits and Vegetables. *Microbes Infect.* **2002,** *4,* 413–423.

Bodaghi, H.; Mostofi, Y.; Oromiehie, A.; Zamani, Z.; Ghanbarzadeh, B.; Costa, C.; Conte, A.; Del Nobile, M. A. Evaluation of the Photocatalytic Antimicrobial Effects of a TiO2 Nanocomposite Food Packaging Film by In Vitro and In Vivo Tests. *LWT-Food Sci. Technol.* **2013,** *50,* 702–706.

Briddon, A. *The Use of Ethylene for Sprout Control;* British Potato Council, Research Review 279, November 2006.

Brody, A. L.; Strupinsky, E.; Kline, L. R. *Active Packaging for Food Applications;* CRC Press: London, 2001.

Burnett, S. L.; Beuchat, L. R. Human Pathogens Associated with Raw Produce and Unpasteurized Juices, and Difficulties in Decontamination. *J. Ind. Microbiol. Biotechnol.* **2000,** *25,* 281–287.

Cagnon, T.; Méry, A.; Chalier, P.; Guillaume, C.; Gontard, N. Fresh Food Packaging Design: A Requirement Driven Approach Applied to Strawberries and Agro-Based Materials. *Innov. Food Sci. Emerg. Technol.* **2013,** *20,* 288–298.

Cameron, A. C.; Talasila, P. C.; Joles, D. W. Predicting Film Permeability Needs for Modified-Atmosphere Packaging of Lightly Processed Fruits and Vegetables. *Hortscience.* **1995,** *30,* 25–34.

Candir, E.; Ozdemir, A. E.; Kamiloglu, O.; Soylu, E. M.; Dilbaz, R.; Ustun, D. Modified Atmosphere Packaging and Ethanol Vapor to Control Decay of 'Red Globe' Table Grapes During Storage. *Postharvest Biol. Technol.* **2012**, *63*, 98–106.

Cha, D. S.; Chinnan, M. S.; Biopolymer-Based Antimicrobial Packaging: A Review. *Crit. Rev. Food Sci. Nutr.* **2004**, *44*, 223–237.

Charles, F.; Anchez, J. S.; Gontard, N. Modeling of Active Modified Atmosphere Packaging of Endives Exposed to Several Postharvest Temperatures. *J. Food Sci.* **2005**, *70*, 443–449.

Charles, F.; Guillaume, C.; Gontard, N. Effect of Passive and Active Modified Atmosphere Packaging on Quality Changes of Fresh Endives. *Postharvest Biol. Technol.* **2008**, *48*, 22–29.

Charles, F.; Sanchez, J.; Gontard, N. Active Modified Atmosphere Packaging of Fresh Fruits and Vegetables: Modeling with Tomatoes and Oxygen Absorber. *J. Food Sci.* **2003**, *68*, 1736–1742.

Charles, F.; Sanchez, J.; Gontard, N. Absorption Kinetics of Oxygen and Carbon Dioxide Scavengers as Part of Active Modified Atmosphere Packaging. *J. Food Eng.* **2006**, *72*, 1–7.

Chervin, C.; Westercamp, P.; Monteils, G. Ethanol Vapors Limit Botrytis Development over the Postharvest Life of Table Grapes. *Postharvest Biol. Technol.* **2005**, *36*, 319–322.

Coma, V. Bioactive Packaging Technologies for Extended Shelf Life of Meat-Based Products. *Meat Sci.* **2008**, *78*, 90–103.

Cruz-Romero, M. C.; Murphy, T.; Morris, M.; Cummins, E.; Kerry, J. P. Antimicrobial Activity of Chitosan, Organic Acids and Nano-Sized Solubilisates for Potential Use in Smart Antimicrobially-Active Packaging for Potential Food Applications. *Food Control.* **2013**, *34*, 393–397.

El-Kazzaz, M.; Sommer, N.; Kader, A. Ethylene Effects on in Vitro and in Vivo Growth of Certain Postharvest Fruit-Infecting Fungi. *Phytopathology.* **1983**, *73*, 998–1001.

Esturk, O.; Ayhan, Z.; Gokkurt, T. Production and Application of Active Packaging Film with Ethylene Adsorber to Increase the Shelf Life of Broccoli. *Packag. Technol. Sci.* **2014**, *27*, 179–191.

European Commission. Commission Regulation (EC) No 450/2009 of 29 May 2009 on Active and Intelligent Materials and Articles Intended to Come into Contact with Food. *Official Journal of the European Union.* 2009. EC No 450/2009.

European Commission. EU Guidance to the Commission Regulation (EC) No 450/2009 of 29 May 2009 on Active and Intelligent Materials and Articles Intended to Come into Contact with Food. 2011. Directorate, E.C.H.a.C.

Fernández, A.; Picouet, P.; Lloret, E. Cellulose-Silver Nanoparticle Hybrid Materials to Control Spoilage-Related Microflora in Absorbent Pads Located in Trays of Fresh-Cut Melon. *Int. J. Food Microbiol.* **2010**, *142*, 222–228.

Ferrocino, I.; La Storia, A.; Torrieri, E.; Musso, S. S.; Mauriello, G.; Villani, F.; Ercolini, D. Antimicrobial Packaging to Retard the Growth of Spoilage Bacteria and to Reduce the Release of Volatile Metabolites in Meat Stored under Vacuum at 1°C. *J. Food Protect.* **2013**, *76*, 52–58.

Gamage, G. R.; Park, H. -J.; Kim, K. M. Effectiveness of Antimicrobial Coated Oriented Polypropylene/Polyethylene Films in Sprout Packaging. *Food Res. Int.* **2009**, *42*, 832–839.

Gomes, C.; Moreira, R. G.; Castell-Perez, E. Radiosensitization of Salmonella spp. and Listeria spp. in Ready-to-Eat Baby Spinach Leaves. *J. Food Sci.* **2011**, *76*, 141–148.

Gontard, N.; Guillaume, C. Packaging and the Shelf Life of Fruits and Vegetables. In *Food Packaging and Shelf Life: A Practical Guide;* Robertson, G. L., Ed.; CRC Press: Boca Raton, FL, 2009.

Goodburn, C.; Wallace, C. A. The Microbiological Efficacy of Decontamination Methodologies for Fresh Produce: A Review. *Food Control* **2013**, *32*, 418–427.

Gorny, J. R. A Summary of CA and MA Requirements and Recommendations for Fresh-Cut (Minimally Processed) Fruits and Vegetables. In *ISHS Acta Horticulturae 600: 8th International CA Conference;* Oosterhaven, J. H. W., Rotterdam, P., Eds.; ISHS: Netherlands; 2003.

Gross, K. C.; Wang, C. Y.; Saltveit, M. E., Eds.; *The Commercial Storage of Fruits, Vegetables, and Florist and Nursery Stocks*; United States Department of Agriculture: Beltsville, MD, 2004.

Gunduz, G. T.; Niemira, B. A.; Gonul, S. A.; Karapinar, M. Antimicrobial Activity of Oregano Oil on Iceberg Lettuce with Different Attachment Conditions. *J. Food Sci.* **2012**, *77*, 412–415.

Gustavsson, J.; Cederberg, C.; Sonesson, U.; Otterdijk, R. V.; Meybeck, A. *Global Food Losses and Food Waste-extent, Causes and Prevention*; Food and Agriculture Organization of the United Nations (FAO): Rome, Italy, 2011.

Han, J. H. Antimicrobial Food Packaging. In *Novel Food Packaging Techniques;* Ahvenainen, R., Ed.; Woodhead Publishing: Cambridge, England, pp 50–70.

Han, J. H. New Technologies in Food Packaging: Overview. In *Innovation in Food Packaging;* Han, J. H., Ed.; Elsevier Ltd.: San Diego, CA, 2005; pp 3–11.

Han, J. H.; Ho, C. H. L.; Rodrigues, E. T. Intelligent Packaging. In *Innovations in Food Packaging* Han, J. H., Ed.; Elsevier Ltd.: San Diego, CA, 2005; pp 138–159.

Hauser, C.; Wunderlich, J. Antimicrobial Packaging Films with a Sorbic Acid Based Coating. *Procedia Food Sci.* **2011**, *1*, 197–202.

Illeperuma, C. K.; Nikapitiya, C. Extension of the Postharvest Life of `Pollock' Avocado Using Modified Atmosphere Packaging. *Fruits* **2002**, *57*, 287–295.

Jideani, V.; Vogt, K. Antimicrobial Packaging for Extending the Shelf Life of Bread – A Review. *Crit. Rev. Food Sci. Nutr.* **2014**, *56*, 1313–1324.

Jin, T.; Zhang, H.; Boyd, G. Incorporation of Preservatives in Polylactic Acid Films for Inactivating Escherichia coli O157:H7 and Extending Microbiological Shelf Life of Strawberry Puree. *J. Food Protect.* **2010**, *73*, 812–818.

Kader, A. A.; Saltveit, M. E. Atmosphere Modification. In *Postharvest Physiology and Pathology of Vegetables;* Bartz, J. A., Brecht, J. K., Eds.; Marcel Dekker Inc.: New York, NY, 2003a; pp 229–246.

Kader, A. A.; Saltveit, M. E. Respiration and Gas Exchange. In *Postharvest Plant Physiology and Pathology of Vegetables* Bartz, J. A., Brecht, J. K., Eds.; Marcel Dekker Inc.: New York, 2003b; pp 7–29.

Kader, A. A.; Zagory, D.; Kerbel, E. L. Modified Atmosphere Packaging of Fruits and Vegetables. *Crit. Rev. Food Sci. Nutr.* **1989**, *28*, 1–30.

Kartal, S.; Aday, M. S.; Caner, C. Use of Microperforated Films and Oxygen Scavengers to Maintain Storage Stability of Fresh Strawberries. *Postharvest Biol. Technol.* **2012**, *71*, 32–40.

Kerry, J. P.; O'Grady, M. N.; Hogan, S. A. Past, Current and Potential Utilization of Active and Intelligent Packaging Systems for Meat and Muscle-Based Products: A Review. *Meat Sci.* **2006**, *74*, 113–130.

Labuza, T. P.; Breene, W. M. Applications of "Active Packaging" for Improvement of Shelf-Life and Nutritional Quality of Fresh and Extended Shelf-Life Foods. *J. Food Process Pres.* **1989**, *13*, 1–69.

Lee, D. S.; Shin, D. H.; Lee, D. U.; Kim, J. C.; Cheigh, H. S. The Use of Physical Carbon Dioxide Absorbents to Control Pressure Buildup and Volume Expansion of Kimchi Packages. *J. Food Eng.* **2001,** *48,* 183–188.

Lee, D. S.; Yam, K. L.; Piergiovanni, L. *Food Packaging Science and Technology*; CRC Press, Taylor and Francis Group, LLC: Boca Raton, FL, 2008.

Li, W. L.; Li, X. H. Zhang, P. P.; Xing, Y. G. Development of Nano-ZnO Coated Food Packaging Film and Its Inhibitory Effect on Escherichia Coli in Vitro and in Actual Tests. In *New Materials and Advanced Materials, Pts 1 and 2;* Jiang, Z. Y., Han, J. T., Liu, X. H. Eds.; Trans Tech Publications Ltd.: Stafa, Zurich, 2011; pp 489–492.

Lu, H.; Zhu, J.; Li, J.; Chen, J. Effectiveness of Active Packaging on Control of *Escherichia coli* O157:H7 and Total Aerobic Bacteria on Iceberg Lettuce. *J. Food Sci.* **2015,** *80,* 1325–1329.

Mahajan, P. V.; Oliveira, F. A. R.; Montanez, J. C.; Frias, J. Development of User-Friendly Software for Design of Modified Atmosphere Packaging for Fresh and Fresh-Cut Produce. *Innov. Food Sci. Emerg. Technol.* **2007,** *8,* 84–92.

Mahajan, P. V.; Rodrigues, F. A. S.; Motel, A.; Leonhard, A. Development of a Moisture Absorber for Packaging of Fresh Mushrooms (*Agaricus bisporus*). *Postharvest Biol. Technol.* **2008,** *48,* 408–414.

Malhotra, B.; Keshwani, A.; Kharkwal, H. Antimicrobial Food Packaging: Potential and Pitfalls. *Front Microbiol.* **2015,** *6,* 1–9.

Martínez-Abad, A.; Ocio, M. J.; Lagarón, J. M.; Sánchez, G. Evaluation of Silver-Infused Polylactide Films for Inactivation of Salmonella and Feline Calicivirus In Vitro and on Fresh-Cut Vegetables. *Int. J. Food Microbiol.* **2013,** *162,* 89–94.

McMillin, K. W. Where is MAP Going? A Review and Future Potential of Modified Atmosphere Packaging for Meat. *Meat Sci.* **2008,** *80,* 43–65.

Montero-Prado, P.; Rodriguez-Lafuente, A.; Nerin, C. Active Label-Based Packaging to Extend the Shelf-Life of "Calanda" Peach Fruit: Changes in Fruit Quality and Enzymatic Activity. *Postharvest Biol. Technol.* **2011,** *60,* 211–219.

Muriel-Galet, V.; Cerisuelo, J. P.; López-Carballo, G.; Aucejo, S.; Gavara, R.; Hernández-Muñoz, P. Evaluation of EVOH-Coated PP Films with Oregano Essential Oil and Citral to Improve the Shelf-Life of Packaged Salad. *Food Control.* **2013,** *30,* 137–143.

Muriel-Galet, V.; Cerisuelo, J. P.; López-Carballo, G.; Lara, M.; Gavara, R.; Hernández-Muñoz, P. Development of Antimicrobial Films for Microbiological Control of Packaged Salad. *Int. J. Food Microbiol.* **2012,** *157,* 195–201.

Olaimat, A. N.; Holley, R. A. Factors Influencing the Microbial Safety of Fresh Produce: A Review. *Food Microbiol.* **2012,** *32,* 1–19.

Oshida, Y.; Takeuchi, T.; Yuyama, M. *Method for Potato Sprout Control;* Mitsubishi Gas Chemical Company: Japan, 2013.

Patrignani, F.; Siroli, L.; Serrazanetti, D. I.; Gardini, F.; Lanciotti, R. Innovative Strategies Based on the Use of Essential Oils and Their Components to Improve Safety, Shelf-Life and Quality of Minimally Processed Fruits and Vegetables. *Trends Food Sci. Technol.* **2015,** *46,* 311–319.

Ponce, A.; Roura, S.; Del Valle, C.; Fritz, R. Characterization of Native Microbial Population of Swiss Chard (Beta Vulgaris, Type Cicla). *LWT-Food Sci. Technol.* **2002,** *35,* 331–337.

Qin, Y.; Liu, D.; Wu, Y.; Yuan, M.; Li, L.; Yang, J. Effect of PLA/PCL/Cinnamaldehyde Antimicrobial Packaging on Physicochemical and Microbial Quality of Button Mushroom (*Agaricus Bisporus*). *Postharvest Biol. Technol.* **2015,** *99,* 73–79.

Quintavalla, S.; Vicini, L. Antimicrobial Food Packaging in Meat Industry. *Meat Sci.* **2002,** *62,* 373–380.

Ramos, B.; Miller, F.; Brandão, T. R.; Teixeira, P.; Silva, C. L. Fresh Fruits and Vegetables-An Overview on Applied Methodologies to Improve Its Quality and Safety. *Innov. Food Sci. Emerg. Technol.* **2013,** *20,* 1–15.

Ray, S.; Jin, T.; Fan, X.; Liu, L.; Yam, K. L. Development of Chlorine Dioxide Releasing Film and Its Application in Decontaminating Fresh Produce. *J. Food Sci.* **2013,** *78,* 276–284.

Realini, C. E.; Marcos, B. Active and Intelligent Packaging Systems for a Modern Society. *Meat Sci.* **2014,** *98,* 404–419.

Restuccia, D.; Spizzirri, U. G.; Parisi, O. I.; Cirillo, G.; Curcio, M.; Iemma, F.; Puoci, F.; Vinci, G.; Picci, N. New EU Regulation Aspects and Global Market of Active and Intelligent Packaging for Food Industry Applications. *Food Control.* **2010,** *21,* 1425–1435.

Robertson, G. L. *Food Packaging Principles and Practice;* CRC Press, Taylor and Francis Group: Boca Raton, FL, 2012.

Rodov, V.; Ben-Yehoshua, S.; Fierman, T.; Fang, D. Modified-Humidity Packaging Reduces Decay of Harvested Red Bell Pepper Fruit. *Hortscience.* **1995,** *30,* 299–302.

Rodriguez-Lafuente, A.; Nerin, C.; Batlle, R. Active Paraffin-Based Paper Packaging for Extending the Shelf Life of Cherry Tomatoes. *J. Agr. Food Chem.* **2010,** *58,* 6780–6786.

Rodríguez, A.; Batlle, R.; Nerín, C. The Use of Natural Essential Oils as Antimicrobial Solutions in Paper Packaging. Part II. *Prog. Org. Coat.* **2007,** *60,* 33–38.

Rooney, M. L. Overview of Active Food Packaging. In *Active Food Packaging;* Rooney, M. L., Ed.; Springer: Dordrecht, Netherlands, 1995; pp 1–37.

Rooney, M. L. Introduction to Active Food Packaging Technologies. In *Innovations in Food Packaging;* Han, J. H., San Diego, Eds.; Elsevier Ltd.: Laguna Hills, CA, 2005.

Roy, S.; Anantheswaran, R. C.; Beelman, R. B. Sorbitol Increases Shelf Life of Fresh Mushrooms Stored in Conventional Packages. *J. Food Sci.* **1995,** *60,* 1254–1259.

Roy, S.; Anantheswaran, R. C.; Beelman, R. B. Modified Atmosphere and Modified Humidity Packaging of Fresh Mushrooms. *J. Food Sci.* **1996,** *61,* 391–397.

Sängerlaub, S.; Böhmer, M.; Stramm, C. Influence of Stretching Ratio and Salt Concentration on the Porosity of Polypropylene Films Containing Sodium Chloride Particles. *J. Appl. Polym. Sci.* **2013,** *129,* 1238–1248.

Scully, A. D.; Horsham, M. A. Active Packaging for Fruits and Vegetables. In *Intelligent and Active Packaging for Fruits and Vegetables;* Wilson, C. L., Ed.; CRC Press: Boca Raton, FL, 2007; pp 57–71.

Seo, H. -S.; Bang, J.; Kim, H.; Beuchat, L. R.; Cho, S. Y.; Ryu, J. -H. Development of an Antimicrobial Sachet Containing Encapsulated Allyl Isothiocyanate to Inactivate Escherichia Coli O157:H7 on Spinach Leaves. *Int. J. Food Microbiol.* **2012,** *159,* 136–143.

Shirazi, A.; Cameron, A. C. Controlling Relative Humidity in Modified Atmosphere Packages of Tomato Fruit. *Hortscience.* **1992,** *27,* 336–339.

Singh, P.; Sängerlaub, S.; Stramm, C.; Langowski, H. -C. Humidity Regulating Packages Containing Sodium Chloride as Active Substance for Packaging of Fresh Raw Agaricus Mushrooms. In *4th International Workshop Cold Chain-Management;* Kreyenschmidt, J. Bonn., Ed.; 2010.

Souza, R.; Peruch, G.; Santos Pires, A. C. d. Oxygen Scavengers: An Approach on Food Preservation. In *Structure and Function of Food Engineering;* Eissa, A. A., Ed.; InTech: Rijeka, Croatia, 2012; pp 21–42. Published Online: http://www.intechopen.com/books/structure-and-function-of-food-engineering:

Sung, S. -Y.; Sin, L. T.; Tee, T. T.; Bee, S. T.; Rahmat, A.; Rahman, W.; Tan, A. C.; Vikhraman, M. Antimicrobial Agents for Food Packaging Applications. *Trends Food Sci. Technol.* **2013**, *33*, 110–123.

Suppakul, P. Natural Extracts in Plastic Food Packaging. In *Multifunctional and Nanoreinforced Polymers for Food Packaging;* Lagarón, J. M., Ed.; Woodhead Publishing: Cambridge, UK, 2011; pp 421–459.

Suppakul, P.; Miltz, J.; Sonneveld, K.; Bigger, S. W. Active Packaging Technologies with an Emphasis on Antimicrobial Packaging and Its Applications. *J. Food Sci.* **2003**, *68*, 408–420.

Takala, P. N.; Salmieri, S.; Boumail, A.; Khan, R. A.; Vu, K. D.; Chauve, G.; Bouchard, J.; Lacroix, M. Antimicrobial Effect and Physicochemical Properties of Bioactive Trilayer Polycaprolactone/Methylcellulose-Based Films on the Growth of Foodborne Pathogens and Total Microbiota in Fresh Broccoli. *J. Food Eng.* **2013**, *116*, 648–655.

Tawakkal, I. S.; Cran, M. J.; Miltz, J.; Bigger, S. W. A Review of Poly (Lactic Acid)-Based Materials for Antimicrobial Packaging. *J. Food Sci.* **2014**, *79*, R1477–R1490.

Tournas, V. Moulds and Yeasts in Fresh and Minimally Processed Vegetables, and Sprouts. *Int. J. Food Microbiol.* **2005**, *99*, 71–77.

Vermeiren, L.; Devlieghere, F.; van Beest, M.; de Kruijf, N.; Debevere, J. Developments in the Active Packaging of Foods. *Trends Food Sci. Technol.* **1999**, *10*, 77–86.

Vermeiren, L.; Heirlings, L.; Devlieghere, F.; Debevere, J. Oxygen, Ethylene and Other Scavengers. *Novel Food Packag. Tech.* **2003**, 22–49.

Vermeulen, A.; Ragaert, P.; De Meulenaer, B.; Devlieghere, F. PredOxyPack®: How to Predict the Impact of the Cold Chain Conditions on the Oxygen Barrier Properties of Packaging. In *5th International Cold-Chain-Management Workshop;* 2013; pp 1–6.

Zagory, D. Ethylene-Removing Packaging. In *Active Food Packaging;* Springer: Boston, MA, 1995; pp 38–54.

Zagory, D.; Kader, A. A. Modified Atmosphere Packaging of Fresh Produce. *Food Technol.* **1988**, *42*, 70–77.

CHAPTER 4

INTELLIGENT PACKAGING APPLICATIONS FOR FRUITS AND VEGETABLES

BAMBANG KUSWANDI[*]

Chemo and Biosensors Group, Faculty of Pharmacy, University of Jember, Jl. Kalimantan 37, Jember 68121, Indonesia

[*]E-mail: b_kuswandi@farmasi.unej.ac.id

CONTENTS

ABSTRACT

Intelligent or smart packaging is the integration of sensor or indicator into the food packaging system that has intelligent functions for monitoring of food quality and safety, such as freshness, leakage, carbon dioxide, oxygen, pH, time or temperature, and pathogens. Thus, this integration is truly integrated and interdisciplinary systems that invoke expertise from the fields of chemistry, biochemistry, physics, and electronics as well as food science and technology. Therefore, this technology is needed as real-time quality control and safety monitoring in terms of consumers, authorities, and food producers. It has great potential in the development of new sensing systems integrated in the food packaging, which are beyond the existing conventional technologies, like control of weight, volume, color, and appearance. In this chapter, the application of intelligent packaging for fruits and vegetables are presented, such as freshness and ripeness monitoring. In the case of fruits, sensors/indicators for ripeness detection have been developed. While for vegetables, sensors/indicators for freshness monitoring have been performed. For ripeness detection, the sensor was constructed to detect volatile compounds (e.g., ethylene, volatile acid, etc., that are produced during the ripeness process). While for freshness detection, the sensor works based on pH change or volatile compounds produced gradually during degradation of vegetable. Furthermore, the sensor responses were found to correlate with pH, sensory evaluation, and bacterial growth patterns in food samples.

4.1   INTRODUCTION

The ability to ship fresh produce to long distances has made a wide variety of fruits and vegetables available to global markets. However, ensuring that produce arrives on store shelves at peak flavor and freshness is a challenge. Produce shipments that are delayed or experienced temperature extremes during transit can result in spoiled or over riped fruits and vegetables, thus increase costs and decrease consumer satisfaction. The new packaging technology can track temperature and ripeness or control the rate of ripening during shipment and on store shelves, thus offers an ideal solution to this problem. In order to meet such demands, new packaging technology such as intelligent/smart packaging has been developed (Smolander, 2003; Yam et al., 2005; Kuswandi et al., 2011a; Wilson, 2007).

Intelligent packaging with intelligent label or indicator provides information on the integrity and the time-temperature history of the food package.

It is helpful in assuring the quality and safety of the packaged food products (Ahvenainen, 2003; Kerry et al., 2006), particularly for perishable foods (Welt [et al.], [2003]; Taoukis & Labuza, 2003; Smolander et al., 2004). This chapter's main focus is on fruits and vegetables. While protecting and preserving food as the principal role of food packaging (Robertson, 2006), it also gives information on food quality, traceability, tamper indication, or safety via its sensor label. Rodrigues and Han (2003) reported the collective list of various commercial intelligent packaging systems and potential applications in fruits and vegetables.

The growing needs for innovation in packaging will drive to progress intelligent packaging, particularly for perishable food products, such as fruits and vegetables. Consumers increasingly need to know what main ingredients or components are in the product and how the product should be stored and used. Intelligent indicators, for instance, are able in communicating directly to the customer via sensor/indicator (i.e., visual information) regarding the condition, quality, and safety of the food product.

This chapter is focusing on the application of intelligent packaging that is related to the integration of sensor technology (i.e., chemical sensor and biosensor). Commonly, they are called sensor labels or indicators on food packages for quality and safety. Their principle, design, construction, and various applications of intelligent packaging are described especially for fruits and vegetables. This is real-time monitoring of ripeness, freshness, quality, and safety of raw and processed products.

4.2 INTELLIGENT LABEL

Regardless of food packaging material, type, and design, practically intelligent/smart label or indicator label is used in intelligent packaging, it is a sensor either chemical sensor or biosensor that sense and inform the status of a food product in terms of its quality (showing the freshness, ripeness, or firmness) or safety (showing either food to be safe or food to be unsafe). In this direction, the intelligent label could be defined as an on-package sensor or indicator that has ability to track the product, sense the environment inside or outside the package and inform the manufacturer, retailer, and consumer regarding the condition of the food product (Kuswandi, 2012) as given in Figure 4.1.

Principally, there are two types of intelligent labels that have been developed in intelligent packaging: (1) direct label and (2) indirect label. The direct label is the sensor label that works based on direct detection of a particular analyte as a marker for food quality or safety. In this case, a variety of different concepts for markers have been presented in the literature for food quality and safety for fruits and vegetables. While indirect label is the sensor label, that works based on indirect detection, for example, based on a reaction system that imitates food degradation inside the package. In indirect labels, these devices are expected to mimic the change of a certain quality parameter of the food product undergoing the same exposure to temperature and surrounding conditions. Some are designed to monitor the evolution of gases and temperature change along the distribution chain, while others are designed to be used in consumer packages (Robertson, 2006). In order to use in food quality and safety monitoring devices, in both labels, the rate of change in the label must correlate well with the rate of deterioration of the food product. The rate of sensor labels change must also be correlated with the temperature variation during transportation and distribution (i.e., time of exposure) (Seldman, 1995).

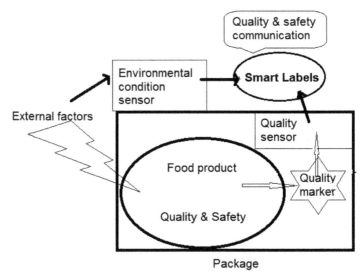

FIGURE 4.1   Principle of intelligent packaging in food incorporated intelligent/smart labels.

One example of the first type of intelligent label is the sensor or indi-cator label that used for direct detection of food freshness. In this case, a variety of different concepts for freshness indicators have been presented in the literature. For example, freshness indicators or sensors concepts have been proposed for total volatile basic nitrogen (Pacquit et al., 2007; Pacquit et al., 2006), CO_2 (Mattila et al., 1990), amines (Wallach & Novicov, 1998), ammonia (Horan, 1998), ethanol (Cameron & Talasila, 1995), hydrogen sulfide (H_2S) (Smolander et al. 1998) and pH (Kuswandi et al., 2012; Kuswandi et al., 2011b). By integrating the sensor or indicator into the food package, the freshness indicators can be realized as visible indicator tags going through a color change in the presence of an analyte or biomarker.

Commonly used indirect label are time–temperature indicators (TTIs). The TTIs presently available on the market have working mechanisms based on different principles, namely, chemical, physical, and biological reactions. For chemical or physical response, usually it is based on chemical reaction or physical change toward time and temperature, such as acid–base reaction, melting, and polymerization. While for biological response, it is based on the change in biological activity, such as microorganism, spores, or enzymes toward time or temperature. The TTIs will change color when exposed to higher than recommended storage temperature and will change as the product reaches the end of its shelf-life. Thus, TTIs based on physical, chemical, or enzymatic activity in the food will give a clear, accurate, and unambiguous indication of product quality, safety, and shelf-life condition.

The intelligent labels can play an important role in indicating the quality and safety of a food product. The most important function is that these labels can monitor and communicate food safety. Furthermore, this becomes extremely important when food is stored in less than optimal conditions such as extreme heat or freezing. In the case of foods that should not be frozen, a smart label would indicate whether the food had been improperly exposed to cold temperatures. Conversely, a smart label could specify whether foods are sensitive to heat and exposed to unnaturally high temperatures. Therefore, various applications of intelligent labels are described in the following sections.

4.2.1 DIRECT INTELLIGENT LABELS

The direct intelligent labels are sensors that directly detect or sense for a particular marker or compounds as an indicator for food quality or safety. In this case, a variety of different intelligent labels have been presented in the literature, such as ripeness indicator, freshness indicator, leak indicator, and pathogen and contaminants indicator. Ripeness indicators are suitable to be used for raw fruits and vegetables, while the rest are used for processed fruits and vegetables. Mainly the sensor labels were presented as color indicator where, the rate of color change in the label correlated well with the rate of deterioration of the food product as a function of temperature variation and time during transportation, distribution, and storage.

Freshness of fruits and vegetables can be determined by gas and volatile compound measurements in the headspace using direct sensor labels. This sensor detects volatile metabolites generated by the produce such as oxygen, carbon dioxide, diacetyl, amines, ethanol and hydrogen sulfide (Ahvenainen, 2003; Kuswandi et al., 2011a; Smolander, 2003; Wilson, 2007; Yam et al., 2005). Ethanol concentration in package headspace can be measured by an enzymatic reaction based on chromogenic substrates and the color indicates the degree of fermentation (Smolander, 2008). Bromothymol blue and methyl red (MR) are chromogenic indicators that can be used as the sensor labels for fruits and vegetables fermentation. They react with carbon dioxide produced by fermentation, and the color density is used to determine the degree of fermentation (Smolander, 2008). Microbial quality indicators based on carbon dioxide production have limited applications in fresh fruits and vegetables because a large amount of carbon dioxide is produced by respiration, which masks the amounts produced by microbial metabolism (Smolander, 2008). Other direct indicator label can also be developed based on changes in the concentration of hydrogen sulfide or organic acids such as *n*-butyrate, L-lactic, D-lactate and acetic acid (Wanihsuksombat, 2010).

4.2.1.1   RIPENESS INDICATOR FOR FRESH FRUITS

It is difficult to know when a fruit has reached its preferred state of ripeness, and this situation always becomes a barrier to purchase or frustrate consumers. One company in New Zealand, Ripesense Ltd., produced *ripeSense*™ (Ripesense, 2015) to eliminate this problem by using a sensor label that reacts to the aromas released by fruit as it ripens. The sensor is initially red and graduates to orange and finally yellow. It is designed to sense the aroma compounds given off by the fruit (Pocas et al., 2008). By matching the color of the sensor with a reference color standard, a customer can choose the fruit with the required ripening stage (Fig. 4.2). Damage and shrinkage are reduced as this sensor significantly reduces damage by consumers as they inspect fruit before purchase and the recyclable sensor pack provides improved hygiene of the product. This sensor has already been applied for pears, and can also be applied for kiwi fruit, melon, mango, avocado, and other stone fruits (Ripesense, 2015).

The number of publications on package indicators for ripeness of fruit is still limited. However, there are some trials to construct indicators for ripeness detection based on ethylene emission during ripening stage, for example, apple (Kim & Shiratori, 2006; Lang & Hubert, 2011). In this case, the color indicator based on molybdenum (Mo) chromophores that change under reaction with ethylene from white to blue because of reduction of Mo(VI) to Mo(V). The sensitivity of molybdenum color change reactions can be varied by composition and pH values (pH 1.4–1.5) of ammonium molybdate solution. The indicator can be combined with color recognition for quantitative measurements of color change.

FIGURE 4.2   Ripeness sensor on fruit package using ripeSense®.

Another approach was developed for on-package color indicator based on bromophenol blue (BPB), and experimental tests have been conducted to detect the ripeness stage of guava (*Psidium guajava* L.) (Kuswandi et al, 2013a). BPB was immobilized onto bacterial cellulose membrane via absorption method. The BPB/cellulose membrane as ripeness indicator works based on pH decrease as the volatile organic acids (e.g., ascorbic acid) produced gradually in the package headspace during ripening stage of guava. Subsequently, the color of the indicator will change from blue to green for over-ripe (juicy) indication (Fig. 4.3). These results show that the color indicator could be used to visualize the state of ripeness of guava, stored at 4 or 28°C. Irrespective of temperature, color changes of the indicators closely reflect the variation of several parameters (pH, soluble solids contents, texture softness, and sensory evaluation) normally used to characterize the ripeness stage of guavas.

FIGURE 4.3   Ripeness sensor for guava packaging (*Source:* Adapted from Kuswandi et al. 2013a).

For non-climacteric fruits, such as strawberries, simple and low-cost on-package color indicator has been fabricated based on MR for ripeness detection of strawberries (*Fragaria vesca* L.) (Kuswandi et al., 2013b). MR was also immobilized onto bacterial cellulose membrane and works based on pH increase as the volatile acids reduced gradually in the package headspace due to enzymatic formation of esters occurring during ripening stages. Subsequently, the color of the indicator will change from yellow to red-purple for over-ripe indication as given in Figure 4.4. The results show that the color indicator could be used to determine the ripeness stage of strawberries, since the correlation between the color changes of indicator toward ripeness stage of strawberries is in similar trend. The MR/cellulose membrane was successfully used as on-package color indicator for on-line ripeness stage monitoring of strawberries in ambient and chillier conditions.

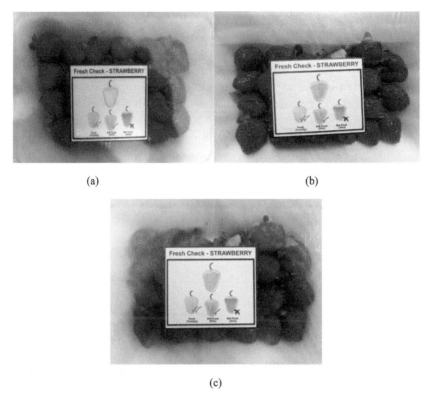

(a)

(b)

(c)

FIGURE 4.4 Ripeness sensor for strawberry, (a) yellow for fresh (crunchy), (b) orange for medium (firm), and (c) purple for not fresh (juicy).

4.2.1.2 FRESHNESS INDICATOR FOR PROCESSED FRUITS AND VEGETABLES

Freshness indicator is commonly used in intelligent packaging for freshness monitoring of fish and meats (Seldman, 1995; Pacquit et al., 2007; Pacquit et al., 2006; Kuswandi, 2012). The similar principle can also be applied in intelligent packaging for fruits and vegetables in particular for cuts and processed fruits and vegetables. Simple and inexpensive sensor labels allow real-time and non-invasive determination of processed fruits and vegetables. These were described in literature based on a pH change. In an enclosed food package, as the processed fruits or vegetable product spoils, a pH increase occurs over time within the headspace, which can be detected with an appropriate pH-indicating sensor. The fundamental characteristic of pH indicator dyes that change color when placed in an acidic or basic environment is the key element of this sensor. Indicators based on color changes due to changes in pH are of great potential as ripeness indicators for vegetable fermentation, such as kimchi (Hong & Park, 2000).

A fast and sensitive detection of spoilage compounds of processed fruits (i.e., dessert) can be achieved by a non-invasive colorimetric method. This colorimetric method was developed using mixed-dye-based food spoilage indicator (Nopwinyuwong et al., 2010). It allows the food product to have an effective shelf-life by permitting dynamic freshness to be monitored visually alongside the best-before date, consequently decreasing margins of error. The expansion of the concept of a colorimetric mixed-dye-based food spoilage indicator to other processed fruits and vegetable products, such as prepared foods, desserts, and fresh-cut fruits and vegetables, could be developed in the future.

4.2.1.3   LEAK INDICATOR FOR CANNED FRUITS AND VEGETABLES

Currently, the canned fruits and vegetables are increasing in the international export market. This industry primarily engaged in canning fruits, vegetables, fruit and vegetable juices; processing ketchup and other tomato sauces; and in producing natural and imitation preserves, jams, and jellies. The primary objective of food processing is the preservation of perishable foods in a stable form that can be stored and shipped to distant markets during all months of the year. Processing also can change foods into new or more consumable forms and make more convenient to prepare food dishes (Woodroof & Luh, 1986).

The goal of the canning process is to destroy any microorganisms in the food and to prevent recontamination. Heat is the most common agent used to destroy microorganisms. Removal of oxygen can be used in conjunction with other methods to prevent the growth of oxygen-requiring microorganisms. Therefore, many processed fruits and vegetables often sold as canned fruits or canned vegetables, mainly stored in modified atmosphere packaging (MAP) (Luh & Woodroof, 1988). MAP and equilibrium MAP are classified as active packaging methods (Shen et al., 2006). In these cases, the atmosphere of package is not air, but consists of a lowered level of O_2 ($\leq 2\%$) and a heightened level of CO_2 (20–80%). Therefore, a leak in MAP means a considerable increase in the O_2 concentration and a decrease in the CO_2 concentration, which in turn, enable aerobic microbial growth to take place. In the worst case, the CO_2 concentration will thus remain high despite leakage and permit microbial growth. Thus, the leak indicators for MAPs are much more important than active packaging, since they become intelligent packaging, and they should rely on the detection of O_2 rather than on the detection of CO_2 (Smolander et al., 1997).

At present, the main application of the commercially available O_2 sensitive MAP indicators is to ensure the proper functioning of O_2 absorption. For example, Mitsubishi Gas Chemical Company (Japan) commercialized their O_2 absorbing sachets under the trade name "Ageless" (Abe, 1994). There are also some other companies producing commercial O_2 indicators to confirm proper O_2 removal by O_2 absorbers (Hurme & Ahvenainen, 1996). Another company Cryovac-Sealed Air Ltd. has developed indicator for checking the types of gas composition (Anonym, 1996).

Usually a typical visual O_2 indicator consists of a redox-dye (e.g., methylene blue (Nakamura et al., 1987; Goto, 1987), 2,6-dichloroindophenol (LeNarvor et al., 1993) or N,N,N',N'-tetramethyl-p-phenylenediamine (Mattila-Sandholm et al., 1998), a reducing compound (e.g., reducing sugars) (Perlman & Linschitz, 1985) and an alkaline compound (e.g., sodium hydroxide (Shirozaki, 1990), potassium hydroxide (Yamamoto, 1992), calcium hydroxide (Yoshikawa et al., 1979), or magnesium hydroxide (Yoshikawa et al., 1982). Oxygen indicator has also been reported in literature, which was based on oxidative enzymes (Gardiol et al., 1996). In addition to these main components, compounds such as a solvent (typically water and/or an alcohol) and bulking agent (e.g., zeolite, silica gel, cellulose materials and polymers) can also be added to the indicator. The indicator can be formulated as a tablet (Nakamura et al., 1987; Goto, 1987), a printed layer (Davies & Garner, 1996; Gardiol et al., 1996) or laminated in a polymer film (Gardiol et al., 1996).

In the case of CO_2-indicators as leak detection, it does not appear to be more reliable leak detection compared to O_2-indicators. This is due to the fact that during the first 1–2 days after the packaging procedure, CO_2 will be dissolved into the product, and then its concentration in the headspace increased followed by a decrease to the final concentration. After this period, a considerable decrease in CO_2 concentration is certainly an evident sign of leakage in a package. Another drawback of CO_2 indicators is related to the production of CO_2 in the microbial metabolism. A leak in a package by decreasing in the CO_2 is often followed by microbial growth, which means increase in the CO_2. In the worst case, the CO_2 will remain constant even in the case of leakage and microbial spoilage.

Few factors need to be considered if O_2 and CO_2 indicators are used as leak indicators of MAP. The very low sensitivity of O_2 indicators is not advantageous as the sensitive indicator might also react with the residual O_2

which is often entrapped in the MAP (typically 0.5–2.0%) (Ahvenainen & Hurme, 1997). Very low sensitivity can also complicate the handling of the indicator, requiring anaerobic conditions during the preparation of the indicator and the food packaging procedure. Furthermore, it has been claimed that the color change of the O_2 indicators used in MAPs containing acidic CO_2 gas is not definite enough (Ahvenainen & Hurme, 1997; Balderson & Whitwood, 1995). The reversibility is undesirable if the indicator is used for leakage control since the O_2 entering the package through the leak will be consumed in the microbial growth (Balderson & Whitwood, 1995). In this case, color will be same as intact packages, even if the product has been spoiled.

4.2.1.4   PATHOGEN AND CONTAMINANTS INDICATOR

Storage and processing technologies have been utilized for centuries to transform these perishable fruits and vegetables into safe, delicious, and stable products. Freezing, canning, and drying are used to transform perishable fruits and vegetables into products that can be consumed around the year and transported safely to consumers all over the world. As a result of processing, respiration is arrested, thereby stopping the consumption of nutritious components, the loss of moisture, and the growth of microorganisms as food pathogen. In addition, it can avoid food contaminants during processing, transportation, distribution, and storage. Most of these sensor platforms are incorporated within devices, and unfortunately, it requires the extraction of analyte in the sample to determine the presence of the target molecule. When considering such systems for food packaging, these are focused on detecting microbial contaminant growth. The challenge for such systems is to integrate within the packaging, to provide an easily distinguished response (most likely a color change), and to make it cheap to manufacture. It is most likely that the presence of microbial contamination will be detected indirectly by measuring changes in gas composition within the package as a result of microbial growth (i.e., using gas sensor as described earlier). The available concepts of package indicators for food contaminants or food pathogens are still very low. Even if the indication of microbial growth by CO_2 is difficult in MAPs, which often already contain a high concentration of CO_2. It is possible to use the increase in CO_2 concentration as a means of determining microbial contamination or pathogen on food, only if the package does not contain CO_2 as protective gas (Mattila et al., 1990).

The color indicators based on reactions caused by microbial metabolites and other concepts for contamination indicators have been proposed in the literature (DeCicco & Keeven, 1995; Kress-Rogers, 1993). The color indicators could be based on a color change of chromogenic substrates of enzymes produced by contaminating microbes (DeCicco & Keeven, 1995), the consumption of certain nutrients in the product or on the detection of microorganism itself (Kress-Rogers, 1993). For this reasons, instead of electrochemical transduction method, optical-based biosensors systems have also been widely developed together with biosensors based on acoustic transduction which is intended to be used mainly for detecting microbial contaminants. These methods have been used for targeting the presence of contaminating microorganisms on food such as *Staphylococcal* enterotoxin A and B, *Salmonella typhimurium, Salmonella* group B, D, and E, *E. coli* and *E. coli 0157:H7* (Kress-Rogers, 1993; Terry et al., 2004).

Alternatively, systems based on caged biomolecules (e.g., fullerenes, liposomes, or nanoporous silica), linked to a colorimetric dye, could be developed for this purpose. These could be provided for stability for the detector molecule and incorporated in a permeable membrane within the main package; however, it requires additional factors (e.g., pre-processing and power). One example is the employment of nano-structured silk as a platform for biosensors. This silk could be incorporated within the package, since the silk is biodegradable and edible (Lawrence et al., 2008). The silk fibrils can be shaped into "lenses" and modified with various biomolecules, which could bound to the targets (such as microbial proteins) and alter their shape, thus resulting in a color change (Lawrence et al., 2008).

Biosensors, such as conducting polymers can also be used for detecting the gases released during microbe metabolism (Ahuja et al., 2007; Retama, 2005). The biosensors are formed by inserting conducting nanoparticles into an insulating matrix, where the change in resistance correlates to the amount of gas released. Such sensors have been developed for detecting foodborne pathogens through quantification of bacterial cultures (Arshak et al., 2007).

4.2.2 INDIRECT INTELLIGENT LABELS

Indirect labels work based on indirect detection, where these devices are expected to mimic the change of a certain quality parameter of the perishable fruits and vegetables undergoing same exposure of temperature. The rate of change of the label must be correlated well with the rate of deterioration of the fruits and vegetables according to the temperature variation over time during transportation, distribution, and storage. Generally, similar with direct labels, the indirect labels will change color when exposed to higher than recommended storage temperature and will change as the product reaches the end of its shelf-life.

4.2.2.1   TEMPERATURE INDICATOR

Since both unprocessed and processed fruits and vegetables must undergo some transport and storage, degradation of some nutrients prior to consumption is expected. Lower temperatures, even in frozen goods, tend to prolong the shelf-life of fruits and vegetables (Kramer, 1982). Therefore, temperature indicator is needed to monitor temperature of fruits and vegetable during transportation, distribution, and storage. For instance, considering self-heating or self-cooling container of fruit juice or syrup, a sensor needs to tell the consumer that it is at the correct temperature and the package becomes "smart" (such packaging is currently commercially available).

The most commonly used temperature indicators are a thermochromic ink dot in order to indicate the product is at the correct serving temperature after refrigeration or microwave heating. Plastic containers of pouring syrup for pancakes can be purchased from the United States and United Kingdom that is labeled with a thermochromic ink dot to indicate that the fruit syrup is at the right temperature after microwave heating. Similar examples can be found on supermarket shelves with orange juice pack labels that incorporate thermochromic-based designs to inform the consumer when a refrigerated orange juice is cold enough to drink as given in Figure 4.5 (Tetrapak, 2015).

FIGURE 4.5   Application of thermochromic ink.

Another example of temperature indicator technology was developed by Smart Lid Systems (Sydney, Australia) (SmartLid, 2015). The smart lid is infused with a color-changing additive, which allows it to change from a coffee bean brown to a bright red color when exposed to an increase in temperature. The change in color starts at 38°C and it reaches full intensity at 45°C. If the red color appears intensely, it indicates that the coffee in the cup is too hot for comfortable drinking (Fig. 4.6). If the lid is cocked and not positioned correctly, the brown color will not be distributed evenly and this indicates that a potential for spillage exists. This color-changing additive is safe in food contact surfaces, since it meets the requirements of the US regulations relating to direct food contact materials additives (SmartLid, 2015).

FIGURE 4.6   Color changing disposable beverage lids showing increasing redness from left to right.

4.2.2.2   TIME AND TEMPERATURE INDICATORS (TTIS)

Variations in temperature are one factor that affects the quality of perishable food products, including fruits and vegetables. Although a product is processed, packed, and shipped at the optimum temperature for prolong shelf-life, during shipment and storage prior to retail sale, quality may be lost due to temperature fluctuations. To ensure that temperature abuse did not occur, TTIs on packages can provide an assurance to the consumer regarding product's quality. TTIs are small measuring devices capable of showing a time-temperature-dependent relationship of irreversible color change (De Johng et al., 2005). Commercially available TTIs are given in Table 4.1.

TABLE 4.1   Some Examples of Commercially Available TTIs.

Product	Company
MonitorMark™	3M™
Timestrip®	Timestrip Plc
Fresh-Check®	LifeLines
CheckPoint®	Vitsab

MonitorMark™ (Three M, 2015) from 3M™ has two versions, one intended for monitoring distribution and threshold indicator; and other intended for consumer information (i.e., sensor label). The former is an abuse indicator, which means that it yields no response unless a predetermined temperature is exceeded. It is based on a special substance having a selected melting point and blue dye. A filmstrip separates the wick from the reservoir that is removed at the activation stage. At this point, the porous wick, white in color, is shown in the window. Upon exposure to a temperature exceeding the critical temperature, the substance melts and begins to diffuse through the porous wick, causing a blue coloring to appear as given in Figure 4.7. There are available indicators with different critical temperatures from −15 to 26°C. The consumer label is a partial-history indicator that changes color when exposed to higher than recommended storage temperature and will also change as the product reaches the end of its shelf-life. The working principle is based on the melting and diffusion of the blue dye as described previously (Three M, 2015).

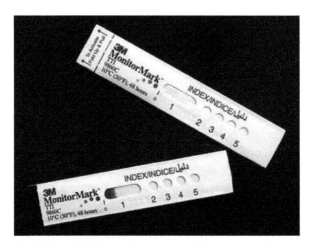

FIGURE 4.7   MonitorMark™ TTIs product from 3M™.

Timestrip® (Timestrip, 2015) is a sensor label that monitors how long a product has been open or how long it has been in use. They can measure elapsed time from minutes up to over a year, in the freezer, refrigerator, at normal ambient or even at elevated temperatures. Inside the Timestrip® is a special porous membrane through which a food-grade liquid diffuses in a consistent and repeatable way. The Timestrip® is activated by squeezing a start button, which moves the liquid into direct contact with the membrane. The laws of physics then take over and the liquid diffuses through the membrane. On the top surface of the Timestrip®, the markers has been printed which communicate all important time since activation, as well as space for branding and other graphics as given in Figure 4.8. Since most applications require the Timestrip® to adhere to a package or a product, they can be chosen from a wide range of adhesive tapes on the underside to suit the specific needs of the customer.

FIGURE 4.8 Timestrips® TTIs product from Timestrip Plc.

Fresh-Check® (Lifelinestechnology, 2015), fresh indicator is supplied as self-adhesive labels, which may be applied to packages of perishable products to ensure consumers at point-of-purchase and at home that the product is still fresh. As given in Figure 4.9, the active center circle of the Fresh-Check darkens irreversibly, faster at higher temperatures, and slower at lower temperatures, so it is easy to see when to use or not to use the food product within the product date codes. As the active center is exposed to temperature over time it gradually changes color to show the freshness of the food product. This full history indicator works based on the color change of a polymer formulated from diacetylene monomers. It consists of a small circle of polymer surrounded by a printed ring for color reference. The polymer, which starts lightly colored, gradually darkens depending on the color that tends to reflect the cumulative exposure to temperature. The polymer color changes at a rate proportional to the rate of food quality loss, thus the higher the temperature, the more rapid the polymer changes its color.

Fresh   Used soon   Should not be used

FIGURE 4.9   Fresh-Check® TTIs product from LifeLines.

CheckPoint® (Vitsab, 2015) is a simple adhesive label attached to food cartons to check for temperature abuse. CheckPoint® monitors a carton or fruits and vegetables package from the processor to the retailer. These labels react with time and temperature in the same way that food product react, and thus give a signal about the state of freshness and remaining shelf-life. This signal is an easy to read color dot as given in Figure 4.10. This is full history indicator based on enzymatic reaction. The device consists of a bubble-like dot containing two compartments: One for the enzyme solution, lipase plus a pH indicating dye and the other for the substrate, consisting primarily of triglycerides. The dot is activated at the beginning of the monitoring period by pressure on the plastic bubble, which breaks the seal between compartments. The ingredients are mixed and as the reaction precede a pH change results in a color change. The dot, initially green in color, becomes progressively yellow as product approaches the end of shelf-life. The reaction is irreversible and will proceed faster as temperature is increased and slows down as the temperature is reduced.

FIGURE 4.10   CheckPont® TTIs product from Vitsab.

4.2.2.3   *RADIO FREQUENCY IDENTIFICATION (RFID)*

As both raw and processed fruits and vegetables must undergo some transport, distribution, and storage, it also needs a sensor label for identification of product that can be used for product traceability. The sensor label that is used for this purpose is RFID that uses radio waves for product identification and traceability. It includes incorporating an RFID tag into the package, from which a proper sensor is used to collect data about the item's status. The data stored in tags are activated by the sensor, which is then transmitted to a reader for decoding and processing by a computer system (Brody et al., 2008; Kuswandi et al., 2011a; Smolander, 2003; Yam et al., 2005). The data used to identify the product (e.g., description of the label content) and its history (e.g., how long the product took to move through the supply chain, its temperature, pressure, humidity, and gas leakage). It can be collected at any point during processing and distribution. The information obtained from data analysis can be used for judgment of the produce status such as traceability in the case of outbreaks of foodborne infections.

A new cold chain monitoring service system using the RFID system was developed to monitor fresh products. This system, which is called X-Track™, composed of an RFID label connected to TTI that continuously monitors and stores data about product temperature and time of exposure. The data are then uploaded via an RFID reader and customer can retrieve them anywhere and anytime (Brody et al., 2008; Pocas et al., 2008).

4.2   CONCLUSION

The current advance of intelligent packaging for fruits and vegetables relies on the development of indicator or sensor technology. The sensor mimics the condition of the product to inform its quality, safety, shelf-life, and usability. For intelligent labels to be used as indicators incorporated in packaging, they need to be suitable with printing technology for mass production, low cost relative to the value of the food product, easy to use, accurate, reliable, simple, and reproducible in their range of operation, environmentally gentle, food contact safe as well as consumer friendly.

The incorporation of an indicator or a sensor in food packaging, as intelligent packaging has made great advances in packaging solutions for fruits and vegetables. These advances have led to improve food quality, safety, shelf-life, and usability. While most packaging innovations have been the result of global trends and consumer preferences, such as ICT (Information and Communication Technology) with their various smart devices. Furthermore, some innovations have stemmed from unexpected sources, such as the emergence of nanosensor technologies, the technology of sensing material in nm size. Undoubtedly, new intelligent packaging development in the near future will be the marriage of these technologies, so that the on-package intelligent labels could communicate with the consumers, not only with the color change, but also via their smartphone in terms of food quality and safety, even more for broader applications, such as tracking, origin, authentication, convenience, and sustainability of food products, especially for fruits and vegetables.

ACKNOWLEDGMENTS

The author gratefully thanks the Higher Education, Ministry of Science, Technology and Higher Education, Republic of Indonesia for supporting this work via the Competency Grant (Hibah Kompetensi 2015/2016).

KEYWORDS

- **intelligent packaging**
- **sensors**
- **indicators**
- **fruits and vegetables**
- **physical response**

REFERENCES

Abe, Y. Active Packaging with Oxygen Absorbers. In *Minimal Processing of Foods*; Ahvenainen, R., Mattila-Sandholm, T., Ohlsson, T., Eds.; VTT Symposium 142; VTT: Finland, 1994; pp 209–223.

Ahuja, T.; Mir, I. A.; Kumar, D.; Rajesh, R. Biomolecular Immobilization on Conducting Polymers for Biosensing Applications. *Biomaterials* **2007,** *28,* 791–805.

Ahvenainen, R. Active and Intelligent Packaging: An Introduction. In *Novel Food Packaging Techniques*; Ahvenainen, R. Ed.; Woodhead Publishing Ltd: Cambridge, 2003; p 5.

Ahvenainen, R.; Hurme, E. Active and Smart Packaging for Meeting Consumer Demands for Quality and Safety. *Food Addit. Contam.* **1997,** *14,* 753–763.

Anonym, Tufflex GS Product Information, Sealed Air (FPD) Limited, Telford, UK, 1996.

Arshak, K.; Adley, C.; Moore, E.; Cunniffe, C.; Campion, M.; Harris, J. Characterisation of Polymer Nanocomposite Sensors for Quantification of Bacterial Cultures. *Sens. Actuat. B.* **2007,** *126,* 226–231.

Balderson, S. N.; Whitwood, R. J. Gas Indicator for a Package, US 5439648, Trigon Industries Limited, Auckland, New Zealand, Aug 8, 1995.

Brody, A. L.; Bugusu, B.; Han, J. H.; Sand, C. K.; McHughet, T. H. Scientific Status Summary: Innovative Food Packaging Solutions. *J. Food Sci.* **2008,** *73,* R107–R116.

Cameron, A. C.; Talasila, T. *Modified-atmosphere Packaging of Fresh Fruits and Vegetables,* IFT Annual Meeting, Book of Abstracts, 1995; p 254.

Davies, E. S.; Garner, C. D. Oxygen Indicating Composition, UK Patent Application GB 2 298273, The Victoria University of Manchester, Manchester, UK, 1996.

DeCicco, B. T.; Keeven, J. K. Detection System for Microbial Contamination in Health-care Products, US 5443987, Aug 22, 1995.

De Johng, A. R.; Boumans, H.; Slaghek, J.; Van Veen, J.; Rijk, R.; Van Zandvoort, M. Active and Intelligent Packaging for Food: Is it the future? *Food Addit. Contam.* **2005,** *22,* 975–979.

Gardiol, A. E.; Hernandez, R. J.; Reinhammar, B.; Harte, B. R. Development of a Gas-phase Oxygen Biosensor Using a Blue Copper Containing Oxidase. *Enz. Microb. Technol.* **1996,** *18,* 347–352.

Goto, M. Oxygen Indicator, JP 62-259059, Mitsubishi Gas Chemical Co., Inc.; Tokyo, Japan, 1987.

Hong, S.; Park, W. Use of Color Indicators as an Active Packaging System for Evaluating Kimchi Fermentation. *J. Food Eng.* **2000,** *46,* 67–72.

Horan, T. J.; Method for Determining Bacterial Contamination in Food Package, US 5753285, 9 May, 1998.

Hurme, E.; Ahvenainen, R. *Active and Smart Packaging of Ready-made Foods,* in Minimal Processing and Ready Made Foods; Ohlsson, T., Ahvenainen, R., Mattila-Sandholm, T. Eds.; SIK: Göteborg, Sweden, 1996; p 169.

Kerry, J. P.; O'Grady, M. N.; Hogan, S. A. Past, Current and Potential Utilisation of Active and Intelligent Packaging Systems for Meat and Muscle-based Products: A Review. *Meat. Sci.* **2006,** *74,* 113–130.

Kim, J. H.; Shiratori, S. Fabrication of Color Film to Detect Ethylene Gas. *Japanese J. Appl. Phys.* **2006,** *45,* 4274–4278.

Kramer, A. Effect of Storage on Nutritive Value of Food. In *Handbook of Nutritive Value of Processed Food;* Rechcigl, M. Ed.; CRC Press: Boca Raton, FL, 1982; p 275.

Kress-Rogers, E. The Marker Concept: Frying Oil Monitor and Meat Freshness Sensor. In *Instrumentation and Sensors for the Food Industry;* Kress-Rogers, E. Ed.; Butterworth-Heinemann: Stoneham, MA, 1993, p 523.

Kuswandi, B. Tell-tale Labels for Food Quality & Safety Control. *Chem. Malaysia* **2012,** *108,* 4–15.

Kuswandi, B.; Wicaksono, Y.; Jayus, Abdullah, A.; Heng, L. Y.; Ahmad, M. Smart Packaging: Sensors for Monitoring of Food Quality and Safety. *Sens. Inst. Food Qual. Safe* **2011a,** *5,* 137–146.

Kuswandi, B. Jayus, Larasati, T. S.; Abdullah, A.; Heng, L. Y. Real-time Monitoring of Shrimp Spoilage Using on-package Sticker Indicator-based on Natural Dye of Curcumin, *Food Anal. Meth.* **2011b,** *5,* 881–889.

Kuswandi, B.; Jayus, Restanty, A.; Abdullah, A.; Heng, L. Y.; Ahmad, M. A Novel Colorimetric Food Package Label for Fish Spoilage Based on Polyaniline Film. *Food Cont.* **2012,** *25,* 184–189.

Kuswandi, B.; Maryska, C.; Jayus, Abdullah, A.; Heng, L. Y. Real Time on-Package Freshness Indicator for Guavas Packaging, *Food Measure.* **2013a,** *7,* 29–39. DOI 10.1007/s11694-013-9136-5, 2013.

Kuswandi, B.; Kinanti, D. P.; Jayus, Abdullah, A.; Heng, L. Y. Simple and Low-Cost Fresh-ness Indicator for Strawberries Packaging, *Acta Manilana*. **2013b**, *61,* 147–159.

Lang, C.; Hubert, T. A Colour Ripeness Indicator for Apples. *Food Bioprocess Technol*. **2011,** *5,* 3244–3249. DOI 10.1007/s1947-011-0694-4

Lawrence, B. D.; Cronin-Golomb, M.; Georgakoudi, I.; Kaplan, D. L.; Omenetto, F. G. Bioac-tive Silk Protein Biomaterial Systems for Optical Devices. *Biomacromolecules*. **2008,** *9,* 1214–1220.

LeNarvor, N.; Hamon, J. R.; Lapinte, C. Dispositif de Détection de Présence et de Disparition d'une Substance Cible et son Utilisation dans les Emballages de Conservation (Detecting the Presence and Disappearance of a Gaseous Target Substance - Using an Indicator Which Forms a Colored Reaction Product with the Substance, and an Antagonist Which Modifies the Color of the Reaction Product), FR 2710751, ATCO, Caen, France, 1993.

Lifelinestechnology. http://www.lifelinestechnology.com (accessed Oct 12, 2015).

Luh, B. S.; Woodroof, J. G., Eds.; *Commercial Vegetable Processing;* Van Nostrand Rein-hold: New York, 2nd ed.; 1988.

Mattila, T.; Tawast, J.; Ahvenainen, R. New Possibilities for Quality Control of Aseptic Pack-ages: Microbiological Spoilage and Seal Defect Detection Using Head-Space Indicators. *Lebensm. Wiss. Technol*. **1990,** *23,* 246–251.

Mattila-Sandholm, T.; Ahvenainen, R.; Hurme, E.; Järvi-Kääriäinen, T. Oxygen Sensitive Colour Indicator for Detecting Leaks in Gas-protected Food Packages, EP 0666977, Tech-nical Research Centre of Finland (VTT), Espoo, Finland, 1998.

Nopwinyuwong, A.; Trevanich, S., Suppakul, P. Development of a Novel Colorimetric Indi-cator Label for Monitoring Freshness of Intermediate-Moisture Dessert Spoilage. *Talanta.* **2010,** *81,* 1126–1132.

Nakamura, H.; Nakazawa, N.; Kawamura, Y. Food Oxidation Indicating Material - Comprises Oxygen Absorption Agent Containing Indicator Composed of Methylene Blue, Reducing Agent and Resin Binder, JP 62-183834, Toppan Printing Co., Ltd., Tokyo, Japan, 1987.

Pacquit, A.; Frisby, J.; Lau, K. T.; McLaughlin, H.; Quilty, B.; Diamond, D. Development of a Smart Packaging for Monitoring of Fish Spoilage. *J. Food Chem*. **2007,** *102,* 466–470.

Pacquit, A.; Lau, K. T.; McLaughlin, H.; Frisby, J.; Quilty, B.; Diamond, D. Development of a Volatile Amine Sensor for the Monitoring of Fish Spoilage. *Talanta*, **2006,** *69,* 515–520.

Perlman, D.; Linschitz, H. Oxygen Indicator for Packaging, US 4526752, 2 July, 1985.

Pocas, M. F. F.; Delgado, T. F.; Oliveira, F. A. R. Smart Packaging Technologies for Fruits and Vegetables. In *Smart Packaging Technologies for Fast Moving Consumer Goods;* Hoboken, N. J., Ed.; John Wiley: Hoboken, NJ, 2008; p 151.

Retama, R. J. Synthesis and Characterization of Semiconducting Polypyrrole/ Polyacryl-amide Microparticles with GOx for Biosensor Applications. *Coll. Surf. A: Physic. Eng. Aspects*. **2005,** *7,* 239–244.

Ripesense. http://www.ripesense.com (accessed Oct 12, 2015).

Robertson, G. *Food Packaging Principles and Practices;* Taylor & Francis: Boca Raton, FL, 2006.

Rodrigues, E. T.; Han, J. H. Intelligent Packaging. In *Encyclopedia of Agricultural, Food, and Biological Engineering*; Heldman, D. R., Ed.; Marcel Dekker: New York, NY, 2003; p 528.

Seldman, J. D. Time-temperature Indicators. In *Active Food Packaging;* Rooney, M. L. Ed.; Chapman & Hall: New York, NY, 1995; p 74.

Shen, Q.; Kong, F.; Wang, Q. Effect of Modified Atmosphere Packaging on the Browning and Lignification of Bamboo Shoots. *J. Food Eng*. **2006,** *77,* 348–354.

Shirozaki, Y. Oxygen Indicator, Rendering Reagent having Oxygen Detecting Capacity to Polyethylene, JP 2-57975, Nippon Kayaku KK, Tokyo, Japan, 1990.

Smartlid. http://www.smartlidsystems.com (accessed Oct 07, 2015).

Smolander, M.; Hurme, E.; Ahvenainen, R.; Siika-aho, M. *Indicators for Modified Atmosphere Packages*; 24th IAPRI Symposium; Rochester, NY, June 1998, 22–24.

Smolander, M. Freshness Indicators for Food Packaging. In *Smart Packaging Technologies for Fast Moving Consumer Goods;* Hoboken, N. J. Ed.; John Wiley: Hoboken, NJ, 2008; p 111.

Smolander, M.; Alakomi, H. L.; Ritvanen, T.; Vainionpa, J.; Ahvenainen, R. Monitoring of the Quality of Modified Atmosphere Packaged Broiler Cuts Stored in Different Temperature Conditions. A. Time-temperature Indicators as Quality-indicating Tools. *Food Cont.* **2004,** *15,* 217–229.

Smolander, M. *The Use of Freshness Indicators in Packaging.* In *Novel Food Packaging Techniques*; Ahvenainen, R., Ed.; Woodhead Publishing Ltd: Cambridge, UK, 2003; p 128.

Smolander, M.; Hurme, E.; Ahvenainen, R. Leak Indicators for Modified-Atmosphere Packages. *Trend. Food Sci. Technol.* **1997,** *8,* 101–106.

Taoukis, P. S.; Labuza, T. P. Time-temperature Indicators (TTIs). In *Novel Food Packaging Techniques*, Ahvenainen, R. Ed.; Woodhead Publishing Ltd: Cambridge, UK, 2003; p 103.

Terry, L. A.; Volpe, G.; Palleschi, G.; Turner, A. P. F. Biosensors for the Microbial Analysis of Food. *Food Sci. Technol.* **2004,** *18,* 22–24.

Tetrapak. http://www.tetrapak.com (accessed Oct 28, 2015).

Three M. http://solutions.3m.com/wps/portal/3M/en_US/Microbiology/FoodSafety/ product-information/product-catalog (accessed Oct 10, 2015), Timestrip. http://www.timestrip.com (accessed Oct 12, 2015).

Vitsab. www.vitsab.com (accessed Oct 12, 2015).

Wallach. D. F. H.; Novikov, A.; Methods and Devices for Detecting Spoilage in Food Products, WO 98/20337, May 14, 1998.

Wanihsuksombat, C.; Hongtrakul, V.; Suppakul, P. Development and Characterization of a Prototype of a Lactic Acid-based Time-temperature Indicator for Monitoring Food Product Quality. *J. Food Eng.* **2010,** *100,* 427–434.

Welt, B. A.; Sage, D. S.; Berger, K. L. Performance Specification of Time-temperature Integrators Designed to Protect Against Botulism in Refrigerated Fresh Foods. *J. Food Sci.* **2003,** *68,* 2–9.

Wilson, C. L., Ed. *Intelligent and Active Packaging for Fruits and Vegetables*; CRC Press: Boca Raton, FL, 2007.

Woodroof, J. G.; Luh B. S., Eds.; *Commercial Fruit Processing;* The Avi Publishing Company: Westport, CT, 1986.

Yam, K. L.; Takhistov, P. T.; Miltz, J. Intelligent Packaging: Concepts and Applications. *J. Food Sci.* **2005,** *70,* 1–10.

Yamamoto, H. Laminated Oxygen Indicator Labels for Food Packages, JP 4151557, Dainippon Printing, 1992.

Yoshikawa, Y.; Nawata, T.; Goto, M.; Fujii, Y. Oxygen Indicator, US 4169811, Mitsubishi Gas Chemical Co. Inc., Tokyo, Japan, 1979.

Yoshikawa, Y.; Nawata, T.; Goto, M.; Kondo, Y. Oxygen Indicator Adapted for Printing or Coating and Oxygen-Indicating Device, US 4349509, Mitsubishi Gas Chemical Co. Inc., Tokyo, Japan, 1982.

CHAPTER 5

SHRINK PACKAGING OF FRUITS AND VEGETABLES

D. V. SUDHAKAR RAO*

Division of Post-Harvest Technology, ICAR-Indian Institute of Horticultural Research, Bengaluru 560089, India

*E-mail: dvsrao@iihr.ernet.in; sudhadvrao@gmail.com

CONTENTS

ABSTRACT

Quantitative and qualitative postharvest losses of fruits and vegetables from harvesting to consumption are still high in developing and underdeveloped countries. Introduction of simple packaging techniques like individual shrink wrapping would help in reduction of these losses in these semi-perishable commodities. Shrink packaging reduces shrinkage without refrigeration or humidity control and it doubles or triples the storage life of the fruits and vegetables as measured by appearance, firmness, shrinkage, weight loss, and other keeping qualities.

5.1   INTRODUCTION

Fresh fruits and vegetables are living commodities and continue to respire and transpire even after harvest. Due to their high respiration and transpiration rates, these become highly perishable. The loss of water from living produce, through transpiration process, is one of the most important determinants of postharvest shelf-life and quality. Loss of water from fruits and vegetables after harvest causes not only shriveling, drying, and softening but also accelerates quality deterioration. High moisture loss from the fruits is the primary factor in deterioration of postharvest quality in tropical and subtropical regions due to high water vapor deficit (Ben-Yehoshua, 1985). Hence, it is always necessary to reduce water loss by reducing the transpiration rate, thus extending the storage life, and maintaining the freshness. The rate of moisture loss is influenced by many physical and morphological factors and it depends on both the types of commodities and the environment. These factors are surface to mass ratio, physical injury, and conditions of storage environment. High humidity in storage rooms helps to reduce moisture losses of many fruits and vegetables. However, in many instances, temperature is controlled, but humidity of the storage is not maintained at the required level.

5.2   SHRINK PACKAGING

Packaging is indispensable for proper postharvest management of fruits and vegetables with high moisture content. Traditional packaging protects the produce from bruising, crushing, and other mechanical damage. Some advanced packaging systems (in addition to protect physical damage) also provide the optimum gas and humidity conditions around the commodity, thus helping to maintain the quality during transit and storage. These also help to reduce moisture loss, prevent re-contamination with spoilage organisms, reduce pilferage, and maintain a sanitary environment during marketing, thus improving sale of produce. Shrink packaging or shrink wrapping is one such packaging technique for better postharvest handling of fresh fruits and vegetables. It is mainly done for individual fruit or vegetable or sometimes as unit packaging.

Shrink wrapping is a packaging technique which greatly reduces the moisture loss and maintains freshness of fresh produce for longer periods. Individual shrink packaging, which may be considered as modified atmosphere packaging (MAP) for individual fruit, involves sealing of fruit in a flexible film, followed by heat shrinking to confine the shape of the fruit. This flexible film shrinks tightly around each piece of produce and acts as a barrier for water loss. Generally, 10–20% reduction in transpiration rate is possible by shrink packaging even under ambient conditions (Ben-Yehoshua, 1985). It also delays the physiological deterioration of fruits and sometimes it is even better than the low-temperature storage. Unlike waxing, the film forms a barrier that markedly increases the resistance to water vapor. Thus, the greatest advantage of shrink packaging is its ability to control moisture loss to a great extent. Such packaging provides protection against abrasion and maintains an attractive appearance of the produce. One of the important advantages of individual shrink packaging is that it prevents secondary infection, which is important for long-term storage. It provides a barrier for the spread of microbial infection from infected lot to the healthy lot. The other advantages of shrinkwrapping are maintenance of firmness, delay in ripening, control of chilling injury (CI) at low temperatures, results extension of storage life, and produce higher consumer appeal (Ben-Yehoshua, 1985).

5.3 INDIVIDUAL SHRINK PACKAGING VERSES MA PACKAGING

In MAP, the produce (usually many pieces of fruit or vegetable) is placed loosely inside a selectively permeable plastic film and sealed, leaving volumes of atmospheric air surrounding the produce. There will be a reduction of O_2 and increase of CO_2 levels around the produce inside the MA package due to natural respiration. In individual shrink wrapping, the film is shrunk tightly around each piece of produce (by shrinkage of the film using hot blown air) to form a tight smooth package of each fruit, making it as another layer of protective cuticle. Another important difference between individual shrink wrapping and MA packaging is the extent of the altered micro atmosphere.

Several factors like respiration produce to pack volume ratio, and storage temperature would affect the gas composition inside the pack. In addition, the film permeability affects the micro atmosphere. A single fruit has more diffusive surface area per respiratory biomass than many fruits sealed in one bag (Ben-Yehoshua, 1985). Thus, the in-package atmosphere necessarily would be altered more in a bag holding more fruits, causing an increased risk of toxic atmosphere that lacks O_2 and has excessive CO_2 and ethylene. This leads to produce off-flavors. Furthermore, the condensation of moisture droplets inside the pack (i.e., bags holding several fruits) is common, thus it promotes the decay. In the case of individual shrink wrapping, it reduces the chance of water condensation within the inner surface of the package, since the temperature gradient between the produce and the package surface is negligible. The extent of decay would also be higher in MA packaging due to secondary infection, in which one rotting fruit infects other fruits in the pack. Whereas, the spoilage due to cross contamination is effectively prevented in individual shrink wrapping.

5.4   METHODS OF SHRINK PACKAGING AND MACHINERY

The shrink packaging process is simple wherein individual fruit or vegetable is loosely sealed first in a suitable polymeric film using any simple film sealer. These packs are then passed through a shrink-tunnel (hot air tunnel) on the conveyer belt, in which they are exposed to hot blown air for a few seconds (Fig. 5.1). In the shrink-tunnel, the loosely sealed film shrinks tightly over the surface of the commodity, thus making it another layer of protective cover (Fig. 5.2).

In the case of unit/tray wrap packaging, the produce is first sealed in a consumer pack of suitable size and then passed through the hot tunnel to form a tight wrap. Different films have different sensitivities to heat shrinkage and the ultimate wrap depends upon the temperature of hot air, the rate of air flow, and the duration that the produce stays inside the hot tunnel. In any case, temperature of tunnel should not be more than 180–200°C and subjecting the produce beyond this temperature may result in tissue injury. Furthermore, the produce should be kept for minimal possible time in the tunnel and the duration of such exposure should not exceed 30 s. A thumb of rule is that the fruit should be maintained for an adequate time inside the tunnel at the lowest temperature, which will provide an effective shrinkage of the film around the produce (Ben-Yehoshua, 1985). The fruit may be cooled after passing through the hot tunnel by rapid ventilation.

Freshly harvested fruits/vegetables

↓

Sort/Grade

↓

Trim the stalks

↓

Wash/Disinfest

↓

Surface dry

↓

Loosely pack and seal

↓

Pass through shrink tunnel (150-170°C) for 10 - 15 sec

↓

Collect and Master pack

FIGURE 5.1   Flowchart for shrink packaging.

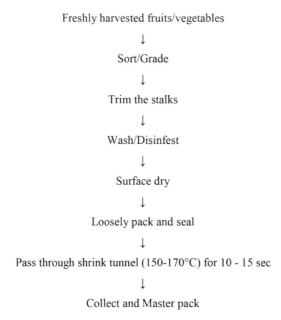

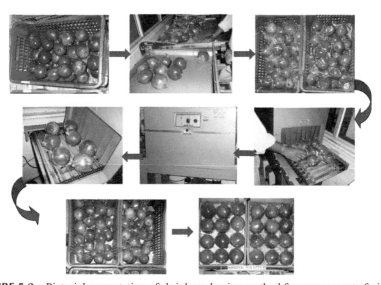

FIGURE 5.2   Pictorial presentation of shrink packaging method for pomegranate fruits.

Besides heat shrinkable polymeric films, the following equipments are required for shrink wrap packaging: (1) Film sealer and (2) Hot air shrink tunnel. Semi or fully automatic L-sealer or more sophisticated automatic form–fill and seal machinery may provide the high-speed sealing needed in packing houses. Some of these form-fill-seal machines can seal 200 fruits per minute (Lins, 1982).

5.5 MODE OF ACTION OF SHRINK PACKAGING

In individual shrink packaging, the film tightly covers the commodity, confining to its shape, leaving only a very less air space between the produce surface and the film. The air in this space immediately becomes saturated with water vapor diffused from the produce. As a result, individual shrink wrapping allows maintaining high humidity around the produce without the risk of condensation (Ben-Yehoshua et al., 1981). This water-saturated atmosphere is one of the controlling factors that delay deterioration of fruit. Due to the close contact between the produce surface and the film, the respiratory heat is efficiently dissipated through conduction rather than the less efficient convection process.

Extension of shelf-life of fresh fruits and vegetables, when they are placed in flexible films, is also attributed to the modification of atmosphere (reduced O_2 and increased CO_2 levels) around the produce due to the natural process of respiration. These conditions also prevail to a certain extent in individual shrink packaging, and thus may be considered as MAP for individual fruit. The increase in resistance to diffusion of these gases, brought about by the film has been counteracted by the inhibition of the respiratory activity, resulting in a new equilibrium level inside the sealed fruit (Ben-Yehoshua, 1983a).

5.5.1 SELECTION OF SUITABLE FILMS FOR SHRINK PACKAGING

Films differ in their permeability to various gases, such as O_2, CO_2, C_2H_4, and water vapor. Therefore, it is possible to select films to create a modified atmosphere within each package and the following points need to be considered:

- Select films having high barrier for the diffusion of water vapor.
- Using thin film of high tensile strength would help in reducing the cost of the film.
- Use films having desirable selective permeability for gases to avoid off-flavor development produced by the fruit over a wide range of temperatures.

- Select films having gloss and more transparency.
- Co-polymers of various combinations can be used to give better result than one polymer.
- Multiple films co-extruded to form one film would give stronger sealing and more attractive appearance after shrinking the film around the fruit.
- Film used should be adapted to the specific requirement of each fruit and packaging machine.

Hence overall, the film for shrink packaging should be low in water and O_2 permeability, high permeability to CO_2, less in thickness but have high tensile strength, should be glossy and transparent, and has the ability to shrink at low temperature (Sudhakar Rao & Gopalakrishna Rao, 2002).

5.5.2 ADVANTAGES

The overall advantages of individual film wrapping as reported by Risse (1989) and other workers include:

- Reduction in weight loss.
- Maintenance of firmness.
- Reduction of deformation.
- Alleviation of CI.
- Reduction of decay from secondary infection.
- Delay in ripening, color development, and senescence.
- Improved sanitation in handling the produce from packing to consumption.
- Creation of modified atmosphere around the fruit thereby extension of storage life.
- Elimination of the need to maintain high humidity during storage/ transit.
- Possible reduction of refrigeration requirement and cost.
- Possibility of substituting air transport of perishables by marine shipment.
- Extending the market season.
- Consumer appeal, in that the whole surface of the produce can be examined prior to purchase.
- Variety and brand identification.

The disadvantages are:

- Acceleration of decay in some cases.
- Development of off-flavors.
- Magnification of defects on produce.
- Cost of film and application.

The response of many fresh fruits and vegetables to a film-wrapped environment was evaluated by Risse (1989). Film wrapping was reported to reduce weight loss, color development, and CI; and maintained firmness and internal quality even under optimum storage conditions. These advantages were generally more pronounced at higher-than-optimum storage temperatures. Some of the disadvantages of film wrapping included the cost of film and machines for application, speed for wrapping, impermeability of some films to commodity treatments (fumigation), possible acceleration of decay, and development of off-flavors. However, he observed that the advantages of film wrapping for high-quality produce at premium prices generally outweigh the disadvantages. The accelerated decay can be controlled by applying chemicals and off-flavor development can be tackled by selecting the right film having optimum gas permeability.

5.6 EFFECTS OF SHRINK PACKAGING ON FRUITS AND VEGETABLES

Shrink packaging reduces various physiological parameters like respiration (Table 5.1), transpiration and ripening sometimes better than low-temperature storage alone. The extent of benefit from shrink packaging depends upon the type of produce, its physiological maturity, and initial quality.

TABLE 5.1   Effect of Shrink Film Wrapping on the Respiration Rate of Different Fruits and Vegetables.

Commodity	Storage temperature	Respiration rate (mg CO_2/kg/h)	
		Non-wrapped	**Shrink-film wrapped**
Pomegranate	25°C	60–70	35–40
Cucumber	25°C	55–65	38–42
	10°C	25–30	15–20
Bell	25°C	60–70	40–50
Pepper	8°C	18–20	14–16

Source: Sudhakar Rao (2003).

5.6.1 REDUCTION OF WEIGHT LOSS

The greatest advantage by individual shrink wrapping is its ability to control moisture loss to a great extent by reducing the transpiration rate (Fig. 5.3). By this packaging technique, the water loss of fruits and vegetables can be reduced 5–20 times (Table 5.2) using less permeable breathable films. Shrink wrapping reduces moisture loss even without refrigeration or humidity control. Fruits having large surface to volume ratio are particularly more susceptible to water loss and this technique has been found to be a boon for extending the storage life of such produce. The money gained by preventing the weight loss will overcome the cost of film application to a greater extent and would be highly economical (Sudhakar Rao & Gopalakrishna Rao, 2002).

FIGURE 5.3   Prevention of moisture loss by shrink packaging in pomegranate cv. Bhagwa.

TABLE 5.2   Effect of Shrink Packaging on the Weight Loss of Different Commodities During Storage.

Commodity	Storage temp. (°C)	Storage period (days)	Weight loss (%)		References
			Shrink packaged	Non-wrapped	
A. Fruits					
Pomegranate	25	25	1.5–2.3	14.0	Nanda et al. (2001)
cv. Ganesh	8	90	< 1.5	20.0	
Pomegranate	8	60	1.8	8.9	Salvatore et al. (2010)
cv. Primosole	20	7	3.0	13	
Pomegranate cv. Wonderful	5	60	2.16	7.05	Nazmy et al. (2012)
Mandarins cv. Kinnow	12–20	60	2.98	> 20	Raghav and Gupta (2000)

TABLE 5.2   *(Continued)*

Commodity	Storage temp. (°C)	Storage period (days)	Weight loss (%)		References
			Shrink packaged	Non-wrapped	
Lemons	21	21	0.8	3.2	Hale et al. (1983)
Mandarins	30–35	21	2–7	> 30	Ladaniya et al. (1997)
cv. Nagpur	6–7	60	< 2	13.3	Sonkar and Ladaniya (1999)
Sweet oranges cv. Mosambi	25 ± 5	40	2.96	25.5	Ladaniya (2003)
Papaya	27–32	10	1.42	20.6	Singh and Sudhakar Rao (2005)
cv. Solo	13	30	3.90	17.8	
Guava cv. Lucknow-49	13–20	14	4	18	Pal et al. (2004)
Mango cv. "Banganapalli" and "Alphonso"	8°C	35	0.5–1.4	5.8–6.9	Sudhakar Rao and Shivashankara (2015)
Apple cv. Royal Delicious	22–28	21	2.3	10.7	Sharma et al. (2010)
	18–22	50	8.4	15.2	Sharma et al. (2013)
Kiwifruit cv. Allison	22–28	18	4.4	18.2	Sharma et al. (2012)
Pear cv. Patharnakh	20–22	18	4.5	8.9	Singh et al. (2012)
Peaches cv. Shan-e-Punjabi	5	21	<0.5	28	Singh et al. (2009)
B. Vegetables					
Capsicum	22–30	10	1.8	16.8	Gopalakrishna Rao and Sudhakar Rao (2002)
	8	35	< 1.0	9.0	
Cucumber	24–26	14	0.64	11.0	Sudhakar Rao et al. (2000)
	10	14	0.22	8.28	
Cucumber	29–33	4	1.1	10.48	Dhall et al. (2012)
cv. Padmini	12	15	0.66	11.1	
Cabbage	22–30	21	< 2.0	24.6	Sudhakar Rao (2003)
	0	80	0.97	19.8	
Egg plant	7	21	0.5	3.6	Risse and Miller (1983)
Muskmelons cv. Honeydew	7.5	21	1.0	5.0	Roger and Stanley (1988)

5.6.1.1   EFFECT OF SHRINK PACKAGING ON WEIGHT LOSS OF FRUITS

In individually shrink-wrapped pomegranates (cv. Ganesh) fruits lost less than 2% weight as compared to 14% in non-wrapped fruits during 25 days of storage at 25°C; while at 8°C, the weight loss was less than 1.5% in wrapped fruits, compared to 20% loss in non-wrapped fruits during 12 weeks storage (Nanda et al., 2001). Salvatore et al. (2010) reported that after 8 weeks of storage at 8°C, weight loss of non-wrapped pomegranates (cv. Primosole) fruit was 8.9% while in shrink-wrapped fruit, it was only 1.8%. At 20°C, weight loss increased to 13% in non-wrapped fruits after 1 week of shelf-life, but it was 3% in wrapped fruits. Polyolefin film wrapping of "Wonderful" pomegranates was also reported (Nazmy et al., 2012) to greatly reduce weight loss during storage at 5°C and subsequent storage under marketing conditions (20 ± 2°C). After 60 days of cold storage, the shrink-wrapped fruits lost only 2.16% weight as compared to 7.05% loss in non-wrapped fruits. After 2 weeks of subsequent storage under marketing conditions, the weight loss increased to 11.82% in non-wrapped fruits and 6.67% in shrink-wrapped fruits.

Kinnow mandarins when shrink packed lost only 2.98% weight after 8 weeks, whereas non-wrapped fruits lost > 20% weight after 3 weeks storage itself under ambient (12–20°C, 63–78% RH conditions) (Raghav & Gupta, 2000). Hale et al. (1982) reported that the weight loss of shrink film wrapped grapefruit was 1.9% during 14–17 days transit at 10°C plus 3 weeks at 15.5°C and 1 week at 21°C compared to 10.6% for waxed grapefruit. Weight loss of shrink film wrapped lemons was less than 0.8% as compared to 3.2% in non-wrapped fruit after 3 weeks storage at 21°C (Hale et al., 1983).

Ladaniya et al. (1997) reported that "Nagpur mandarins" shrink wrapped with different heat shrinkable films (PE film and Cryovac BDF 2001, D-955 films) had a weight loss of only 2–7% after 3 weeks of storage under ambient conditions (30–35°C and 25–35% RH) compared to > 30% loss in non-wrapped fruits. Under refrigerated (6–7°C and 90–95% RH) conditions, "Nagpur mandarins" shrink wrapped with low density polyethylene (LDPE) film had a weight loss of <2% after 60 days of storage + 2 days holding period at ambient conditions compared to 13.3% loss in non-wrapped fruits (Sonkar & Ladaniya, 1999). Sweet oranges (cv. Mosambi) shrink wrapped with Cryovac D-955 film and stored at 25 ± 5°C and 40–45% RH for 40 days had a weight loss of only 2.96% as compared to 25.5% weight loss in non-wrapped fruits (Ladaniya, 2003).

Shrink packed papaya fruits (cv. Solo) lost only 1.42% of weight after 10 days of storage at ambient conditions (27–32°C and 50–55% RH), when as compared to 20.6% in non-wrapped fruits. The weight loss of shrink-wrapped fruits was 3.9 and 3.8% when stored at 13 and 7°C for 30 days as compared to 17.8 and 16.7% loss in non-wrapped fruits, respectively (Singh & Sudhakar Rao, 2005). Individual shrink wrapping of guava fruits (cv. Lucknow-49) with 9 μm linear low density polyethylene (LLDPE) film reduced the weight loss about four times when compared to non-wrapped fruits during 2 weeks storage at ambient (13.5–20.5°C and 55–80% RH) conditions (Pal et al., 2004).

Shrink film wrapping of mangoes (cv. Tommy Atkins) with 60-gauge RD 106 film greatly reduced the weight loss stored at 12°C. After 1 week of storage, non-wrapped fruits lost about 1.8% of their initial weight compared with 0.2% in wrapped fruits. After ripening at 21°C, non-wrapped fruits lost 6.6% weight, whereas the wrapped fruits lost only 0.6% (McCollum et al., 1992). Weight loss of film-wrapped mangoes was 0.2% compared to 7% in non-wrapped mangoes stored for 2 weeks at 12°C plus ripening to the soft-ripe stage at 21°C (Miller et al., 1983). At suboptimal temperature of 8°C, shrink wrapping significantly reduced weight loss during 5 weeks storage in two cultivars ("Banganapalli" and "Alphonso") of mango fruits. Non-wrapped "Banganapalli" and "Alphonso" fruits lost 5.75 and 6.87% of weight whereas, shrink wrapped fruits lost only 0.53 and 1.41%, respectively (Sudhakar Rao & Shivashankara, 2015).

Shrink wrapping of apples (cv. Royal Delicious) with a Cryovac (9 μm) film was found to be most effective in reducing the weight loss from 10.7 (non-wrapped) to 2.3% (wrapped) during 3 weeks of storage under ambient conditions (22–28°C and 52–68% RH) and from 15.2 (non-wrapped) to 8.4% (wrapped) during 7 weeks of storage in zero energy cool chamber (Sharma et al., 2010 and 2013). Shrink wrapped kiwifruits (cv. Allison) recorded the least weight loss of 4.4% after 18 days of storage when compared to 18.2% loss in non-wrapped fruits under ambient conditions of 22–28°C and 62–68% RH (Sharma et al., 2012). Physiologically, mature pear fruits packed in shrink film registered the lowest weight loss of 4.52%, whereas the non-wrapped fruits recorded the highest weight loss of 8.97% stored for 18 days under supermarket conditions of 20–22°C and 80–85% RH (Singh et al., 2012). Singh et al. (2009) reported that weight loss of shrink-wrapped peaches (cv. Shan–e–Punjab) was found to be only 0.7% after 6 weeks of storage in cool chamber (5 ± 1°C, 90–95% RH), when compared to 28% loss in non-wrapped fruits after 3 weeks of storage itself. Shrink-wrapped fruits did not show any shriveling symptoms, whereas unwrapped fruits shriveled due to this excessive moisture loss.

5.6.1.2 EFFECT OF SHRINK PACKAGING ON WEIGHT LOSS OF VEGETABLES

Weight loss of shrink packaged peppers was less than 0.3% compared to 3% for nonwrapped fruits after 3 weeks of storage at 7°C plus 1 week at 15.5°C (Miller et al., 1986). Green bell peppers lost less than 1% weight even after 5 weeks of storage at 8°C when shrink-wrapped, compared to 9% in non-wrapped peppers within 3 weeks of storage at the same temperature (Gopalakrishna Rao & Sudhakar Rao, 2002). Individual shrink packaging of cabbage heads extended the shelf-life by 8–10 days at ambient temperature and 100 days at 0°C with minimum loss in quality. The weight loss of shrink-wrapped cabbage was < 2% during 3 weeks of storage at ambient temperature as compared to 25% in non-wrapped fruits. At 0°C storage, the weight loss was less than 1% in wrapped cabbage heads, as compared to 20% in non-wrapped heads after 80 days of storage (Sudhakar Rao & Gopalakrishna Rao, 2002).

Sudhakar Rao et al. (2000) reported that the weight loss of cucumbers shrink wrapped with different polymeric films (D-955, BDF-2001, PE-film) was less than 1% after 22 days of storage at different cold storage temperatures (10, 15, and 20°C) when compared to 5.6– 8.28% weight loss in non-wrapped fruits at the end of 14 days itself under similar storage conditions. At room temperature (24–26°C) shrink-wrapped cucumbers lost less than 1% of weight during 14 days of storage as compared to 11% loss in non-wrapped fruits. Weight loss of wrapped cucumbers was 0.4% during 3 weeks storage at 7°C plus 3 days at 21°C, compared to 1.3% for waxed cucumbers and 4.2% for non-wrapped cucumbers (Risse et al., 1985a).

Dhall et al. (2012) also reported that the weight loss was significantly less in all shrink-wrapped (60 gauge Cryovac D955 film) cucumber fruits at both ambient and low-temperature storage conditions. The weight loss was 10.48% in non-wrapped fruits after 4 days of storage at ambient conditions (29–33°C, 65–70% RH), whereas shrink packaging resulted in the lowest weight loss of 1.1% even after 6 days of storage. At 12°C, individual shrink-wrapped cucumbers recorded a minimum of 0.66% weight loss as compared to 11.1% in case of unwrapped cucumbers at the end of 15 days storage.

The weight loss of shrink-wrapped eggplant was 0.5% during 3 weeks storage at 7°C plus 1 week at 15.5°C compared to 3.6% for paper tissue

wrapped fruits (Risse & Miller, 1983). Weight loss of shrink-wrapped tomatoes was less than 0.8% compared to 4.6% in non-wrapped fruits after 3
weeks of storage at 13°C plus 1 week at 21°C (Risse et al., 1985b). Individually wrapped Honeydew muskmelons with polyvinyl chloride (PVC) shrink
film ripened slower than non-wrapped fruit during 21 days low-temperature
storage (2.5 or 7.5°C) and additional 2–3 days subsequent holding at 20°C.
The fresh weight loss was about 1% in wrapped melons, but 5% in non-
wrapped fruit, regardless of storage temperature (Roger & Stanley, 1988).

5.6.2 MAINTENANCE OF FIRMNESS

Shrink packaging prevents desiccation, shriveling, softening, and maintains
firmness and texture of different commodities mainly by preventing moisture loss. Due to low water vapor transmission rate (WVTR) of films, shrink
wrapping reduces the transpiration rate and thus maintains higher firmness.
Therefore, the packed produce remains fresh for longer periods.

5.6.2.1 EFFECT OF SHRINK PACKAGING ON THE FIRMNESS OF FRUITS

The firmness of pomegranate fruit (cv. Ganesh) was maintained throughout
the storage period both at ambient and low-temperature storage when they
were shrink-wrapped with different films (Nanda et al., 2001). Firmness
was maintained significantly better when the fruits were wrapped using
BDF-2001 film with very low WVTR as compared to other films both
at ambient and low-temperature storage. The non-wrapped fruits became
tough, desiccated, and less firm after 1, 5, and 7 weeks of storage at 25, 15,
and 8°C, respectively. Whereas shrink wrapped fruits were firm and fresh
even after 3, 10, and 12 weeks of storage at 25, 15, and 8°C, respectively.

Nazmy et al. (2012) reported that polyolefin wrapping of "Wonderful"
pomegranates maintained the fruit firmness during storage at 5°C and subsequent storage under marketing conditions (20 ± 2°C). After 60 days of cold
storage, the fruit firmness was 102 N in shrink-wrapped fruits as compared
to 78.5 N in non-wrapped fruits. After 2 weeks of subsequent storage under
marketing conditions, a sharp decrease in fruit firmness was noticed in
control fruits (58.8 N) as compared to shrink-wrapped fruits (94.1 N).

Singh and Sudhakar Rao (2005) reported better maintenance of papaya
fruit firmness when they were shrink-wrapped and stored at low temperatures

as compared to non-wrapped fruits. The firmness of shrink-wrapped fruits was 9.8 and 11.9 kg/cm^2 when stored at 13 and 7°C for 30 days as compared to 2.0 and 7.7 kg/cm^2 in non-wrapped fruits, respectively. However, when shrink-wrapped fruits were unwrapped and ripened at 20°C for 1 week, there was no significant difference in the firmness between the treatments.

In case of apples, the non-wrapped fruits had higher firmness force (26.5 N) over those wrapped in shrink film (18.6 N) at the end of 6 weeks of storage under ambient (22–28°C and 52–68% RH) conditions (Sharma et al., 2010). Higher firmness (i.e., force) in non-wrapped apples was reported and this is due to greater loss of water from fruits resulting in shriveling, thereby providing higher force for puncturing the fruits. The decrease in firmness force in apples wrapped in Cryovac films during storage was due to the fact that there was the development of mealiness in the fruit with the advancement in storage period. Similar results were also observed in the apples stored in zero energy cool chambers (18–22°C, 82–88% RH) for 60 days (Sharma et al., 2013). Shrink packaged peach fruits were firmer than the non-wrapped fruits after 6 weeks of storage in cool chamber (5 ± 1°C, 90–95% RH). The skin hardness, pulping energy, and firmness force values decreased from 4.06, 7.46, and 3.14 to 1.72, 3.65, and 0.9 N, respectively, in case of shrink-wrapped fruit, whereas the same values decreased to 1.31, 1.36, and 0.61 N, respectively, in case of non-wrapped fruits (Singh et al., 2009).

Firmness of shrink-wrapped grapefruit was higher than non-wrapped fruits during 4 weeks storage at 21°C (Hale et al., 1982) and deformation of the fruit was 5% for wrapped fruits, compared to 20.3% for non-wrapped fruit after 35 days in transit at 10°C (Hale et al., 1981). The firmness of the shrink-wrapped kinnow mandarin fruits remained almost constant during 8 weeks of storage under ambient (12–20°C, 63–78% RH) conditions (Raghav & Gupta, 2000). The firmness decreased from 78.5 to 68.7 N at the end of 8 weeks storage in shrink packaged fruits, whereas it decreased to 19.6 N by the end of 3 weeks itself in non-wrapped fruits.

The softening of guava was very much influenced by shrink wrapping treatment and it was observed that shrink wrapping delayed the softening significantly during storage at ambient and cool chamber conditions (Pal et al., 2004). The non-wrapped fruits had a firmness of 10 N against 28 N of shrink-wrapped fruits after 2 weeks of storage at ambient (13.5–20.5°C and 55–80% RH) conditions and the farmer became very soft within 7 days. Similarly, in cool chamber (8–12°C and 88–90% RH), shrink-wrapped fruits maintained higher firmness (30 N) compared to non-wrapped fruits (18 N) after 18 days of storage. Singh et al. (2012) observed that pear fruits packed in shrink film maintained the highest firmness (52.67 N) when compared to non-wrapped fruits which registered the lowest firmness (34.9 N) stored for 18 days under supermarket conditions of 20–22°C and 80–85% RH. The loss in firmness thus leads to excessive softening and shriveling of fruits.

5.6.2.2 EFFECT OF SHRINK PACKAGING ON THE FIRMNESS OF VEGETABLES

Shrink packaged cucumbers maintained their firmness and freshness at both ambient and cold storage conditions as compared to non-wrapped cucumber (Dhall et al., 2012). The loss of firmness (softening) was maximum (from 12.8 to 8.6 N) in non-wrapped cucumbers, whereas the loss was minimum (from 12.8 to 10.44 N) in shrink-wrapped fruits after 12 days storage at 12 ± 1°C and 90–95% RH. The greater loss of firmness made the non-wrapped cucumbers unmarketable after 9 days of storage. At ambient conditions (29–33°C, 65–70% RH) also the loss in firmness was maximum (from 12.8 to 7.51 N) in non-wrapped fruits after 4 days of storage as compared to shrink-wrapped fruits (12.8 to 9.44 N) even after 6 days of storage.

Sudhakar Rao et al. (2000) observed better maintenance of firmness in shrink-packaged cucumbers during storage at both low temperatures (10, 15, and 20°C) and ambient conditions. After 16 days of storage, the fruit firmness decreased from the initial value of 9.18 kg/cm^2 to 8.39, 8.78, and 7.68 at 10, 15, and 20°C, respectively, whereas the same value decreased to 7.31, 7.30, and 6.41 kg/cm^2, respectively, in case of non-wrapped fruits. Firmness rating of film wrapped eggplant was 4.6 compared to 2.9 for paper tissue wrapped ones (Risse & Miller, 1983) after 3 weeks of storage at 7°C plus 1 week at 15.5°C. Similarly, firmness rating was higher (4.8) for shrink-wrapped peppers compared to non-wrapped (4.2) fruits after 3 weeks of storage at 7°C plus 1 week at 15.5°C (Miller et al., 1986). Firmness of tomatoes after 3 weeks storage at 13°C plus 1 week at 21°C was higher (21 N) compared to non-wrapped (15 N) tomatoes (Risse et al., 1985b).

Gopalakrishna Rao and Sudhakar Rao (2002) reported better maintenance of firmness in shrink packaged capsicum during storage at both low temperature (8°C) and ambient conditions (Fig. 5.4). After 5 weeks of storage at 8°C, the fruit firmness decreased very little from the initial value of 6.5–6.4 kg in case of wrapped fruits, whereas the firmness value decreased to 3.99 kg in case of non-wrapped fruits. At ambient temperature (22–30°C), the firmness decreased to 5.22 and 3.47 in wrapped and non-wrapped fruits, respectively.

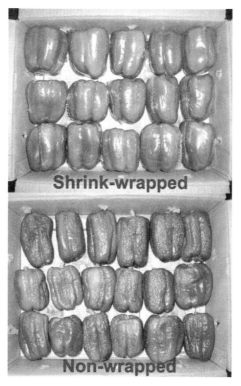

FIGURE 5.4   Maintenance of firmness by shrink packaging in green bell pepper.

5.6.3   EXTENSION OF STORAGE LIFE

Shrink packaging extends the shelf-life of fruits and vegetables mainly by preventing the moisture loss, maintaining firmness, reducing the respiration rate, and sometimes delaying ripening in climacteric fruits. It can extend the storage life of fresh produce both at ambient and low-temperature conditions. Shrink packaging also helps in alleviation of CI at suboptimal temperatures, thereby further extending the storage life.

5.6.3.1   EFFECT OF SHRINK PACKAGING ON THE STORAGE LIFE AND QUALITY OF FRUITS AND VEGETABLES

Apples: Sharma et al. (2010 and 2013) studied the effect of individual shrink packaging on the shelf-life and quality of apples (cv. Royal Delicious) under zero energy cool chamber and ambient conditions. Among different films tried viz., Cryovac (9 µm), polyolefin (13 µm), and LDPE (25 µ), best results were obtained with Cryovac (9 µm) film, which exhibited least weight loss and decay loss and higher juice recovery and TSS over other films or control under both storage conditions. These studies indicated that apples could be very well packed in Cryovac heat-shrinkable films and stored for 45 days in zero energy cool chamber (22–28°C and 52–68% RH) and 35 days under ambient conditions (22–28°C and 52–68% RH) with least weight or decay loss and without any adverse effect on fruit quality.

Shrink-wrapped apples (cv. Starkrimson) had acceptable qualities during 38 weeks storage at 26°C and 40–42% RH with good yellow color development in less mature apples due to accumulation of ethylene (Heaton et al., 1990). Anzueto and Rizvi (1985) also reported the usefulness of shrink wrapping of cold-stored apples (cv. Red Delicious) to extend the life of apples during retail marketing. When shrink wrapped, cold stored apples (up to 4 months) had a shelf-life extension of 3–4 weeks at 21°C over non-packaged fruits.

Citrus fruits: Shrink packaging has been extensively used for citrus fruits in the United States, Australia, Florida, Spain, and China, and good results were obtained in several other parts of the world. Shrink wrapping of lemons with 15 µm high-density polyethylene (D-950) film and stored at ambient temperature had a shelf-life (75–90 days) similar to that of refrigerated non-wrapped lemons owing to a marked reduction in weight loss. Unrefrigerated non-wrapped lemons became unsalable in less than 30 days. Wrapped lemons kept in a cold room maintained good commercial quality for 120 days (Testoni & Grassi, 1983).

Individual wrapping of "Nagpur mandarins" in heat shrinkable films (PE film and Cryovac BDF 2001, D-955 films) extended the shelf-life up to 3 weeks at ambient (30–35°C and 25–35% RH) conditions (Ladaniya et al., 1997) and the storage life could be further extended to 60 days under refrigerated (6–7°C and 90–95% RH) conditions (Sonkar & Ladaniya, 1999). The wrapped fruits were firm and fresh without losing their natural appearance and flavor. Cryovac (BDF 2001, D-955) films had better gloss, strength, and clarity as compared to PE film. Deformation was less in film wrapped fruits as compared to non-wrapped ones. Fungicide (Bavistin) treatment before wrapping minimized decay, and individually wrapped fruits had less decay as compared to tray wrapped fruits. Golshan-Tafty and Shahbake (2004) investigated the effects of shrink wrapping, curing, and thiabendazole (TBZ) treatment on the storage life of orange cultivars, namely Valencia, Mars early, and Jiroft. A combination of shrink wrapping and TBZ treatment not only reduced decay but also prevented weight loss and was effective in maintaining fruit quality during 3 months of storage at 10°C and 85–90% RH.

Shahbake (1999) investigated the effects of combined treatments of shrink wrapping, curing, quarantine heat disinfestation treatment, and MAP on the storage life of Washington Navel and Valencia oranges. Shrink wrapping and curing significantly reduced *Penicillium* sp. in all treatments, whereas MAP significantly increased stem end rot (*Phomopsis citri*) in Washington Navel oranges, but had no effect on Valencia oranges. Generally, the combined treatments of shrink wrapping, curing, quarantine heat disinfestation treatment, and MAP adversely affected flavor, peel color, and increased ethanol concentrations in both cultivars. In the absence of MAP, however, there were no adverse effects on fruit quality, flavor, and overall acceptability in both cultivars after 4 and 8 weeks storage.

Ladaniya (2003) reported that "Mosambi" sweet oranges shrink wrapped with PE film and Cryovac (BDF 2001 and D-955) films and stored at $25 \pm 5°C$ and 40–45% RH were fresh and turgid even after 40 days as compared with shriveled and lusterless non-wrapped fruits. Kinnow mandarin fruits shrink packaged individually with 60 gauge polyolefin films were found to be better in firmness, appearance, flavor, and overall eating quality than the non-wrapped fruits. Under ambient conditions (12–20°C, 63–78% RH), the shelf-life of wrapped fruit was 8 weeks as compared to 2 weeks in non-wrapped fruits (Raghav & Gupta, 2000).

Custard apple: Shrink wrapping of atemoya fruits (cvs. African Pride and Gefner) reduced weight loss and delayed ripening and also reduced the number of fruit splits, whereas waxing increased the number of splits (Paull, 1996).

Guava: Individually shrink packaged guava fruits (cv. Lucknow-49) with 9 µm LLDPE film could be successfully stored up to 12 days at ambient temperatures (13.5–20.5°C and 55–80% RH) and 18 days in evaporative cool chamber (8–12°C and 88–90% RH) with negligible loss in vitamin C content when compared to 25–30% loss in vitamin C content in case of non-wrapped fruits within a week after harvest (Pal et al., 2004). Delay in senescence and metabolic activities as supported by less changes in soluble solids, sugars, acidity, respiration, and ethylene evaluation rate was also observed in individual shrink-wrapped fruits stored in cool chamber. Individually, wrapped fruits had a better sensory score over Sta-fresh under both the storage conditions.

Kiwifruit: Sharma et al. (2012) evaluated the effect of different types of heat shrinkable films on shelf-life and quality of kiwifruits (cv. Allison) stored under ambient conditions. All the heat-shrinkable films delayed ripening of kiwifruit, but the effect rendered by Cryovac film was much more significant. Among different films tried viz., Cryovac (9 µm), polyolefin (13 µm) and LDPE (25 µm), the best results were obtained with Cryovac (9 µm) film, which exhibited least PLW and decay loss and fruits with good total soluble solids (TSS) and higher ascorbic acid content over other films or control. The shrink-wrapped kiwifruits could be stored up to 18 days, that is, a week extension over non-wrapped fruits under ambient conditions (22–28°C and 62–68% RH).

Mangoes: The effect of individual shrink film (60 gauge RD 106 film) wrapping of mangoes (cv. Tommy Atkins) on shelf-life and fruit quality (stored at 12 and 21°C) were evaluated by McCollum et al. (1992). Although shrink film wrapping reduced the weight loss of mangoes, the wrapped fruit had more decay and lower fruit quality when compared with non-wrapped fruit and concluded that individual shrink film wrapping of mango fruit was not a beneficial treatment for improvement of shelf-life. Individual shrink wrapping of mango fruits (cv. Dashehari) with perforated (5%) heat shrinkable Cryovac D 995 film (15 µm) was also found to be not helpful in extending the shelf-life under ambient temperature over those wrapped individually in perforated LDPE bags (Periyathambi et al., 2013). However, Thanaa and Rehab (2011) reported that shrink film wrapping was most effective in keeping storability of two mango cultivars (Hindi Be-Besennara and Alphonse) at 8°C with maintenance of physical and chemical parameters. Shrink film had no decay till 24 days under three storage temperatures of 8, 10, and 13°C.

Sudhakar Rao and Shivasankar (2015) also reported that shrink wrapping of two cultivars of mangoes ("Alphonso" and "Banganapalli") with D-955 film (15 µm) extended their storage life to 5 weeks at 8°C followed by 1 week ripening under ambient conditions. Both varieties of mangoes showed good surface yellow color development, maintained optimum fruit firmness, higher TSS, sugars, and carotenoid contents, when unwrapped and ripened under ambient conditions (Fig. 5.5).

Papaya: Singh and Sudhakar Rao (2005) reported that shrink wrapping of mature green papaya (Cv. Solo) fruits with Cryovac D-955 film could extend the shelf-life for 15 days at ambient conditions (27–32°C) over 7 days in non-wrapped fruits. At 13°C, they could be stored for 30 days without any CI in unripe conditions and the fruits after unwrapping ripened normally in 7 days at 20°C and 75–80% RH.

Pears: The application of shrink and cling film proved quite effective in prolonging the shelf-life and maintaining the quality of pear fruits (cv.

Patharnakh) under supermarket conditions (20–22°C and 80–85% RH) up to 18 days. However, under ordinary retail market conditions (30–32°C and 60–65% RH), fruit quality was affected with film wrapping due to buildup of high water vapor condensation and abnormal gaseous atmosphere inside the package as a result of high temperature (Singh et al., 2012).

FIGURE 5.5   Shrink-wrapped Alphonso mangoes stored for 5 weeks at 8°C.

Peaches: Individual heat shrink packaging of peaches (cv. Shan–e–Punjab) with 20 μm LLDPE film after postharvest treatment with carbendazim (500 ppm) was found to be quite useful for extending the shelf-life up to 42 days in cool chamber (5 ± 1°C, 90–95% RH) with better quality retention as compared to 7 days in non-wrapped fruits (Singh et al., 2009).

Pomegranate: Shrink packaging was found to be highly effective in maintaining the quality and extending the storage life of pomegranate fruits. Individual shrink wrapping of pomegranate (cv. Ganesh) fruits maintained freshness, firmness, and extended storage life both at ambient and low-temperature storage. The shrink-wrapped fruits could be stored for 12 weeks at 8°C, 9 weeks at 15°C, and 4 weeks at 25°C without much deterioration in quality when compared to 7, 5, and 1 week at, respectively, under similar storage conditions (Nanda et al., 2001). The non-wrapped fruits were dull, desiccated, tough, deformed, and discolored beyond 1 week of storage at RT resulting in unmarketable fruits. The wrapped fruits were fresh with bright yellowish rind color.

Salvatore et al. (2010) assessed the effectiveness of individual film packaging alone or in combination with fludioxonil, on reducing the occurrence of husk scald, weight loss, and decay of pomegranates (cv. Primosole) fruits. Wrapping the fruits with a polyolephinic heat-shrinkable film almost completely inhibited weight loss and husk scald and preserved fruit freshness for 12 weeks at 8°C. Fludioxonil, both alone and in combination with wrapping, effectively controlled mold development, resulting in 50–67% less decay than control fruit after 12 weeks at 8°C plus 1-week shelf-life.

The effect of film wrapping and calcium chloride treatments on post-harvest quality of "Wonderful" pomegranates was studied by Nazmy et al. (2012). Fruits treated with 2% calcium chloride combined with film wrapping using heat shrinkable polyolefin film (BDF-2001) retained maximum firmness, peel thickness, L- ascorbic acid, and sensory quality for 2 months at $5 \pm 1°C$ with $85 \pm 5\%$ RH and 2 weeks subsequent storage under marketing conditions ($20 \pm 2°C$).

Cabbage: Individual shrink packaging of cabbage with 25-μm polyolefin film (D-955) extended the shelf-life to 3 weeks at ambient temperature and 100 days at 0°C with minimum loss in quality (Sudhakar Rao & Gopalakrishna Rao, 2002).

Cucumbers: Shrink wrapping of individual cucumbers with polyethylene (PE) film extended the storage life up to 24 days at 10°C with an additional shelf-life of 2 days at ambient temperature (Sudhakar Rao et al., 2000). However, the optimum storage temperature for cucumbers when shrink-wrapped with other films (Cryovac BDF-2001 and D-955) was found to be 15°C where it could be stored for only 16 days. Wrinkling/shriveling of stalk ends and rotting restricted the storage life at 10°C when cucumbers were wrapped with these films. Dhall et al. (2012) also reported that individually shrink wrapped cucumber can be stored well up to 15 days at 12°C and for 5 days at ambient conditions (29–33°C, 65–70% RH) with maximum retention of green color, minimum losses in weight, and firmness and very good sensory quality attributes when compared to 9 and 2 days, respectively, in case of non-wrapped fruits.

Firmness and color development changes were slight, but better for shrink-wrapped cucumber than for non-wrapped ones during 3 weeks storage at 7°C plus 3 days at 21°C (Risse et al., 1985). Cucumbers dipped in an imazalil solution (1000 ppm) before wrapping or wrapped with film-coated imazalil (12,000 ppm) significantly reduced decay. Shrink wrapped cucumbers did not produce any volatiles (acetaldehyde, methanol, or ethanol), whereas waxed cucumbers, either wrapped or non-wrapped produced volatiles which resulted in off-flavor development (Risse et al., 1987).

Melons: Honeydew muskmelons (Cucumis melo L.) shrink wrapped with PVC shrink film ripened slower than non-wrapped fruit during 21 days storage at 2.5 or 7.5°C and subsequent holding at 20°C, after which time 70% of the wrapped melons were rated eating ripe, but 62% of the non-wrapped fruits were overripe. The concentration of CO_2 measured in the cavity of wrapped melons stored for 21 days was 5.6% at 2.5°C, 9.1% at 7.5°C, but only 1.1 and 1.5% in the non-wrapped held at 2.5 or 7.5°C, respectively. Decay incidence was about equal in wrapped and non-wrapped melons after storage at 2.5°C, but was greater for wrapped than non-wrapped melons after storage at 7.5°C. Soluble solids content was same (12.5%) in wrapped and non-wrapped melons stored at either temperature (Roger & Stanley, 1988).

Ripe green and yellow muskmelon (cv. Tam Uvalde) fruits shrink film wrapped in 12.7 μm high-density PE film were compared with non-wrapped fruits during 21 days of storage at 4°C and 90–95% RH (Collins et al., 1990). After 21 days of storage, both yellow and green shrink-wrapped melons had a better appearance, less surface mold, and less vein tract browning than non-wrapped melons. However, the flavor and taste of shrink-wrapped fruits were significantly inferior to those of non-wrapped fruits. Wrapped green melons were rated poorer in taste and flavor than wrapped yellow and non-wrapped melons after 14 days of storage. These results indicate that shrink wrapping may enhance undesirable flavor changes in muskmelon during storage.

Peppers: Individual shrink packaging with PE film was found to be advantageous over D-955 film in extending the storage life of capsicum both at ambient and low-temperature storage. The storage life could be extended to 12 days a RT and 35 days at 8°C (+5 days at RT) as compared to 5 and 14 days, respectively, in non-wrapped fruits (Gopalakrishna Rao & Sudhakar Rao, 2002). Color development (yellowing) of green peppers was also less in film wrapped than non-wrapped fruits. Individual shrink wrapping extended the storage life mainly by preventing the moisture loss, maintaining firmness, and by reducing the respiration rate. The main disadvantage found was the enhancement of spoilage if initial fruits used for wrapping were not sound and healthy.

Tomato: Harvesting maturity played an important role in extending the storage life of shrink packaged tomato fruits. Individual shrink wrapped tomato fruits harvested at mature green stage had a 10 days longer shelf-life than non-wrapped fruits stored at 18°C. However, shrink wrapping did not extend the shelf-life when tomatoes were harvested at pink stage though helped in greater water retention. At both 13 and 18°C storage, wrapped tomatoes (pink and mature green stage) were firmer than control and had lower titratable acidity and soluble solids, but no differences were noted in monosaccharide or citric acid levels (Hulbert & Bhowmik, 1987). Double-wrapping of mature green tomato fruits with heat-shrinkable plastic film delayed the onset of climacteric respiration and increased the shelf-life by 10 days at 13°C and 85–90% RH with superior eating quality (Bhowmik & Hulbert, 1989).

Zucchini: Shrink packaging of individual zucchini fruits with a selective film (micro-perforated LDPE of 18-μm thickness) could induce cold tolerance in two cultivars (Sinatra and Natura) by considerably reducing CI and weight loss and maintained firmness during the storage at 4°C. This improvement in fruit quality parameters was associated with a reduction in the production of ethylene and a down regulation of ethylene biosynthesis genes CpACS1 and CpACO1, together with a reduction in the respiration rate of fruit and the inhibition of oxidative stress and oxidative damage processes (Zoraida et al., 2015).

5.6.4 DECAY CONTROL

Prevention of secondary infection is an important factor for produce meant for long-term storage or shipment. Individual shrink wrapping prevents the contamination of adjacent fruit by sporulation and drip from rotten ones. The spoiled fruit can be discarded easily without destroying the commercial appeal of the whole container. However, the decay of the fruit would be faster than the unwrapped if initially infected fruits are used for wrapping. Film wrapping has a varied effect on decay development depending on the type of decay organism and the specific produce item. The use of fungicides coated onto films may provide a partial control for some types of decay.

Shrink wrapping also enhances the effectiveness of fungicide by slowing down the dissipation rate of these volatile fungicides (as film acts as a physical barrier) when fungicide treated fruits are shrink wrapped. Individual shrink wrapping thus produces a micro atmosphere that can be enriched with a suitable volatile fungicide so that the sealed enclosure forms a fumigation chamber to control decay over a prolonged period. Recently, several films have been produced that contain fungicides, which are released gradually. The film thus acts as a reservoir of the fungicide for several weeks without contaminating the fruit with excessive toxic residue (Ben-Yehoshua, 1985). However, in case of film wrapped mangoes, the decay was higher when the fruits were stored for 2 weeks at 12°C followed by ripening at 21°C (Miller et al., 1983).

5.6.5 ALLEVIATION OF CHILLING INJURY

Many tropical and sub-tropical fruits and vegetables are susceptible to low temperatures stress know as CI when stored below their optimum temperatures. Individual shrink packaging has been reported to reduce CI of various fruits and vegetables and thus it could be combined with low-temperature storage to extend storage life of chilling sensitive fruits and vegetables. The effect of shrink wrapping in reducing CI may be attributed to the modified and water saturated atmosphere produced by the seal-packaging (Ben-Yehoshua et al., 1983b). Individual shrink wrapping reduced CI of mango (Alphonso and Banganapalli cultivars) at 8°C and extended the storage life to 5 weeks (Sudhakar Rao & Shivasankar, 2015). Similarly shrink wrapped papaya fruits could be stored for 1 month at 13°C without any CI (Singh & Sudhakar Rao, 2005). Individual shrink packaging significantly reduced CI of red bell pepper fruit (cv. Selika) during 21 days of cold storage at 2°C plus 3 days at 20°C while maintaining the quality (Zoran et al., 2012). The wrapped fruit ripened normally (carotenoid content increased) during shelf-life period, when shifted to 20°C after unwrapping.

Film wrapped lemons had less CI and decay than non-wrapped lemons after 3 weeks storage at 1°C plus 2 weeks at 21°C (Hale et al., 1983). Individually wrapped Honeydew muskmelons with PVC exhibited 30% less CI symptoms (reddish-brown to dark-brown surface discolorations and dry sunken areas of skin) than non-wrapped fruit stored for 21 days at 2.5°C, followed by additional 2–3 days subsequent holding at 20°C (Roger & Stanley, 1988).

5.6.6 REDUCING BLEMISHES

Shrink packaging has contributed to the healing of wounds caused during harvesting and has reduced markedly the incidence of blemishes on various fruits (Ben-Yehoshua, 1978). The percentage of spoilage of grapefruit (Golomb et al., 1984) due to mechanical harvest was reduced drastically by shrink wrapping.

5.6.7 DELAYING RIPENING AND FRUIT SENESCENCE

Shrink film-wrapping delays changes related to senescence of fruit, such as deterioration of membrane integrity and softening (Ben-Yehoshua et al., 1983b). Another important effect of shrink wrapping is a considerable delay in the development of undesirable advanced color (senescence-related yellowing) in citrus, green bell pepper, cucumber, and other fruits (Sudhakar Rao & Gopalakrishna Rao, 2002). Individually, sealed lemons maintained a yellow color for 10 months at 14°C with a shelf-life of additional 2 weeks whereas the non-wrapped fruits developed an overripe brownish yellow color within 6 months (Ben-Yehoshua et al., 1979). Shrink packaging of mature green papaya (Cv. Solo) fruits with Cryovac D-955 film delayed ripening and extend the shelf-life for 15 days at ambient conditions (27–32°C) over 7 days in non-wrapped fruits (Singh & Sudhakar Rao, 2005).

5.7 CONCLUSIONS

Quantitative and qualitative postharvest losses of fruits and vegetables from harvesting to consumption are still high in developing and under-developed countries. Introduction of simple packaging techniques like individual shrink wrapping would help in reduction of these losses in these semi-perishable commodities. Shrink packaging reduces shrinkage without refrigeration or humidity control and it doubles or triples the storage life of the fruits and vegetables as measured by appearance, firmness, shrinkage, weight loss, and other keeping qualities. In the absence of cold chain facilities, shrink wrapping may enable storage without refrigeration with possible extension of storage life at lower cost than by conventional means. The money gained by preventing the weight loss to a greater extent will overcome the cost of film application and would be economical. Seal packing would also help to substitute air transport by marine shipment (long-term transport) for some commodities. Since CI is reduced by individual shrink wrapping, considerable extension of storage life of CI-susceptible cultivars or the mixing of otherwise incompatible cultivars in storage is possible. Shrink packaging is more effective than conventional practice of waxing as it prevents shrinkage and weight loss 5–10 times more than that of waxing. This technique could be used effectively for reduction of weight loss and maintenance of freshness of high-value fruits and vegetables during storage, long-distance transport, and during marketing at supermarkets and other retail outlets. Thus, this technique will be useful for commercial growers, wholesale dealers, exporters, and retailers for extending the storage life by reducing shriveling, weight loss, and maintenance of fresh quality. It can also be adopted by a group of small farmers as a joint activity or by co-operative societies of farmers mainly for distant marketing and export purpose.

KEYWORDS

- shrink wrapping
- low-density polyethylene
- high-density polyethylene
- fruits
- vegetables

REFERENCES

Anzueto, C. R.; Rizvi, S. S. H. Individual Packaging of Apples for Shelf-life Extension. *J. Food. Sci.* **1985,** *50* (4), 897–900.

Ben-Yehoshua, S. *Delaying Deterioration of Individual Citrus Fruits by Seal Packaging in Film of HDPE. I. General Effects*, Proceedings of the International Society for Citriculture, Griffith, Australia, 1978, pp 110–115.

Ben-Yehoshua, S. Individual Seal Packaging of Fruits and Vegetables in Plastic Film—A New Postharvest Technique. *Hortic. Sci.* **1985,** *20* (1), 32–37.

Ben-Yehoshua, S.; Shapiro, B.; Chen, Z. E.; Lurie, S. Mode of Action of Plastic Film in Extending Life of Lemon and Bell Pepper Fruit by Alleviation of Water Stress. *Plant Physiol.* **1983b,** *73,* 87–93.

Ben-Yehoshua, S. Extending Life of Fruit by Individual Seal Packaging in Plastic Film-status and Prospects. *Plasticulture.* **1983a,** *58,* 45–58.

Ben-Yehoshua, S.; Kobiler, I.; Shapiro, B. Some Physiological Effects of Delaying Deterioration of Citrus Fruit Individual Seal Packaging in High Density Polyethylene. *J. Am. Soc. Hort. Sci.* **1979,** *104* (6), 868–872.

Ben-Yehoshua, S.; Kobiler, L.; Shapiro, B. Effect of Cooling Versus Seal-Packaging with High Density Polyethylene on Keeping Quality of Various Citrus Cultivars. *J. Am. Soc Hort. Sci.* **1981,** *106,* 536–54.

Bhowmik, S. R.; Hulbert, G. J. Effect of Individual Shrink-wrapping on Shelf Life, Eating Quality and Respiration Rate of Tomatoes. *Lebensm. Wiss. Technol.* **1989,** *22* (3), 119–123.

Collins, J. K.; Bruton, B. D.; PerkinsVeazie, P. Organoleptic Evaluation of Shrink-wrapped Muskmelon. *Hort. Sci.* **1990,** *25* (11), 1409–1412.

Dhall, R. K.; Sharma, S. R.; Mahajan, B. V. C. Effect of Shrink Wrap Packaging for Maintaining Quality of Cucumber during Storage. *J. Food. Sci. Technol.* **2012,** *49* (4), 495–499.

Golomb, A.; Ben-Yehoshua, S.; Sarig, Y. High Density Polyethylene Wrap Improves Wound Healing and Lengthens Shelf Life of Mechanically Harvested Grapefruits. *J. Am. Soc. Hort. Sci.* **1984,** *109,* 155–159.

Golshan-Tafty, A.; Shahbake, M. A. Effect of Chemical and Physical Treatments on Storage Life of Valencia, Marsearly and Jiroft Local Oranges Varieties. *Iran. J. Agr. Sci.* **2004,** *35* (3), 713–720.

Gopalakrishna Rao, K. P.; Sudhakar Rao, D. V. *Extension of Storage Life of Fruits and Vege-tables by Individual Seal Packaging*. Proceedings of National Workshop on Post-Harvest Management of Horticultural Produce, *2002*, pp 333–342.

Hale, P. W.; Davis, P. L.; Marousky, F. J.; Bongers, A. J. Evaluation of a Heat-shrinkable Polymer Film to Maintain Quality of Florida Grapefruit during Export. *Citrus Veg. Mag.* **1982**, *46,* 39–48.

Hale, P. W.; Hatton, T. T.; Albrigo, L. G. Exporting Individual Packaged Grapefruit in Bulk Bins and Non-packaged Grapefruit in Bulk Bins with Film Liners. *J. Food Dist. Res.* **1981**, *12,* 9–18.

Hale, P. W.; Miller, W. R.; Davis, P. L. Wrapping of Florida Bears Lemons in a Heat-shrink-able Polymer Film. *Citrus Veg. Mag.* **1983**, *46* (49), 52–55.

Heaton, E. K.; Dobson, J. W.; Lane, R. P.; Beuchat, L. R. Evaluation of Shrink-wrap Pack-aging for Maintaining Quality of Apples. *J. Food. Prot.* **1990**, *53* (7), 598–599.

Hulbert, G. J.; Bhowmik, S. R. Quality of Fungicide Treated and Individually Shrink-wrapped Tomatoes. *J. Food Sci.* **1987**, *52* (5), 1293–1297.

Ladaniya, M. S. Shelf-life of Seal-packed Mosambi Sweet Orange Fruits in Heat Shrinkable and Stretchable Films. *Haryana J. Hort. Sci.* **2003**, *32,* 50–53.

Ladaniya, M. S.; Sonkar, R. K.; Dass, H. C. Evaluation of Heat Shrinkable Film Wrapping of 'Nagpur' Mandarin (*Citrus reticulata*, Blanco) for Storage. *J. Food Sci. Technol.* **1997**, *34,* 324–337.

Lins, D. Practical Aspects of Film Wrapping. Packinghouse News Letter no. 130:1–5, Univer-sity of Florida: Gainseville, FL, 1982.

McCollum, T. G.; Salvatore, D.; Miller, W. R.; McDonald, R. E. Individual Shrink Wrapping of Mangoes. *Proc. Fla. State Hort. Soc.* **1992**, *105,* 103–105.

Miller, W. R.; Hale, P. W.; Spalding, D. H.; Davis, P. Quality and Decay of Mango Fruit Wrapped in Heat-shrinkable Film. *Hort. Sci.* **1983**, *18,* 957–858.

Miller, W. R.; Risse, L. A.; McDonald, R. E. Deterioration of Individually Wrapped and Non-wrapped Bell Peppers during Long-term Storage. *Trop. Sci.* **1986**, *26,* 1–8.

Nanda, S.; Sudhakar Rao, D. V.; Shantha Krishnamurthy. Effects of Shrink Film Wrapping and Storage Temperature on the Shelf Life and Quality of Pomegranate Fruits cv. Ganesh. *Postharvest Biol. Technol.* **2001**, *22,* 61–69.

Nazmy, A. A.; Samah, I. N.; Hassan, M. K. Effects of Polyolefin Film Wrapping and Calcium Chloride Treatments on Postharvest Quality of "Wonderful" Pomegranate Fruits. *J. Hortic. Sci. Ornamental Plants.* **2012**, *4* (1), 7–17.

Pal, R. K.; Ahmad, M. S.; Roy, S. K.; Singh, M. Influence of Storage Environment, Surface Coating, and Individual Shrink Wrapping on Quality Assurance of Guava (*Psidium guajava* L.) Fruits. *Plant Food Hum. Nutr.* **2004**, *59,* 67–72.

Paull, R. E. Postharvest Atemoya Fruit Splitting during Ripening. *Postharvest Biol. Technol.* **1996**, *8* (4), 329–334.

Periyathambi, R.; Navprem, S.; Harminder, K.; Jawandha, S. K. Effect of Post-harvest Treatments on the Ambient Storage of Mango cv. Dashehari. *Progr. Hortic.* **2013**, *45* (1), 104–109.

Roger, E. R.; Stanley, R. R. Effects of Shrink Film Wrap on Internal Gas Concentrations, Chilling Injury, and Ripening of Honeydew Melons. *J. Food Qual.* **1988**, *11* (3), 175–182.

Raghav, P. K.; Gupta, A. K. Quality and Shelf Life of Individually Shrink Wrapped Kinnow Fruits. *J. Food Sci. Technol.* **2000**, *37* (6), 613–616.

Risse, L. A. Individual Film Wrapping of Florida Fresh Fruit and Vegetables. *Acta Hort.* **1989**, *258,* 263–270.

Risse, L. A.; Miller, W. R.; Chun, D. Effect of Film Wrapping, Waxing and Imazalil in Weight Loss and Decay of Florida Cucumbers. *Proc. Fla. State Hort. Soc.* **1985a,** *98,* 189–191.

Risse, L. A.; Miller, W. R.; Ben Yehoshua, S. Weight Loss, Firmness, Colour and Decay Development of Individually Film Wrapped Tomatoes. *Tropical Sci.* **1985b,** *25,* 117–121.

Risse, L. A.; Chun, D.; McDonald, R. E.; Miller, W. R. Volatile Production and Decay during Storage of Cucumber Waxed, Imazalil-treated and Film Wrapped. *Hort. Sci.* **1987,** *22,* 274–276.

Risse, L. A.; Miller, W. R. Film Wrapping and Decay of Eggplant. *Proc. Fla. State Hort. Soc.* **1983,** *96,* 350–352.

Salvatore, D.; Amedeo, P.; Mario, S.; Alberto, C.; Eugenio, T.; Stefano, L. M. Influence of Film Wrapping and Fludioxonil Application on Quality of Pomegranate Fruit. *Postharvest Biol. Technol.* **2010,** *55,* 121–128.

Shahbake, M. A. Effects of Heat Disinfestation Treatment and Modified Atmosphere Packaging on the Storage Life of Washington Navel and Valencia Oranges. *Iran. J. Agr. Sci.* **1999,** *30* (1), 93–102.

Sharma, R. R.; Pal, R. K.; Singh, D.; Samuel, D. V. K.; Kar, A.; Asrey, R. Storage Life and Fruit Quality of Individually Shrink-wrapped Apples (*Malus domestica*) in Zero Energy Cool Chamber. *Indian J. Agr. Sci.* **2010,** *80* (4), 338–41.

Sharma, R. R.; Pal, R. K.; Vishal Rana. Effect of Heat Shrinkable Films on Storability of Kiwifruits under Ambient Conditions. *Indian J. Hort.* **2012,** *69* (3), 404–408.

Sharma, R. R.; Pal, R. K.; Singh, D.; Samuel, D. V. K.; Sethi, S.; Kumar, A. Evaluation of Heat Shrinkable Films for Shelf Life, and Quality of Individually Wrapped Royal Delicious Apples under Ambient Conditions. *J. Food Sci. Technol.* **2013,** *50* (3), 590–594.

Singh, J.; Mahajan, B. V. C.; Dhillon, W. S. Effect of Packaging Films on Shelf Life of Individually Wrapped Pear Fruits. *J. Res. Punjab Agric. Univ.* **2012,** *49* (1 & 2), 35–39.

Singh, D.; Mondal, G.; Sharma, R. R. Effect of Individual Shrink Wrapping on Spoilage and Quality of Peaches during Storage. *J. Agr. Eng.* **2009,** *46* (2), 22–25.

Singh, S. P.; Sudhakar Rao, D. V. Quality Assurance of Papaya by Shrink Film Wrapping during Storage and Ripening. *J. Food Sci. Technol.* **2005,** *42* (6), 523–525.

Sonkar, R. K.; Ladaniya, M. S. Individual Film Wrapping of Nagpur Mandarin (*Citrus reticulate* Blanco) with Heat-shrinkable and Stretch-cling Films for Refrigerated Storage. *J. Food Sci. Technol.* **1999,** *36* (3), 273–276.

Sudhakar Rao, D. V. Quality Assurance of Fruits and Vegetables through Advanced Packaging System. In *Value Addition in Horticultural Crops. Delhi Garden Magazine;* Agri-Horticultural Society: New Delhi, India, 2003, Vol. 41, pp 188–196.

Sudhakar Rao, D. V.; Gopalakrishna Rao, K. P. Individual Shrink Wrapping of Fruits and Vegetables. *Packag. India.* **2002,** *34* (5), 27–32.

Sudhakar Rao, D. V.; Gopalakrishna Rao, K. P.; Shantha, K. Extension of Shelf Life of Cucumber by Modified Atmosphere Packaging (MAP) and Shrink Wrapping. *Indian Food Packer.* **2000,** *54,* 65–71.

Sudhakar Rao, D. V.; Shivasankar, K. S. Individual Shrink Wrapping Extends the Storage Life and Maintains the Antioxidants of Mango (cvs. 'Alphonso' and 'Banganapalli') Stored at 8°C. *J. Food Sci. Technol.* **2015,** *52* (7), 4351–4359.

Testoni, A.; Grassi, M. Individual Shrink Wrapping with Plastic Film: A New Technology for Citrus Fruits. *Annali-dell'Istituto-Sperimentale-per-la-Valorizzazione-Tecnologica-dei-Prodotti-Agricoli.* **1983,** *14,* 49–56.

Thanaa, M. E.; Rehab, M. A. Effect of Some Post Harvest Treatments under Different Low Temperature on Two Mango Cultivars. *Aust. J. Basic Appl. Sci.* **2011,** *5* (10), 1164–1174.

Zoran, S. I.; Radmila, T.; Radoš, P.; Sharon, A.; Yaacov, P.; Elazar, F. Effect of Heat Treatment and Individual Shrink Packaging on Quality and Nutritional Value of Bell Pepper Stored at Suboptimal Temperature. *Int. J. Food Sci. Technol.* **2012,** *47* (1), 83–90.

Zoraida, M.; Cecilia, M.; Susana, M.; Alicia, G.; María del Mar Rebolloso-Fuentes, Dolores, G., Juan Luis, V.; Manuel, J. Individual Shrink Wrapping of Zucchini Fruit Improves Post-harvest Chilling Tolerance Associated with a Reduction in Ethylene Production and Oxidative Stress Metabolites. *PLoS One* **2015,** *10,* e0133058. DOI:10.1371/journal.pone.0133058

CHAPTER 6

ANTIMICROBIAL PACKAGING: BASIC CONCEPTS AND APPLICATIONS IN FRESH AND FRESH-CUT FRUITS AND VEGETABLES

BASHARAT YOUSUF[1*], OVAIS SHAFIQ QADRI[1,2], and ABHAYA KUMAR SRIVASTAVA[1]

[1]*Department of Post-Harvest Engineering and Technology, Faculty of Agricultural Sciences, Aligarh Muslim University, Aligarh, India*

[2]*Department of Bioengineering, Integral University, Lucknow, India*

Corresponding author. E-mail: yousufbasharat@gmail.com

CONTENTS

ABSTRACT

Among different reasons of deterioration of food, spoilage due to microbial growth is a predominant factor leading to deterioration and subsequently rendering the food unsafe for consumption. In other words, microbial food spoilage largely accounts to food loss. Further, since the microbiological quality of any food has a direct impact on the public health, protection of food from pathogenic, and spoilage microorganisms is of critical importance. Various strategies have been used over time to prevent or limit the growth of microorganisms in food products. Packaging is effectively exploited to develop strategies to prevent food spoilage and growth of pathogenic microorganisms. In addition to other functions, food packaging can serve as an effective means of protecting foods from different hazards including the microbial hazards.

6.1   INTRODUCTION

The deterioration of fresh or processed foods is mainly caused by interaction of food with a variety of factors including water or gases, contamination with bacteria, yeast or fungi, or due to infestation by insects and rodents (Makwana et al., 2015). Among different reasons of deterioration of food, spoilage due to microbial growth is a predominant factor leading to deterioration and subsequently rendering the food unsafe for consumption. In other words, microbial food spoilage largely accounts to food loss. Further, since the microbiological quality of any food has a direct impact on the public health, protection of food from pathogenic, and spoilage microorganisms is of critical importance. Various strategies have been used over time to prevent or limit the growth of microorganisms in food products. Packaging is effectively exploited to develop strategies to prevent food spoilage and growth of pathogenic microorganisms. In addition to other functions, food packaging can serve as an effective means of protecting foods from different hazards including the microbial hazards. Antimicrobial agents may be either directly incorporated into the food or packaging material, which maintains a sustained release over time to protect quality and safety, and ensures extended shelf-life. A relatively new concept known as antimicrobial packaging is an emerging area of interest and rapid developments have been made in the recent past.

Antimicrobial packaging, one of many applications or forms of active packaging is a relatively novel technique wherein an antimicrobial agent is incorporated into the packaging material, which suppresses the growth and subsequently the multiplication of target spoilage and pathogenic microorganisms. The spoilage or pathogenic microorganisms contaminating the food may either be killed or inhibited by using antimicrobial packaging. Food pathogens potentially harmful to humans can be dealt with multifunctional antimicrobial packaging systems, which can provide enhanced safety (Sung et al., 2013). The antimicrobial packaging limits the growth rate of microbes by extension in their lag phase and decreases the live microbial counts. Antimicrobial packaging can thus reduce food losses and increase the shelf-life of foods (Zhang et al., 2015). The potential of antimicrobial agents for prevention of foods from microbes has led to a large-scale scientific research. Quality and safety are main attributes associated with a food product and today's consumers are increasingly concerned about their health, making the safety aspect more critical. Therefore, the goal of food processors is to maximize the safety to increase the consumer acceptance. Though in all foods, quality and safety are necessary aspects but they become of critical importance in case of fresh-cut or other minimally processed fruits and vegetables since, they are usually prepared by trimming, peeling, washing, cutting the fruits, or vegetables. These preparatory steps often lead to increased risk of microbial growth, which may either cause loss of products due to spoilage microorganisms or threat to public health due to pathogenic microorganisms. In short, the fresh-cut produce being extremely sensitive, it is a huge challenge to maintain quality and safety efficiently for a longer period.

In context with the above discussion, in this chapter, we will discuss on food packaging with special emphasis on antimicrobial packaging for fresh and fresh-cut fruits and vegetables.

6.2 PACKAGING STRATEGIES AND CONCEPT OF ACTIVE PACKAGING

Packaging as defined by Coles and Kirwan (2011) is "coordinated system of preparing goods for transport, distribution, storage, retailing and end use, a means of ensuring safe delivery to the ultimate consumer in sound condition at optimum cost and a techno-commercial function aimed at optimizing the costs of delivery while maximizing sales." Packaging has major role in food supply chain and it performs many important functions. Four main functions include:

- Protection
- Containment
- Information
- Ease of use.

A desired packaging should fulfill most of the requirements, which may include nominal cost, consumer attractiveness, compatibility with the food, ease of use (for both manufacturer and consumer), environment friendly and recyclability in addition to maintaining quality and safety. Practically a perfect packaging, having all the requisite characteristics, may not be possible and, generally a packaging, which delivers most of the characteristic features without any compromise in the safety and the quality of the food is selected (Ahvenainen, 2003).

Numerous terms for novel packaging methods have been extensively reported in the literature including active, smart, interactive, clever, or intelligent packaging, but they are mostly undefined. Most of the researchers have classified the recent advanced packaging into two broad groups; active packaging and intelligent packaging. According to Robertson (2013), "active packaging is defined as packaging in which subsidiary constituents have been deliberately included in or on either the packaging material or the package headspace to enhance the performance of the package system." These subsidiary constituents usually counter the possible factors of deterioration of the food product and the use of such constituents depend upon the type of food product they are to be used, for example, ethylene scavengers used in the packaging of respiring fruits. Table 6.1 presents the information regarding the use of active packaging systems in various foods. Modified atmosphere packaging, reported, as active packaging by many authors, if practiced in a normal way is not active packaging as per the above definition, although modified atmosphere packaging (MAP) may be included in active packaging system if it is innovated to actively modify the gas composition within the package.

The packaging that helps us monitor the food quality while it is being transported or stored is intelligent packaging. Indicators are used within the packaging that typically undergoes permanent color change making it easier for the consumer to check the quality. A sensor RipeSense™, developed in New Zealand, helps the consumers to detect the quality of fruits as per their taste. RipeSense™ is sensitive to the complex volatile mixture emitted by the fruits as they ripen and changes color as per the concentration of volatiles. The volatile mixture generally composed of alcohols, esters, aldehyde, ketones, and lactones gives the particular aroma to the fruit and as the fruit

TABLE 6.1 Different Active Packaging Concepts with Some Successfully Tried Examples.

Active agent	Chemical nature	Purpose	Successfully tried with	References
O_2-scavenger	Iron-based, metal/acid, metal (e.g., platinum) catalyst, and ascorbate/metallic salts	Growth inhibition of aerobic microorganisms, prevention of oxidative deterioration of foods	Catfish, seerfish, and fios-de-ovos (egg product)	Mohan et al. (2008); Mohan et al. (2009); Suppakul et al. (2016)
CO_2-scavenger/emitter	Iron oxide/calcium hydroxide, ferrous carbonate/metal halide, calcium oxide/activated charcoal and ascorbate/sodium bicarbonate	Growth inhibition of gram-negative bacteria and moulds, delayed senescence, preserved sensory score, and chemical quality attributes	Strawberry, pear eggplant, soy paste, and red pepper paste	Aday et al. (2011); Nugraha et al. (2015); Veasna et al. (2012); Jang et al. (2000)
C_2H_4-scavenger	Potassium permanganate, activated carbon, and activated clays/zeolites	Extend the shelf-life of climacteric fruits by delaying the ripening process	Date fruit and banana	Mortazavi et al. (2015); Yadav et al. (2015)
Moisture scavenger	Poly(vinyl acetate) blanket, activated clays, and minerals and silica gel	Prevent the moisture uptake by low moisture foods.	Banana	Yadav et al. (2015)
Antimicrobials	Organic acids, spice and herb extracts, BHA/BHT antioxidants, antibiotics, chlorine dioxide, and sulfur dioxide	Growth inhibition spoilage and pathogenic microorganisms	Chicken meat, persimmon, and strawberry	Mulla et al. (2016); Sanchis et al. (2016); Duran et al. (2016); Robertson (2013)
Flavor releasing film	Cellulose triacetate, acetylated paper, citric acid, ferrous salt/ascorbate, and activated carbon/clays/zeolites	Minimization of flavor scalping Masking off-odors Improving the flavor of food	Strawberry	Almenar et al. (2009)

softens the aroma becomes stronger. The sensor changes color from red when the fruit is unripe through orange and finally yellow for the fully ripe fruit (Mills, 2011).

Fruits and vegetables have long been one of the important domains of the packaging industry. While numerous packaging systems have been developed and are in use for these products, extensive research is presently focused on the packaging of fresh-cut produce. Active and intelligent packaging present a great scope in development of packaging for fresh-cut produce, which are highly susceptible to microbial damage and auto quality deterioration.

6.3   BASIC CONCEPTS OF ANTIMICROBIAL PACKAGING

Food products are by their nature prone to microbial attack and contamination of food after processing is considered as one of the important cause that results in foodborne illness, a major public health concern (Mangalassary et al., 2007). Post processing the food is either stored or consumed and control of microorganisms during this phase is necessary in order to preserve the aesthetic as well as nutritional quality. Any food material after being processed is packed till its delivery to the consumer, so packaging can be used as a delivery mechanism of antimicrobials to limit the growth of microorganisms (Mangalassary, 2012). Since microorganisms vary widely in their physiology, the effect of antimicrobial agents against different microorganisms also varies. Table 6.2 presents some common antimicrobials used in food packaging and their target microorganisms. Incorporation of antimicrobials in the packaging system and their controlled release throughout the shelf-life of the product presents a promising active packaging mechanism that will help in ensuring safety and improving shelf-life of food products. Researchers have incorporated different antimicrobial agents including bacteriocins, enzymes, spices, chitosan, organic acids and others, depicting varied rate of success. The basic methods of developing antimicrobial packaging system as reported include incorporation of antimicrobial agents directly into the packaging; use of a sachet which is attached to the packaging containing bioactive antimicrobial substances that are released during storage because of their volatile nature; coating of the packaging film with a layer of antimicrobial matrix and edible coating incorporated with suitable antimicrobials (Appendini & Hotchkiss, 2002). The antimicrobial agents that are volatile in nature diffuse within the package and restrict the growth of microorganisms but the non-volatile antimicrobial agents must

TABLE 6.2 Some Common Antimicrobials Used in Food Packaging and Their Target Microorganisms.

Antimicrobials	Packaging materials	Foods	Microorganisms	References
Benzoic acid	Polyethylene, low-density polyethylene	Cheese	Total bacteria, molds	Weng and Hotchkiss (1993)
Sorbates	Low-density polyethylene, methylcellulose/chitosan, starch/glycerol	Caviar, cheese	Yeast, mold	Heshmati et al. (2013)
Essential oils	Gelatin–chitosan-based edible films	Fish	*Pseudomonas fluorescens, Shewanella putrefaciens, Photobacterium phosphoreum, Listeria innocua, Escherichia coli, and Lactobacillus acidophilus*	Gómez-Estaca et al. (2010)
Chlorine dioxide	Sachet	Chicken, fresh meat	Total bacteria	Cooksey (2005)
Chitosan	Chitosan/paper	Strawberry	*Escherichia coli*	Hernández-Muñoz et al. (2008)
Grapefruit seed extract	Low-density polyethylene/nylon	Ground beef	Aerobes and coliforms	Ha et al. (2001)
Horseradish extract	Paper, polyethylene, polyethylene terephthalate pouch	Ground beef	*Escherichia coli*	Nadarajah et al. (2005); Muthukumarasamy et al. (2003)

come in contact with the food surface. Any packaging system should be able to ensure a continuous diffusion of the anti-microbial agent especially non-volatile agent at a desired rate to attain maximum efficiency of the system (Cooksey, 2005).

Most of fruits and vegetables are nutritionally rich and present favorable conditions for growth of different microorganisms. The microorganisms can spoil fruits and vegetables at any stage of development but during the ripening and senescence of agricultural produce, there is maximum susceptibility. Once the fruit is detached from the parent plant the defense mechanism starts weakening and the microorganisms if present in vicinity can easily infect once the conditions become favorable (Sommer et al., 1992). Fresh-cut produce, the novel and convenient form of fruits and vegetables presently dominating the market worldwide, are at even higher risk of microbial spoilage. Since fresh-cut fruits are not usually subjected to any extreme processing other than basic preparatory steps like washing cutting, slicing, peeling pitting, and others, they present an open invitation to the microorganisms as their outer protective layer gets damaged in the process. The packaging used for fruits and vegetables (whole or minimally processed) present a great scope for the incorporation of antimicrobial agents in order to increase the shelf-life of such products and ensure safety of consumers.

6.4   ANTIMICROBIAL PACKAGING FOR FRESH AND FRESH-CUT AGRICULTURAL PRODUCE

Microbial contamination occurs in almost all foods but fruits and vegetables especially minimally processed products can serve as an excellent platform for the growth and multiplication of microorganisms. In fact, the risk of microbial growth increases many fold in fresh-cut or minimally processed produce and they may act as potential sources of foodborne diseases. A large number of attempts have been made to develop antimicrobial packaging systems for safeguarding the microbiological quality, thereby ensuring safety and extension in shelf-life of different foods including fresh and fresh-cut fruits and vegetables.

Of different strategies, application of antimicrobial coatings and films is most widely investigated for fresh and fresh-cut fruits and vegetables. Here, we discuss about the antimicrobial packaging for fresh and fresh-cut fruits and vegetables with emphasis on use of antimicrobial coatings and films.

6.4.1 ANTIMICROBIAL COATINGS/FILMS AND BIOPOLYMER-BASED PACKAGING

The antimicrobial features of food packaging materials can be achieved by different strategies. Coatings and films with inherent antimicrobial properties or an antimicrobial agent being incorporated in them serve as an effective way of packaging and it basically, involves the concept of active packaging or one can say that antimicrobial packaging is a form of active packaging (Appendini & Hotchkiss, 2002). These films/coatings may be categorized into three types based on the nature of material, namely, polysaccharides, protein, or lipid-based coatings (Qadri et al., 2016). Another category can be the combination of hydrocolloid and lipid commonly called as composite coatings. These coatings or films reduce the microbial growth on surface of food thereby maintaining the quality and extending the shelf-life. Films and coating are being extensively used for fruits and vegetables. Antimicrobial packaging is a promising form of food packaging, particularly for fresh and minimally processed produce. The contamination of such products mostly occurs on surface, therefore, various antimicrobial coatings and films have been used to delay spoilage and improve safety by limiting the growth of both spoilage and pathogenic microorganisms. Recently, they have been in a wide use for quality and shelf-life enhancement of fresh-cut fruits and vegetables. Fresh-cut industry is emerging globally and its demand is growing at a rapid pace due to the fast lifestyle adopted by the people who demand use of convenience foods.

Ideally, biodegradable packaging materials may be obtained from renewable biological resources, often known as biopolymers. Biopolymers are polymeric biomolecules containing monomeric units that are covalently bonded to form larger structures having good mechanical and barrier properties and most importantly are biodegradable. They act as efficient alternatives to non-biodegradable plastic packaging materials and can be employed as vehicles for incorporating various food additives including antimicrobial components into the packaging, which makes the food safer and extends the storage life of food products. Figure 6.1 categorizes the biopolymers into different classes based on their origin. Biopolymers from natural sources have been most commonly used for coating of fresh and fresh-cut fruits and vegetables.

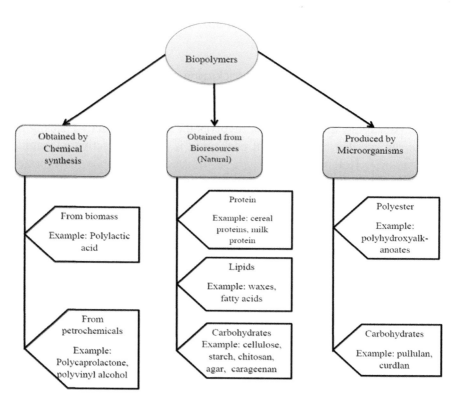

FIGURE 6.1   Classification of biopolymers based on their origin.

6.4.1.1   BIOPOLYMERS WITH INHERENT ANTIMICROBIAL PROPERTIES

Some polymers used as films or coatings possess inherent antimicrobial activity. The second category is where we incorporate antimicrobial agents into the films or coatings for food packaging, which is discussed in next sub-section. Examples of biopolymers or biomaterials are polysaccharides, proteins, lipids, and their derivatives obtained either from animal or plant origin. The best example of biopolymers with inherent antimicrobial activity is chitosan, which is the most common and widely used polymer. Chitosan is one of a few natural cationic polysaccharides derived from crustaceans or various fungi and food applications of chitin, chitosan, and their derivatives have been well reported by a number of researchers (Cha & Chinnan, 2004). Chitosan is the second most abundant polysaccharide found in nature after cellulose and is nontoxic, biodegradable, biofunctional, and biocompatible.

A number of studies have been conducted wherein chitosan and its derivatives have been investigated as an effective material for the formulation of edible packaging for fresh and minimally processed fruits and vegetables. Similarly, many other biopolymers possess inherent antimicrobial activity. Some examples of coating fruits with polymers having inherent antimicrobial properties are given in Table 6.3.

TABLE 6.3 Recent Studies on Edible Coatings/Films with Inherent Antimicrobial Activity Used on Fresh and Fresh-cut Fruits.

S. No.	Biopolymer	Fruit on which coating/ packaging is applied	References
1	Chitosan-based coating	Fresh-cut lotus root	Xing et al. (2010)
2	Hydroxypropyl methylcellulose-lipid edible coatings	Cherry tomato fruit	Fagundes et al. (2013)
	Edible coatings enriched with chitosan	Strawberries	Gol et al. (2013)
3	Aloe vera gel coating	Ready-to-eat pomegranate arils	Martinez-Romeroa et al. (2013)
4	Alginate–chitosan coating	Fresh-cut melon	Poverenov et al. (2014)
5	Aloe vera	kiwifruit slices	Benítez et al. (2014)

6.4.1.2 *INCORPORATION OF ANTIMICROBIAL AGENTS IN BIOPOLYMERS*

In the preceding section, we discussed and emphasized on bio-based polymers possessing inherent or inbuilt antimicrobial properties. However, there are some coatings or film-forming materials including some polymers, which perform better in terms of barrier and other mechanical properties, however, lack inherent antimicrobial nature. For such materials, the antimicrobial agents can be incorporated into them for better performance. For this purpose, numbers of antimicrobial agents have already been investigated and used different food products including fresh and fresh-cut fruits and vegetables. It is not only antimicrobial agents which are being incorporated into packaging materials but presently there is general trend of incorporating various active agents into packaging materials rendering them capable of maintaining the quality and safety of packaged foods. Table 6.4 different antimicrobial agents incorporated into the biopolymers for coating of fresh and fresh-cut fruits and vegetables.

TABLE 6.4 Application of Biopolymers Incorporated with Antimicrobial Agents on Fresh and Fresh-cut Fruits and Vegetables.

S. No.	Biopolymer	Antimicrobial agent or materials containing antimicrobial components	Fruit on which coating/packaging is applied	References
1	Apple puree-alginate edible coating	Lemongrass, oregano oil and vanillin	Fresh-cut apples	Rojas-Grau et al. (2007)
2	Multi-layered antimicrobial edible coating	Trans-cinnamaldehyde	Fresh-cut cantaloupe	Martinon et al. (2014)
3	Alginate-based coating	Malic acid and essential oils of cinnamon, palmarosa and lemongrass	Fresh-cut melon	Raybaudi-Massilia et al. (2008)
4	Edible coating based on calcium caseinate and whey protein isolate	Extracts of *Quillaja saponaria*	Fresh strawberries	Zuniga et al. (2012)
5	Alginate-based edible coating	Lemongrass essential oil	Fresh-cut pineapple	Azarakhsh et al. (2014)
6	Multilayered edible coating	Trans-cinnamaldehyde	Fresh-cut pineapple	Mantilla et al. (2013)
7	Chitosan matrix	β-cyclodextrin and trans-cinnamaldehyde	Minimally processed melon	Moreira et al. (2014)
8	Multilayered edible coating made of chitosan and pectin	β-cyclodextrin and trans-cinnamaldehyde complex	Fresh-cut papaya	Brasil et al. (2012)
9	Alginate-based edible coating	Beta-cyclodextrin and microencapsulated trans-cinnamaldehyde	Fresh-cut watermelon	Sipahi et al. (2013)
10	Alginate-based edible coating	Cinnamon oil and rosemary oil	Fresh-cut apple	Chiabrando and Giacalone (2015)
11	Alginate coating	Malic acid	Mango	Salinas-Roca et al. (2016)

Here again, in antimicrobial packaging either synthetic antimicrobial components or natural antimicrobial agents may be incorporated into the packaging polymers. The synthetic chemicals may include organic acids, metals, alcohols while as natural bio-preservatives may include enzymes or bacteriocins. Due to the health consciousness and preference of consumers toward natural food additives over chemical ones, more research is focused to exploit natural materials as effective antimicrobial agents. Recently, a number of natural materials having antimicrobial properties, such as essential oils from various sources have been incorporated in the coatings and films used for fresh and fresh-cut fruits and vegetables. United States Food and Drug Administration categorizes essential oils are as GRAS (generally recognized as safe) and their use as food grade additives is allowed in the United States (Makwana et al., 2015). Antimicrobial activity of these essential oils is attributed mostly to terpenoids and phenolic compounds present. The advances in development and use of biopolymeric matrices and incorporating essential oil as antimicrobial barriers in food as an alternative to synthetic additives has increased the attention toward the use of films and coatings in food preservation. Table 6.5 shows that a number of studies have been conducted to investigate antimicrobial activity of edible films as affected by addition of various essential oils.

TABLE 6.5   Various Studies Conducted to Investigate Antimicrobial Activity of Edible Films as Affected by Addition of Various Essential Oils.

S. No	Essential oil	Polymer matrix in which oil is incorporated	Tested against microorganism	References
1	Oregano essential oil	Chitosan	*Staphylococcus aureus* and *Listeria* monocytogenes	Hosseini et al. (2015)
2	*Eucalyptus globulus* essential oil	Chitosan	*Escherichia coli, Staphylococcus aureus, Pseudomonas aeruginosa, Candida albicans,* and *Candida parapsilosis*	Hafsa et al. (2016)
3	Thyme, lemongrass, and sage essential oil	Sodium alginate	*Escherichia coli*	Acevedo-Fani et al. (2015)
4	Cinnamon, clove, and oregano essential oil	Starch–gelatin blend films	*Colletotrichum gloeosporioides* and *Fusarium oxysporum*	Acosta et al. (2016)
5	Lime essential oil	Pectin-based edible films	*Escherichia coli, Salmonella typhimurium, Bacillus cereus, Staphylococcus aureus,* and *Listeria monocytogenes*	Aldana et al. (2015)
6	Basil or thyme essential oil	Chitosan	*Aspergillus niger, Botrytis cinerea,* and *Rhizopus stolonifer*	Perdones et al. (2016)

TABLE 6.5   *(Continued)*

S. No	Essential oil	Polymer matrix in which oil is incorporated	Tested against microorganism	References
7	*Zataria multiflora* essential oil	Carboxymethyl cellulose films	*Staphylococcus aureus, Bacillus cereus, Escherichia coli, Pseudomonas aeruginosa,* and *Salmonella typhimurium*	Dashipour et al. (2015)
8	Citrus essential oils	Chitosan and methylcellulose edible films	*Listeria monocytogenes*	Randazzo et al. (2016)
9	*Zataria multiflora* essential oil	Zein edible film	*Listeria monocytogenes* and *E. coli O157: H7*	Moradi et al. (2016)
10	Ginger, finger root, and plai essential oil	Methylcellulose-based nanocomposite films	*Staphylococcus aureus* and *Escherichia coli*	Klangmuang and Sothornvit (2016)
11	Oregano, cinnamon, and lemongrass essential oils	Amaranth, chitosan, or starch edible films	*Aspergillus niger and Penicillium digitatum*	Avila-Sosa et al. (2012)
12	Oregano essential oil	Cellulose acetate	*Alternaria alternata, Geotrichum candidum,* and *Rhizopus stolonifer*	Pola et al. (2016)
13	Carvacrol essential oils	Chitosan films	*Escherichia coli* and *Listeria innocua*	Tastan et al. (2016)

In our work, we have investigated gums from various sources as coatings for various fresh-cut fruits. Examples include psyllium seed gum (Yousuf & Srivastava, 2015), flaxseed gum, and okra gum. We also incorporated some natural antimicrobial agents into the coatings such as lemongrass essential oil. Figure 6.2 shows gum (dried) extracted from flax seeds, and pomegranate arils coated with edible coating of flaxseed gum incorporated with lemongrass oil as antimicrobial agent then stored in polypropylene trays under refrigeration. Some researchers have also incorporated essential oils derived from various natural sources, such as rosemary, thyme, and peppermint into edible coatings owing to their antimicrobial properties against several foodborne pathogens.

(a)

(b)

(c)

FIGURE 6.2 (a) Gum (dried) extracted from flax seeds. (b and c) Pomegranate arils coated with edible coating of flaxseed gum incorporated with lemongrass oil as antimicrobial agent then stored in polypropylene trays under refrigeration (work carried out at Food Processing Laboratory, Department of Post-Harvest Engineering and Technology, Aligarh Muslim University, India).

6.5   PRIMARY FUNCTIONS AND FACTORS AFFECTING USE OF ANTIMICROBIALS IN FRESH AND FRESH-CUT FRUITS AND VEGETABLES

Some primary functions of an antimicrobial agent used in packaging of foods including fresh and fresh-cut fruits and vegetables are:

- assurance of safety to consumers;
- maintenance of quality;
- extension in shelf-life.

There are number of antimicrobial agents and in general, there may be one or many potential concerns of using antimicrobial packaging/agents.

- One of the main issues is that the polymer or the antimicrobial agent may impart its own attributes and mask the attributes originally associated with food.
- The antimicrobial agent incorporated in edible coating or film should necessary be safe for human consumption.
- Economic viability of the antimicrobial agent.
- Stability of the antimicrobial agent.
- Possible adverse reactions of the antimicrobial compounds with the packaging material/polymer and food product.

6.6   NANOTECHNOLOGY IN FOOD PACKAGING

Antimicrobial packaging coupled with the concept of nanotechnology is an emerging area of interest. In recent times, nanomaterials have received a great attention due to their unique impact on a wide range of industries including food industries. Use of antimicrobials and nanomaterials in packaging materials has renewed and rejuvenated the ideas and strategies of food packaging. Nanotechnology can have a significant role in the development of food packaging. In fact, the advancement of nanotechnology may enable the antimicrobial packaging as a successful application for packaging of a variety of foods (Makwana et al., 2015). Nanotechnology involves the characterization, production, and use of material that has dimension of about 1–100 nm. Though nanotechnology finds its application in almost every sector of the food industry, but research and developments in food packaging using nanotechnology is one of the most promising areas and it can potentially provide solutions to food packaging challenges (Chaudhry et al., 2008).

The main applications of nano-antimicrobials for food safety may be outlined as:

- Prevention of biofouling
- Nano-based antimicrobial packaging systems
- Nanosensors for microbial detection

And the potential applications of any given nanoparticle depend on many factors such as:

- Material type
- Particle shape
- Usage concentration
- Microbial sensitivity toward a particular nanoparticle

6.6.1 METAL-BASED NANO-COATINGS

Silver (Ag) and titania (TiO_2) among the metal and metal nanocomposites are considered to be most important (Li et al., 2016). Silver nanoparticles incorporated into biopolymer films, such as chitosan and starch is most widely studied nanocomposite for food packaging. Several studies have been conducted which attempted to explain antibacterial mechanism of silver particles. Silver being heavy metal may sometimes cause adverse health effects upon migration into food. Nonetheless, silver nanoparticles are commonly used as photocatalytic antibacterials in packaging systems and have gained attention due to photocatalysis, antibacterial activity, and presence of nanocrystals (Zhao et al., 2012). Certain other metal oxides possessing excellent antimicrobial activity have been exploited for the formulation of antimicrobial packaging films.

Much of the scientific research has been directed toward the development and characterization of active composites based on the addition of silver as antimicrobial agent into polymer matrices. Addition or incorporation of nanofillers like silicate, clay, and titanium dioxide (TiO_2) into polymers may not only improve its mechanical or barrier properties but also provide beneficial functions such as antimicrobial, biosensor, and oxygen scavenger (De Azeredo, 2009).

Biopolymers like cellulose are regarded as good carriers of silver nanoparticles. The antimicrobial activity of cellulose-silver nanoparticle materials was investigated during storage of minimally processed melon by Fernandez et al. (2010). Significant effect of silver was observed against all the potential spoilage microorganisms and they suggested that these cellulose-silver nanoparticles could potentially be used to increase aseptic conditions during the storage of fresh-cut products.

Various other bio-nanocomposites have also been recognized to have good antimicrobial activity and can be helpful in reducing post-processing contamination and subsequently extending shelf-life and ensuring food safety. Bio-nanocomposite is a multiphase material which comprises of two or more constituents which are continuous phase or matrix particularly biopolymer and discontinuous nano-dimensional phase or nanofiller (Othman, 2014). For instance, effectiveness of several chitosan-based nanocomposites containing silver nanoparticles has been investigated against different microorganisms. The nano-sized fillers also act as reinforcement to improve the mechanical and barrier properties of the matrix.

6.6.2 NANOEMULSION-BASED COATINGS

Generally, essential oils are directly incorporated into biopolymer matrices for the development of edible films or coatings. This direct incorporation offers the advantage of being simple and versatile exercise. However, there may be certain drawbacks, such as reduction in the loading capability of essential oil into the coating or films (Tastan et al., 2016). Use of nanoemulsions may enhance the efficiency of essential oil-based antimicrobial coating. Despite some known merits of using nanoemulsions over the direct incorporation, no studies exist in literature wherein the effect of essential oil nanoemulsion loading on the film forming properties of the biopolymer has been discussed. Moreover, the role of nanoemulsion formulation and properties on the resulting antimicrobial activity of the coatings or films is yet to be elaborated.

Nanoemulsion may be defined as thermodynamically stable isotropically clear dispersion of two immiscible liquids, such as oil and water, stabilized by an interfacial film of surfactant molecules (Aqil et al., 2016). Nanoemulsions have different physicochemical properties than conventional emulsions owing to their small droplet size. Nanoemulsions have potential applications in food systems. The dispersed nanoemulsions are imperative vehicles for the delivery of water-insoluble or sparingly soluble antimicrobials, such as plant essential oils (Ghosh et al., 2014). Emulsification can be achieved by various high-energy emulsification methods, such as homogenization, microfluidization, and sonication. Here, we will discuss nanoemulsion coatings with particular reference to coatings applied to fresh-cut fruits and vegetables.

Edible coatings are most commonly used for quality maintenance and shelf-life extension of fresh and fresh-cut fruits and vegetables. In addition to other benefits, edible coatings serve as excellent carriers of active agents, such as antimicrobials. Moreover, nanoemulsion-based edible coatings may enhance the effective delivery of such active agents. For instance, Salvia-Trujillo et al. (2015) studied the effect of nanoemulsion-based edible coating containing lemongrass essential oil on safety and quality parameters of fresh-cut apples. Nanoemulsion-based coating inhibited the microbial flora of apples. Some recent studies on use of nanoemulsions-based edible coatings for fresh and fresh-cut fruits and vegetables are shown in Table 6.6.

TABLE 6.6   Nanoemulsions-based Edible Coatings Used for Fresh and Fresh-cut Fruits and Vegetables.

S. No.	Nanoemulsion coating	Fruit	References
1	Tocopherol/mucilage nanoemulsion	Fresh-cut Red Delicious apples	Zambrano-Zaragoza et al. (2014a)
2	Nanoemulsion-based edible coating containing lemongrass essential oil	Fresh-cut apples	Salvia-Trujillo et al. (2015)
3	Nanocoatings with α-tocopherol and xanthan gum	Apples	Zambrano-Zaragoza et al. (2014b)
4	Nanoemulsion coatings of lemongrass oil	Grape berry	Kim et al. (2014)
5	Nanoemulsion carnaubashellac wax coating containing lemongrass oil	Apples	Jo et al. (2014)
6	Lemon, mandarin, oregano or clove essential oils-based nanoemulsion edible coatings	Rucola	Sessa et al. (2015)
7	Nanoemulsions containing lemongrass oil	Plums	Kim et al. (2013)
8	Chitosan-loaded nanoemulsions	Various tropical fruits	Zahid et al. (2012)
9	Tocopherol nanocapsules/xanthan gum coatings	Fresh-cut Apples	Galindo-Perez et al. (2015)
10	Oregano oil nanoemulsion	Fresh lettuce	Bhargava et al. (2015)
11	Chitosan-based coating containing nanoemulsion of essential oils	Green beans	Severino et al. (2015)

6.7   FUTURE RESEARCH PROSPECTIVE AND CONCLUSION

Antimicrobial packaging provides an effective means of reducing or preventing contamination by pathogens or spoilage microorganisms. Antimicrobial packaging is gaining popularity in the food industry due to its promising potential to maintain the quality and safety of different foods. In addition, it may also prove beneficial in extending the shelf-life of certain foods, such as fresh and minimally processed agricultural produce. Over the last decade, there have been some great developments wherein researchers have successfully found some novel polymers, antimicrobial agents, and tested them for regulatory concerns. However, there is further need of research to explore the area more efficiently. The diversity of the sources from which biopolymers and antimicrobial agents can be obtained also warrants the scope for further research. Though the use of antimicrobial packaging is promising, however, there are some limitations, which need to be addressed and overcome to commercialize this technology at large scale. Some limiting factors may be high cost, regulatory concerns of some biopolymers or antimicrobial agents for use in foods, stability issues of antimicrobial agents, coatings/films, or antimicrobial agent imparting its own quality attributes on a food system while masking the attributes particular to a food product. For instance, essential oils used as antimicrobial agents in coating of fresh and fresh-cut fruits may impart their own flavor on these products. These issues thus warrant further research and development.

ACKNOWLEDGMENTS

Support from Department of Post-Harvest Engineering and Technology, Aligarh Muslim University, India is highly acknowledged and appreciated.

KEYWORDS

- **antimicrobial packaging**
- **edible coatings**
- **active packaging**
- **nano-packaging**
- **biopolymers**

REFERENCES

Acevedo-Fani, A.; Salvia-Trujillo, L.; Rojas-Graü, M. A.; Martín-Belloso, O. Edible Films from Essential-Oil-Loaded Nanoemulsions: Physicochemical Characterization and Antimicrobial Properties. *Food Hydrocoll.* **2015,** *47,* 168–177.

Acosta, S.; Chiralt, A.; Santamarina, P.; Rosello, J.; Gonzalez-Martínez, C.; Chafer M. Antifungal Films Based on Starch-Gelatin Blend, Containing Essential Oils. *Food Hydrocoll.* **2016,** *61,* 233–240.

Aday, M. S.; Caner, C.; Rahval, F. Effect of Oxygen and Carbon Dioxide Absorbers on Strawberry Quality. *Postharvest Biol. Technol.* **2011,** *62,* 179–187.

Ahvenainen, R. *Novel Food Packaging Techniques;* Woodhead Publishing Limited CRC Press: England, 2003.

Aldana, D. S.; Andrade-Ochoa, S.; Aguilar, C. N.; Contreras-Esquivel, J. C.; Nevarez-Moorill, G. V. Antibacterial Activity of Pectic-Based Edible Films Incorporated with Mexican Lime Essential Oil. *Food Control.* **2015,** *50,* 907–912.

Almenar, E.; Catala, R.; Hernandez-Muñoz, P.; Gavara, R. Optimization of an Active Package for Wild Strawberries Based on the Release of 2-Nonanone. *Food Sci. Technol.* **2009,** *42,* 587–593.

Appendini, P.; Hotchkiss, J. H. Review of Antimicrobial Packaging. *Innov. Food Sci. Emerg. Technol.* **2002,** *3,* 113–126.

Aqil, M.; Kamran, M.; Ahad, A.; Imam, S. S. Development of Clove Oil Based Nanoemulsion of Olmesartan for Transdermal Delivery: Box-Behnken Design Optimization and Pharmacokinetic Evaluation. *J. Mol. Liq.* **2016,** *214,* 238–248.

Avila-Sosa, R.; Palou, E.; Munguía, M. T. J.; Nevárez-Moorillón, G. V.; Cruz, A. R. N.; López-Malo, A. Antifungal Activity by Vapor Contact of Essential Oils Added to Amaranth, Chitosan, or Starch Edible Films. *Int. J. Food Microbiol.* **2012,** *153,* 66–72.

Azarakhsh, N.; Osman, A.; Ghazali, H. M.; Tan, C. P.; Adzahan, N. M. Lemongrass Essential Oil Incorporated into Alginate-Based Edible Coating for Shelf-Life Extension and Quality Retention of Fresh-Cut Pineapple. *Postharvest Biol. Technol.* **2014,** *88,* 1–7.

Benítez, S.; Achaerandio, I.; Pujolà, M.; Sepulcre, F. Aloe Vera as an Alternative to Traditional Edible Coatings Used in Fresh-Cut Fruits: A Case of Study with Kiwifruit Slices. *Food Sci. Technol.* **2014,** *61,* 184–193.

Bhargava, K.; Conti, D. S.; Rocha, S. R. P.; Zhang, Y. Application of an Oregano Oil Nanoemulsion to the Control of Foodborne Bacteria on Fresh Lettuce. *Food Microbiol.* **2015,** *47,* 69–73.

Brasil, I. M.; Gomes, C.; Puerta-Gomez, A.; Castell-Perez, M. E.; Moreira, R. G. Polysaccharide-Based Multilayered Antimicrobial Edible Coating Enhances Quality of Fresh-Cut Papaya. *Food Sci. Technol.* **2012,** *47,* 39–45.

Cha, D. S.; Chinnan, M. S. Biopolymer-Based Antimicrobial Packaging: A Review. *Crit. Rev. Food Sci. Nutr.* **2004,** *44,* 223–237.

Chaudhry, Q.; Scotter, M.; Blackburn, J.; Ross, B.; Boxall, A.; Castle, N.; Aitken, R.; Watkins, R. Applications and Implications of Nanotechnologies for the Food Sector. *J. Food Addit. Contam.* **2008,** *25,* 241–258.

Chiabrando, V.; Giacalone, G. Effect of Essential Oils Incorporated into an Alginate-Based Edible Coating on Fresh-Cut Apple Quality during Storage. *Qual. Assur. Saf. Crop. Foods.* **2015,** *7,* 251–259.

Coles, R.; Kirwan, M. *Food and Beverage Packaging Technology;* Wiley-Blackwell: Oxford, 2011.

Cooksey, K. Effectiveness of Antimicrobial Food Packaging Materials. *Food Addit. Contam.* **2005,** *22,* 980–987.

Dashipour, A.; Razavilar, V.; Hosseini, H.; Shojaee-Aliabadi, S.; German, J. B.; Ghanati, K.; Khakpour, M.; Khaksar, R. Antioxidant and Antimicrobial Carboxymethyl Cellulose Films Containing Zatariamultiflora Essential Oil. *Int. J. Biol. Macromol.* **2015,** *72,* 606–613.

De Azeredo, H. M. Nanocomposites for Food Packaging Applications. *Food Res. Int.* **2009,** *42,* 1240–1253.

Duran, M.; Aday, M. S.; Zorba, N. N. D.; Temizkan, R.; Büyükcan, M. B.; Caner, C. Potential of Antimicrobial Active Packaging "Containing Natamycin, Nisin, Pomegranate and Grape Seed Extract in Chitosan Coating" to Extend Shelf Life of Fresh Strawberry. *Food Bioprod. Process.* **2016,** *98,* 354–363.

Fagundes, C.; Perez-Gago, M. B.; Monteiro, A. R.; Palou, L. Antifungal Activity of Food Additives in Vitro and as Ingredients of Hydroxypropyl Methylcellulose-Lipid Edible Coatings against Botrytis Cinerea and Alternaria Alternate on Cherry Tomato Fruit. *Int. J. Food Microbiol.* **2013,** *166,* 391–398.

Fernandez, A.; Picouet, P.; Lloret, E. Cellulose-Silver Nanoparticle Hybrid Materials to Control Spoilage-Related Microflora in Absorbent Pads Located in Trays of Fresh-Cut Melon. *Int. J. Food Microbiol.* **2010,** *142,* 222–228.

Galindo-Perez, M. J.; Quintanar-Guerrero, D.; Mercado-Silva, E.; Real-Sandoval, S. A.; Zambrano-Zaragoza, M. L. The Effects of Tocopherol Nanocapsules/Xanthan Gum Coatings on the Preservation of Fresh-Cut Apples: Evaluation of Phenol Metabolism. *Food Bioprocess Technol.* **2015,** *8,* 1791–1799.

Ghosh, V.; Mukherjee, A.; Chandrasekaran, N. Eugenol-Loaded Antimicrobial Nanoemulsion Preserves Fruit Juice against, Microbial Spoilage. *Colloids. Surf. B. Biointerfaces.* **2014,** *114,* 392–397.

Gol, N. B.; Patel, P. R.; Rao, T. V. R. Improvement of Quality and Shelf-Life of Strawberries with Edible Coatings Enriched with Chitosan. *Postharvest Biol. Technol.* **2013,** *85,* 185–195.

Gómez-Estaca, J.; López deLacey, A.; López-Caballero, M. E.; Gómez-Guillén, M. C.; Montero, P. Biodegradable Gelatin–Chitosan Films Incorporated with Essential Oils as Antimicrobial Agents for Fish Preservation. *Food Microbiol.* **2010,** *27,* 889–896.

Ha, J. U.; Kim, Y. M.; Lee, D. S. Multilayered Antimicrobial Polyethylene Films Applied to the Packaging of Ground Beef. *Packag. Technol. Sci.* **2001,** *14,* 55–62.

Hafsa, J.; Smach, M.; Khedher, M. R. B.; Charfeddine, B.; Limem, K.; Majdoub, H; Rouatbi, S. Physical, Antioxidant And Antimicrobial Properties of Chitosan Films Containing Eucalyptus Globulus Essential Oil. *Food Sci. Technol.* **2016,** *68,* 356–364.

Hernández-Muñoz, P.; Almenar, E.; Valle, V.; Del Velez, D.; Gavara, R. Effect of Chitosan Coating Combined with Postharvest Calcium Treatment on Strawberry (*Fragariaananassa*) Quality during Refrigerated Storage. *Food Chem.* **2008,** *110* (2), 428–435.

Heshmati, M. K.; Hamdami, N.; Shahedi, M.; Nasirpour, A. Impact of Zatariamultiflora Essential Oil, Nisin, Potassium Sorbate and LDPE Packaging Containing Nano-ZnO on Shelf Life of Caviar. *Food Sci. Technol. Res.* **2013,** *19,* 749–758.

Hosseini, S. F.; Rezaei, M.; Zandi, M.; Farahmandghavi, F. Bio-Based Composite Edible Films Containing *Origanum vulgare* L. Essential Oil. *Ind. Crops Prod.* **2015,** *67,* 403–413.

Jang, J. D.; Hwang, Y. I.; Lee, D. S. Effect of Packaging Conditions on the Quality Changes of Fermented Soy Paste and Red Pepper Paste. *J. Korean Soc. Packag. Sci. Technol.* **2000,** *6,* 31–40.

Jo, W. S.; Song, H. Y.; Song, N. B.; Lee, J. H.; Min, S. C.; Song, K. B. Quality and Microbial Safety of "Fuji" Apples Coated with Carnauba Shellac Wax Containing Lemongrass Oil. *Food Sci. Technol.* **2014,** *55,* 490–497.

Kim, I. H.; Lee, H.; Kim, J. E.; Song, K. B.; Lee, Y. S.; Chung, D. S.; Min, S. C. Plum Coatings Of Lemongrass Oil-Incorporating Carnauba Wax-Based Nanoemulsion. *J. Food Sci.* **2013,** *78,* 1551–1559.

Kim, I. H.; Oh, Y. A.; Lee, H.; Song, K. B.; Min, S. C. Grape Berry Coatings of Lemongrass Oil-Incorporating Nanoemulsion. *Food Sci. Technol.* **2014,** *58,* 1–10.

Klangmuang, P.; Sothornvit, R. Barrier Properties, Mechanical Properties and Antimicrobial Activity of Hydroxypropyl Methylcellulose-Based Nanocomposite Films Incorporated with Thai Essential Oils. *Food Hydrocoll.* **2016,** *61,* 609–616.

Li, L.; Zhao, C.; Zhang, Y.; Yao, J.; Yang, W.; Hu, Q.; Wang, C.; Cao, C. Effect of Stable Antimicrobial Nano-Silver Packaging on Inhibiting Mildew and Storage of Rice. *Food Chem.* **2016,** *215,* 477–482.

Makwana, S.; Choudhary, R.; Kohli, P. Advances in Antimicrobial Food Packaging with Nanotechnology and Natural Antimicrobials. *Int. J. Food Sci. Nutr. Eng.* **2015,** *5,* 169–175.

Mangalassary, S. Antimicrobial Food Packaging to Enhance Food Safety: Current Developments and Future Challenges. *J. Food Process Technol.* **2012,** *3,* 3–5.

Mangalassary, S.; Han, I.; Rieck, J.; Acton, J.; Jiang, X.; Sheldon, B.; Dawson, P. L. Effect of Combining Nisin and/or Lysozyme with in-Package Pasteurization on Thermal Inactivation of *Listeria Monocytogenes* in Ready-to-Eat Turkey Bologna. *J. Food Prot.* **2007,** *70,* 2503–2511.

Mantilla, N.; Castell-Perez, M. E.; Gomes, C.; Moreira, R. G. Multilayered Antimicrobial Edible Coating and Its Effect on Quality and Shelf-Life of Fresh-Cut Pineapple (*Ananascomosus*). *Food Sci. Technol.* **2013,** *51,* 37–43.

Martinez-Romeroa, D.; Castillo, S.; Guillena, F.; Díaz-Mulaa, H. M.; Zapataa, P. J.; Valeroa, D.; Serranob, M. Aloe Vera Gel Coating Maintains Quality and Safety of Ready-To-Eat Pomegranate Arils. *Postharvest Biol. Technol.* **2013,** *86,* 107–112.

Martinon, M. E.; Moreira, R. G.; Castell-Perez, M. E.; Gomes, C. Development of a Multilayered Antimicrobial Edible Coating for Shelf-life Extension of Fresh-Cut Cantaloupe (*Cucumismelo* L.) Stored at 4°C. *Food Sci. Technol.* **2014,** *56,* 341–350.

Mills, A. Intelligent Inks. In *Kirk-Othmer Encyclopedia of Chemical Technology;* Wiley: New York, NY, 2011; pp 1–15.

Mohan, C. O.; Ravishankar, C. N.; Srinivasagopal, T. K. Effect of O_2 Scavenger on the Shelf-Life of Catfish (Pangasiussutchi) Steaks during Chilled Storage. *J. Sci. Food Agric.* **2008,** *88,* 442–448.

Mohan, C. O.; Ravishankar, C. N.; Srinivasa Gopal, T. K.; Ashok Kumar, K.; Lalitha, K. V. Biogenic Amines Formation in Seer Fish (*Scomberomoruscommerson*) Steaks Packed with O_2 Scavenger during Chilled Storage. *Food Res. Int.* **2009,** *42,* 411–416.

Moradi, M.; Tajik, H.; Rohani, S. M. R.; Mahmoudian, A. Antioxidant and Antimicrobial Effects of Zein Edible Film Impregnated with *Zataria multiflora* Boiss Essential Oil and Monolaurin. *Food Sci. Technol.* **2016,** *72,* 37–43.

Moreira, S. P.; Carvalho, W. M.; Alexandrino, A. C.; Paula, H. C. B.; Rodrigues, M. C. P.; Figueiredo, R. W.; Maia, G. A.; Figueiredo, E. M. A. T.; Brasil, I. M. Freshness Retention of Minimally Processed Melon Using Different Packages and Multilayered Edible Coating Containing Microencapsulated Essential Oil. *Int. J. Food Sci. Technol.* **2014,** *49,* 1–9.

Mortazavi, S. M. H.; Karami, Z.; Ahmad, M. Use of Ethylene Scavenger Sachet in Modified Atmosphere Packaging to Maintain Storage Stability of Khalal Date Fruit. *Int. J. Postharvest Technol. Innov.* **2015,** *5,* 52–63.

Mulla, M.; Ahmed, J.; Al-Attar, H.; Castro-Aguirre, E.; Arfat, Y. A.; Auras, R. Antimicrobial Efficacy of Clove Essential Oil Infused into Chemically Modified LLDPE Film for Chicken Meat Packaging. *Food Control.* **2016,** *73,* 663–671. Doi: http://doi.org/10.1016/j.foodcont.2016.09.018

Muthukumarasamy, P.; Han J. H.; Holley, R. A. Bactericidal Effects of *Lactobacillus reuteri* and Allyl Isothiocyanate on *Escherichia coli* O157:H7 in Refrigerated Ground Beef. *J. Food Prot.* **2003,** *66,* 2038–2044.

Nadarajah, D.; Han, J. H.; Holley, R. A. Inactivation of *Escherichia coli* O157:H7 in Packaged Ground Beef by Allyl Isothiocyanate. *Int. J. Food Microbiol.* **2005,** *99,* 269–279.

Nugraha, B.; Bintoro, N.; Murayama, H. Influence of CO_2 and C_2H_4 Adsorbents to the Symptoms of Internal Browning on the Packaged "Silver Bell" Pear (*Pyruscommunis L.*). *Agric. Agric. Sci. Procedia.* **2015,** *3,* 127–131.

Othman, S. H. Bio-nanocomposite Materials for Food Packaging Applications: Types of Biopolymer and Nano-sized Filler. *Agric. Agric. Sci. Procedia.* **2014,** *2,* 296–303.

Perdones, A.; Chiralt, A.; Vargas, M. Properties of Film-Forming Dispersions and Films Based on Chitosan Containing Basil or Thyme Essential Oil. *Food Hydrocoll.* **2016,** *57,* 271–279.

Pola, C. C.; Medeiros, E. A. A.; Pereira, O. L.; Souza, V. G. L.; Otoni, C. G.; Camillotoa, G. P.; Soares, N. F. F. Cellulose Acetate Active Films Incorporated with Oregano (*Origanumvulgare*) Essential Oil and Organophilic Montmorillonite Clay Control the Growth of Phytopathogenic Fungi. *Food Packaging Shelf Life.* **2016,** *9,* 69–78.

Poverenov, E.; Danino, S.; Horev, B.; Granit, R.; Vinokur, Y.; Rodov, V. Layer-By-Layer Electrostatic Deposition of Edible Coating on Fresh Cut Melon Model: Anticipated And Unexpected Effects of Alginate–Chitosan Combination. *Food Bioprocess Technol.* **2014,** *7,* 1424–1432.

Qadri, O. S.; Yousuf, B.; Srivastava, A. K. Fresh-Cut Produce: Advances in Preserving Quality and Ensuring Safety. In *Postharvest Management of Horticultural Crops: Practices for Quality Preservation*; Siddiqi, M. W, Ali, A., Eds.; Apple Academic Press: Waretown, NJ, 2016; pp 265–290.

Randazzo, W.; Jimenez-Belenguer, A.; Settanni, L.; Perdones, A.; Moschetti, M.; Palazzolo, E.; Guarrasi, V.; Vargas, M.; Germana, M. A.; Moschetti, G. Antilisterial Effect of Citrus Essential Oils and their Performance in Edible Film Formulations. *Food Control.* **2016,** *59,* 750–758.

Raybaudi-Massilia, R. M.; Mosqueda-Melgar, J.; Martín-Belloso, O. Edible Alginate-Based Coating as Carrier of Antimicrobials to Improve Shelf-Life and Safety of Fresh-Cut Melon. *Int. J. Food Microbiol.* **2008,** *121,* 313–327.

Robertson, G. L. *Food Packaging, Principles and Practice*, 3rd Ed.; Marcel Dekker: New York, 2013.

Rojas-Grau, M. A.; Raybaudi-Massilia, R. M.; Soliva-Fortuny, R. C.; Avena-Bustillos, R. J.; McHughb, T. H.; Martin-Belloso, O. Apple Puree-Alginate Edible Coating as Carrier of Antimicrobial Agents to Prolong Shelf-Life of Fresh-Cut Apples. *Postharvest Biol. Technol.* **2007,** *45,* 254–264.

Salinas-Roca, B.; Soliva-Fortuny, R.; Welti-Chanes, J.; Martín-Belloso, O. Combined Effect of Pulsed Light, Edible Coating and Malic Acid Dipping to Improve Fresh-Cut Mango Safety and Quality. *Food Control.* **2016,** *66,* 190–197.

Salvia-Trujillo, L.; Rojas-Grau, M. A.; Soliva-Fortuny, R.; Martin-Belloso, O. Use of Antimicrobial Nanoemulsions as Edible Coatings: Impact on Safety and Quality Attributes of Fresh-Cut Fuji Apples. *Postharvest Biol. Technol.* **2015**, *105,* 8–16.

Sanchís, E.; Ghidelli, C.; Sheth, C. C.; Mateos, M.; Palou, L.; Pérez-Gago, M. B. Integration of Antimicrobial Pectin-Based Edible Coating and Active Modified Atmosphere Packaging to Preserve the Quality and Microbial Safety of Fresh-Cut Persimmon (Diospyros Kaki Thunb. Cv. Rojobrillante). *J. Sci. Food Agric.* **2016**, *97,* 252–260. Doi:10.1002/jsfa.7722

Sessa, M.; Ferraria, G.; Donsi, F. Novel Edible Coating Containing Essential Oil Nanoemulsions to Prolong the Shelf Life of Vegetable Products. *Chem. Eng. Trans.* **2015**, *43,* 55–60.

Severino, R.; Ferrari, G.; Vu, K. D.; Donsì, F.; Salmieri, S.; Lacroix, M. Antimicrobial Effects of Modified Chitosan Based Coating Containing Nanoemulsion of Essential Oils, Modified Atmosphere Packaging and Gamma Irradiation Against *Escherichia coli* O157:H7 and *Salmonella Typhimurium* on Green Beans. *Food Control.* **2015**, *50,* 215–222.

Sipahi, R. E.; Castell-Perez, M. E.; Moreira, R. G.; Gomes, C.; Castillo, A. Improved Multi-layered Antimicrobial Alginate-Based Edible Coating Extends the Shelf Life of Fresh-Cut Watermelon (*Citrulluslanatus*). *Food Sci. Technol.* **2013**, *51,* 9–15.

Sommer, N. F.; Fortlagae, R. J.; Edwards, D. C. Postharvest Diseases of Selected Commodities. In *Postharvest Technology of Horticultural Crops*; Kader, A., Ed.; University of California Division of Agriculture and Natural Resources, Publications: Davis, CA, 1992; pp 117–160.

Sung, S. Y.; Sin, L. T.; Tee, T. T.; Bee, S. T.; Rahmat, A. R.; Rahman, W. A.; Tan, A. C.; Vikhraman, M. Antimicrobial Agents for Food Packaging Applications. *Trends Food Sci. Technol.* **2013**, *33,* 110–123.

Suppakul, P.; Thanathammathorn, T.; Samerasut, O.; Khankaew, S. Shelf Life Extension of "Fios De Ovos," an Intermediate-Moisture Egg-Based Dessert, by Active and Modified Atmosphere Packaging. *Food Control.* **2016**, *70,* 58–63.

Tastan, O.; Ferrari, G.; Baysal, T.; Donsì, F. Understanding the Effect of Formulation on Functionality of Modified Chitosan Films Containing Carvacrol Nanoemulsions. *Food Hydrocoll.* **2016**, *61,* 756–771.

Veasna, H.; Hwang, Y. S.; Choi, J. M.; Ahn, Y. J.; Lim, B. S.; Chun, J. P. 1- Methyl Cyclopropene and Carbon Dioxide Absorber Reduce Chilling Injury of Eggplant (*Solanum melongena* L.) during MAP Storage. *J. Bio-Environ. Control.* **2012**, *21,* 50–56.

Weng, Y. M.; Hotchkiss J. H. Anhydrides as Antimycotic Agents Added to Polyethylene Films for Food Packaging. *Packag. Technol. Sci.* **1993**, *6,* 123–128.

Xing, Y.; Li, X.; Xu, Q.; Jiang, Y.; Yun, J.; Li, W. Effects of chitosan-Based Coating and Modified Atmosphere Packaging (MAP) on Browning and Shelf Life of Fresh-Cut Lotus Root (*Nelumbo nucifera* Gaerth). *Innov. Food Sci. Emerg. Technol.* **2010**, *11,* 684–689.

Yadav, R.; Aradhita, R.; Khatkar, B. S. A Study of Active Packaging (AP) Technology on Banana Fruit. *J. Basic Appl. Eng. Res.* **2015**, *2,* 1518–1522.

Yousuf, B.; Srivastava, A. K. Psyllium (*Plantago*) Gum as an Effective Edible Coating to Improve Quality and Shelf Life of Fresh-Cut Papaya (*Carica papaya*). *Int. J. Biol. Biomol. Agric. Food Biotechnol. Eng.* **2015**, *9,* 702–707.

Zahid, N.; Ali, A.; Manickam, S.; Siddiqui, Y.; Maqbool, M. Potential Of Chitosan-Loaded Nanoemulsions to Control Different *Colletotrichum* Spp. and Maintain Quality of Tropical Fruits during Cold Storage. *J. Appl. Microbiol.* **2012**, *113,* 925–939.

Zambrano-Zaragoza, M. L.; Gutierrez-Cortez, E.; Real, A. D.; Gonzalez-Reza, R. M. Galindo-Perez, M. J.; Quintanar-Guerrero, D. Fresh-cut Red Delicious Apples Coating

Using Tocopherol/Mucilage Nanoemulsion: Effect of Coating on Polyphenol Oxidase and Pectin Methylesterase Activities. *Food Res. Int.* **2014a,** *62,* 974–983.

Zambrano-Zaragoza, M. L.; Mercado-Silva, E.; Real, A. D.; Gutierrez-Cortez, E.; Cornejo-Villegas, M. A.; Quintanar-Guerrero, D. The Effect of Nano-Coatings with A-Tocopherol and Xanthan Gum on Shelf-Life and Browning Index of Fresh-Cut "Red Delidous" Apples. *Innov. Food Sci. Emerg. Technol.* **2014b,** *22,* 188–196.

Zhang, H.; Hortal, M.; Dobon, A.; Bermudez, J. M.; Lara-Uedo, M. The Effect of Active Packaging on Minimizing Food Losses: Life Cycle Assessment (LCA) of Essential Oil Component-Enabled Packaging for Fresh Beet. *Packag. Technol. Sci.* **2015,** *28,* 761–774.

Zhao, L.; Li, F.; Chen, G.; Fang, Y.; An, X.; Zheng, Y.; Xin, Z.; Zhang, M.; Yang, Y.; Hu, Q. Effect of Nanocomposite-based Packaging on Preservation Quality of Green Tea. *Int. J. Food Sci. Technol.* **2012,** *47,* 572–578.

Zuniga, G. E.; Junqueira-Gonc-alves, M. P.; Pizarro, M.; Contreras, R.; Tapia, A.; Silva, S. Effect of Ionizing Energy on Extracts of *Quillaja saponaria* to be Used as an Antimicrobial Agent on Irradiated Edible Coating for Fresh Strawberries. *Radiat. Phys. Chem.* **2012,** *81,* 64–69.

CHAPTER 7

EDIBLE COATINGS AND THEIR EFFECT ON POSTHARVEST FRUIT QUALITY

K. PRASAD[1], M. W. SIDDIQUI[2*], R. R. SHARMA[1],
ABHAY KUMAR GAURAV[3], PALLAVI NEHA[4], and NIRMAL KUMAR[1]

[1]Division of Food Science and Postharvest Technology, Indian Agricultural Research Institute, New Delhi, India
[2]Department of Food Science and Postharvest Technology, Bihar Agricultural University, Sabour, Bhagalpur 813210, Bihar, India
[3]Division of Floriculture and Landscaping, Indian Agricultural Research Institute, New Delhi, India
[4]Division of Postharvest Technology, Indian Institute of Horticultural Research, Bengaluru, India

*Corresponding author. E-mail: wasim_serene@yahoo.com

CONTENTS

ABSTRACT

Edible coating is a healthy and environment friendly approach. It is directly consumable and does not leave any residues. This technology not only retains the quality of fresh fruits but also extends their overall shelf-life. Edible coating alters various physical, physiological and biochemical aspects of fruit growth, to carry out its effect. Alteration and control over respiration rate, moisture retention, retaining glossiness, firmness, are few to be listed from the long list of its functions. Intelligent selection and use of edible coating helps us to incorporate other natural preservatives in it, which in turn offers resistance to fruits against various quality-deteriorating factors. Although some lacunas still exists in the technology with respect to gas exchange, off flavor, functional barriers, etc. These lacunas can be rectified to a maximum extent by the knowledge of crop-based response of various edible coatings and is an important future thrust. The success of edible coating depends on its ability for enhancing food safety along with retaining nutritional and sensory attributes. The chapter deals with the prop-erties, types, classification, formulations, available products forms, advan-tages, disadvantages, mode of action, and phenomenon involved. Along with this a detailed overview on action of edible coating with the reports of its physical, physiological, and biochemical effect on fruit crops has been provided. This also includes the recent botanicals and plant products tested successfully for their effect on fruit crops, when used as edible coating. Thus this chapter provides an overview for the researchers for evolving and modifying the edible coatings for their enhanced effects on fruit crops and to explore this emerging and expanding green technology of edible coating for humankind.

7.1 INTRODUCTION

Fruit quality maintenance after harvest is the need of the hour (Prasad & Sharma, 2016). Use of postharvest treatments for this purpose now becomes mandatory from the traders and growers point of view (Prasad et al., 2016). Edible coatings are important for retaining postharvest quality through many ways. Edible coatings denote the application of commercial food grade waxes to restore and protect the loss of natural glossiness during the post-harvest period. Coating or waxing reduces shriveling, wilting, and respira-tion rate of fruits and enhances the gloss and cosmetic appearance of fruits (El-Anany et al., 2009; Dhal, 2013). The use of food grade wax coating on fruits is safe and is approved for application on fresh fruits and vegetables (PFA, 2008).

Edible coatings are thin layers of edible material applied to the surface of the product in addition to or as a replacement for natural protective waxy coatings to provide a barrier to moisture, oxygen, and solute movement for the food (Olivas et al. 2008; Smith et al., 1987; McHugh & Senesi, 2000: Raghav et al., 2016; Rupak & Suman, 2016). They are applied in the form of dipping, spraying, or brushing on the food surface in order to create a modified atmosphere (Krochta & Mulder-Johnston, 1997; McHugh & Senesi, 2000; Olivas et al., 2008). Edible coatings maintain the quality of fruits and vegetables by forming an edible film over the produce. The film acts as a partial barrier to different gases like O_2 and CO_2, water vapor and other chemical compounds, which creates a modified atmosphere around the fruits and vegetables, thus decreasing the respiration rate and the water loss, and preserving its texture and flavor. The main advantage of edible films over traditional synthetics is that they can be consumed with the packaged products. There is no package to dispose of even if the films are not consumed they could still contribute to the reduction of environmental pollution. They do not add unfavorable properties to the foodstuff. Edible coatings and films do not replace traditional packaging materials but provide an additional support to fruits and vegetables for their preservation.

7.2 PROPERTIES OF EDIBLE COATINGS

The properties of edible coating depend primarily on molecular structure rather than molecular size and chemical constitution. Specific requirements for edible films and coatings are (Arvanitoyannis & Gorris, 1999; Bourtoom, 2008; Dhall, 2013; Raghav et al., 2016; Rupak & Suman, 2016):

- The coating should be water-resistant so that it remains intact and covers a product adequately when applied.
- It should not deplete O_2 or build up excessive CO_2. A minimum of 1–3% oxygen is required around a commodity to avoid a shift from aerobic to anaerobic respiration.
- It should reduce water vapor permeability.
- It should improve appearance, maintain structural integrity, improve mechanical handling properties, carry active agents (antioxidants, vitamins, etc.) and retain volatile flavor compounds.

- It should provide biochemical and microbial surface stability while protecting against contamination, pest infestation, microbe proliferation, and other types of decay.
- It should melt above 40°C without decomposition.
- It should be easily emulsifiable, non-sticky, or should not be tacky, and have efficient drying performance.
- It should never interfere with the quality of fresh fruit or vegetable and not impart undesirable odor.
- It should have low viscosity and be economical for large quantity.
- It should be translucent to opaque for good visibility but not like glass and should be capable of tolerating slight pressure.

7.3　TYPES OF EDIBLE COATINGS AND COMMERCIAL AVAILABILITY

There are various types of edible coatings available in the markets; they have their own merits and demerits. They may be composed of hydrocolloids (polysaccharides or proteins), hydrophobic compounds (lipids or waxes), or of a combination of both (composite coatings). Sometimes, additives like antimicrobials, texture enhancers, nutrients, plasticisers, and emulsifiers are also added to improve the usefulness of the film in certain environments. The content and composition of the coating materials determine its properties with regard to gases like O_2, CO_2, water vapor, and other chemicals from the fruits and vegetable to the environment. Any compound added to the coating has to be safe to eat and generally recognized as safe (Raghav et al., 2016; Rupak & Suman, 2016). Classification and commercial forms are given in the Tables 7.1 and 7.2.

7.4　ADVANTAGES AND DISADVANTAGES OF EDIBLE COATINGS

7.4.1　ADVANTAGES OF EDIBLE COATINGS

It provides various advantages:

- The biggest advantage is that; it can be consumed with the packaged products.
- As these are edible, there are no materials to discard; even if the films are not consumed they are expected to degrade more readily

TABLE 7.1   Classification of Edible Films and Coatings.

Sl No.	Major types	Sub-types	Source and characteristics
1	Polysaccharides	Cellulose	Cellulose is the most abundantly occurring polysaccharides found on earth. Methylcellulose, HPMC, HPC, and CMC films possess a good film-forming characteristic, are generally odorless and tasteless, flexible and of moderate strength, transparent, resistant to oil and fats, water-soluble, and moderately permeable to moisture and oxygen transmission (Krochta & Mulder-Johnston, 1997).
		Starch	Starch is the storage polysaccharide of cereals, legumes, and tubers, widely available as raw material and suitable for a variety of industrial uses. It contains amylose and amylopectin. Starches are good oxygen barrier, used for coating fruits and vegetables characterized by high respiration rates.
		Chitosan	Chitosan is derived from chitin; it is an edible polymer, isolated from crustacean animal shells. It is a natural product which is non-toxic in nature.
		Gums	Gums are soluble in water. They include exudate gums (gum Arabic), extractive gums (locust bean and guar) and microbial fermentation gums (xanthan and gellan).
		Alginate	Alginate is extracted from brown algae, which are sodium salts of alginic acid. Alginate contains excellent barrier for moisture and water vapor.
		Pectin	Pectin is naturally found in plants mainly in fruits and vegetables like guava, apple, etc. Pectin is good for low moisture fruits and vegetables but is not a good moisture barrier.
		Carrageenan	Carrageenan is extracted from red seaweeds. It is a complex mixture of several water-soluble galactose polymers.
		Agar	Agar gums are obtained from red seaweeds, which is galactose polymer. Agar is hydrophilic colloidal in nature it contains a mixture of agro pectin and agarose.
		Dextrins	Derivative of starch with a smaller molecular size. Coatings provide a better water vapor resistance
2	Proteins	Casein	Casein is a milk-derived protein. Casein is commonly used in the preparation of emulsion because it is amphipathic in nature and contains hydrophilic and hydrophobic ends. Casein forms transparent, flavorless, and flexible films. Generally used in caseinate form.

TABLE 7.1   *(Continued)*

Sl No.	Major types	Sub-types	Source and characteristics
		Whey proteins	Whey protein is a by-product of the cheese production. It produces transparent, flavorless, and flexible films, similar to caseinate films. Whey protein-based films possess excellent oxygen barrier properties and also are good grease barriers.
		Zein	Zein proteins are obtained from maize and are insoluble in water. It is dissolved in a solvent. Characterized by excellent barrier property to water vapor. Corn-Zein protein is effective to prevent color change, firmness, weight loss and it increases the shelf life of fruits and vegetables, it has good barrier property to O_2.
		Gluten	Gluten is a water-insoluble protein of wheat flour, which is composed of a mixture of polypeptide Molecules. It contains gliadin and glutenin. Edible films can be formed by drying aqueous ethanol solution of wheat gluten.
		Gelatin and collagen	Gelatin is obtained from the fibrous insoluble protein, collagen, which is widely found in nature as the major constituent of skin, bones, and connective tissue. Gelatin is produced via either partial acid or alkaline hydrolysis of collagen. Gelatin is used to encapsulate low moisture or oil phase food ingredients which provide protection against oxygen and light.
		Soy protein	Isolated from soybean. Most of the protein in soybeans is insoluble in water but soluble in dilute neutral salt solutions (Kinsella & Phillips, 1979). Soy protein coatings generally have poor moisture resistance and water vapor barrier properties whilst they are potent oxygen barriers.
		Albumen	Egg protein. Commonly available and cheap.
3	Lipids	Carnauba wax	Carnauba wax is derived from the palm tree leaves (*Copernicia cerifera*). Carnauba wax has a very high melting point and is used as an additive to other waxes to increase toughness and luster.
		Bee wax	Beeswax is made by honeybees.
		Paraffin wax	Paraffin wax is derived from distillate fractions of crude petroleum and consists of a mixture of solid hydrocarbons. Paraffin wax is used on raw fruits and vegetables.

TABLE 7.1   *(Continued)*

Sl No.	Major types	Sub-types	Source and characteristics
		Minerals oil	Mineral oil is made of a mixture of liquid paraffin and naphthenic hydrocarbon. Mineral oil is commonly used for coating fruits and vegetables and as a food release agent (Hernandez, 1994).
		Fatty acids and monoglycerides	Fatty acids and monoglycerides are used for coatings mainly as emulsifiers and dispersing agents. Fatty acids are generally extracted from vegetable oils, while monoglycerides are prepared by trans-esterification of glycerol and triglycerol (Hernandez, 1991). The acetylated monoglyceride displays the unique characteristic of solidifying from the molten state into a flexible, wax-like solid.
		Resins	Resin coatings are effective in reducing water loss but are least permeable to gases resulting in anaerobic respiration and flavor changes thus in poor quality. Shellac resin is a secretion by the insect Laccifer lacca and is composed of a complex mixture of aliphatic alicyclic hydroxyl acid polymers. This resin is soluble in alcohols and in alkaline solutions. Shellac is not generally regarded as safe substance as it is only permitted as an indirect food additive in food coatings and adhesives.
4	Composites	Composites and bilayer	Composite and bilayer coatings are the edible coatings of the future. These coatings may be heterogeneous in nature, consisting of a blend of polysaccharides, proteins, and/or lipids. This approach enables one to utilize the distinct functional characteristics of each class of film used (Kester & Fennema, 1986).
		Conglomerates	A mixture of several components in one layer. Components are like (1) lipid component; (2) a protein component; (3) a gelled plant gum; (4) an emulsifier; and optionally (5) Buffering agent and/or a plasticizer.
5	Additives	Plasticizers	Glycerol, acetylated monoglyceride, polyethylene glycol, and sucrose.
		Antimicrobial compounds	The use of edible coatings as carriers of antimicrobial compounds is another potential alternative to enhance the safety of produce (e.g., nisin).
		Enzymes	Lysozyme, peroxidase, and lactoperoxidase.

TABLE 7.1 *(Continued)*

Sl No.	Major types	Sub-types	Source and characteristics
		Essential oils	Cinnamon, oregano, lemongrass, clove, rosemary, tea tree, thyme, and bergamot.
		Nitrites and sulfites	
		Synthetic antioxidants	Butylated hydroxyanisole, butylated hydroxytoluene, propyl gallate, octyl gallate, dodecyl gallate, and ethoxyquin.
		Natural antioxidants	Tocopherols, tocotrienols, ascorbic acid, citric acid, and carotenoids.
6	Herbal coating	Aloe gel	This gel is isolated from aloe vera plant. Aloe vera gel is a polysaccharide matrix rich in active compounds. This gel is tasteless, colorless, and odorless. This natural product is a safe and environment friendly alternative to synthetic preservatives such as sulfur dioxide.
		Neem, lemongrass, rosemary, tulsi, and turmeric	These herbs have antimicrobial properties; it consists of vitamins, antioxidants, and essential minerals. These days herbal edible coating is getting higher importance.

TABLE 7.2 Commercially Available Edible Coatings.

Sl No.	Trade name	Composition
1.	FreshSeal™ (Agricoat Industries Limited, Berkshire, United Kingdom)	Polyvinyl alcohol, starch, and surfactant
2.	Fry shield	Calcium pectinate
3.	Nature seal (Agricoat Industries Limited, Berkshire, United Kingdom)	Calcium ascorbate
4.	Nutri-save	N,O-carboxymethyl chitosan
5.	Opta Glaze	Wheat gluten
6.	Seal gum, spray gum	Calcium acetate
7.	Semperfresh™ (Agricoat Industries Limited, Berkshire, United Kingdom)	Sucrose esters
8.	Z Coat	Corn protein

TABLE 7.2   *(Continued)*

Sl No.	Trade name	Composition
9.	Prolong	Mixture of sucrose fatty acid esters, sodium CMC, and mono and diglycerides
10.	Tal prolong	Mixture of sucrose fatty acid esters, sodium CMC, and mono and diglycerides
11.	Nature-seal™	Cellulose-based edible coating
12.	Zein	Corn zein protein
13.	Brilloshine	Sucrose esters and wax
14.	Nu-coatFlo, Ban-seel	Sucrose esters of fatty acids and sodium salt of CMC
15.	Citrashine	Sucrose ester and wax
16.	Sta-Fresh Series (Bornnet Corporation Company Ltd.)	Shellac/carnauba/resin
17.	Natural Shine™ Series (Pace International, Branch Road, Wapato, WA)	Vegetable wax, carnauba, and shellac
18.	PrimaFresh® Series (Pace International, Branch Road, Wapato, WA)	Vegetable wax, carnauba, and shellac
19.	Shield-Brite® Series (Pace International, Branch Road, Wapato, WA)	Carnauba and shellac
20.	Peach, Nectarine & Plum Lustr® (Decco Cerexagri Inc., S. California Ave, Monrovia, CA)	Vegetable oil and mineral oil
21.	Syncera Series (Paramelt B.V., Netherlands)	Carnauba wax, shellac resin, and polyethylene

than synthetic and thus contribute to the reduction of environmental pollution.

- It enhances the organoleptic properties of fruits and vegetables by giving additional shine to the surface of the fruits besides act as flavoring agent and sweetener.
- Reduces fresh weight loss and keeps the fruit firm, so that its fresh look can be maintained.
- The coating reduces the respiration rate as well as senescence by delaying and reducing ethylene production (partial barrier to gas exchange).
- It also prevents fruits and vegetables against different storage disorders including chilling injuries mainly in tropical crops.

- It also provides a carrier for different postharvest chemical treatments like anti-microbial compounds.
- Encapsulation of different aroma compounds, antioxidants, pigments, ions with edible coats can reduce the browning reactions and enhance the shelf-life of fruits and vegetables.
- Different nutritional substances such as vitamins and minerals can also be added to enrich the product. They enhance the nutritional composition of fruits and vegetables without affecting its quality.
- Reduces the use of synthetic packaging material.
- These can be also be used for individual packaging of small products such as pears, beans, nuts, and strawberries as practical solutions.

7.4.2 DISADVANTAGES OF EDIBLE COATINGS

Edible coatings also have some disadvantages. They are:

- Thick coatings could restrict the respiratory gas exchange, causing the product to accumulate high levels of ethanol and develops off-flavor.
- Edible coatings have good gas barrier properties which cause anaerobic respiration due to this normal ripening process is disturbed in fruits and vegetables.
- Coating formulations that provide adequate gas exchange are often not good barriers to water vapor. Poor water vapor barrier could be a potential source of microbial spoilage.
- Antioxidants compounds also have some disadvantages like vulnerability to high temperature and light, high volatility, limited solubility and unpleasant flavor; these characteristics result in a loss of functionality, and hence limiting their application.

7.5 EFFECTS OF EDIBLE COATINGS IN FRUIT CROPS

The fruit edible coatings affect the physical, physiological, and biochemical characteristics of fruits. Edible coatings are utilized in several fruit crops. Understanding of the results of the experiments carried out by researchers of various countries increases the knowledge of better use of edible coatings. Crop-specific use of edible coatings will support the better postharvest management and handling of fruit crops.

7.6 PHYSIOLOGICAL IMPACT

7.6.1 EDIBLE COATINGS AFFECT PHYSIOLOGICAL LOSS IN WEIGHT IN VARIOUS FRUIT CROPS

Physiological loss in weight (PLW) is an important parameter which determines the freshness of a fruit. The percent of PLW in general increased with the advancement of the storage period rather slowly in the beginning, but at a faster pace as the storage period advanced.

7.6.1.1 MANGO

Two types of fruit coatings were tested by Baldwin et al. (1999) for their effect on external and internal mango fruit atmospheres and quality factors during simulated commercial storage at 10 or 15°C with 90–99% RH followed by simulated marketing conditions of 20°C with 56% RH. One coating was polysaccharide-based while the other had carnauba wax as the main ingredient. The carnauba wax coating significantly reduced water loss compared to uncoated and polysaccharide-coating treatments.

7.6.1.2 GRAPES

Valverde et al. (2005) treated table grapes (cv. Crimson seedless) with aloe vera gel (1:3; aloe vera gel: distilled water) and stored at 1°C for 21 days. It was observed that, the loss in weight of aloe vera gel treated fruits was only 8% compared to 15% in case of control. When grapes were treated with aloe vera gel before harvest, similar results were reported during postharvest storage (Castillo et al., 2010).

7.6.1.3 BANANA

Malmiri et al. (2011) worked on banana cv. Berangan. They coated fruits with chitosan (0.5–2.5%, w/w) with glycerol (0–2% w/v) and Tween 80 (0.1% w/v) and stored at 26°C and 40–50% RH for 10 days. Water can increase the rate of several reactions in fruits such as browning, vitamin degradation, and enzyme activity; enhance the rate of microorganism's growth and cause texture changes. The study shows the weight loss of coated banana as function of chitosan and glycerol concentrations. As clearly observed, at low concentrations of glycerol (less than 0.2%), an increase in the concentration of chitosan, significantly decreased the weight loss. Han et al. (2004) reported that edible coating significantly reduced weight loss of fresh strawberries (Puget Reliance) and red raspberries (Tullmeen) during storage at 2°C and 88% RH compared to the control.

7.6.1.4   BLUEBERRIES

Duan et al. (2011) worked on fresh blueberries cv. Duke and Elliott used Semperfresh, acid-soluble chitosan, water-soluble chitosan, water-soluble chitosan+sodium alginate, applied concentrations were 2% (w/v) acid soluble chitosan, 3% water-soluble chitosan, 1% Semperfresh. It has been reported that Semperfresh coating significantly decreased weight loss.

7.6.1.5   CHERRY

Yaman and Bayoundurlc (2002) worked on sweet cherry and applied coating and its concentration as Semperfresh@10, 20 g/L, stored at ambient temperature (30 ± 3°C) and at humidity (40–50% RH), and stored at cold storage (0°C) and at humidity (95–98% RH). They found that Semperfresh@ 20 g/L was most effective to reduce the weight loss. The primary mechanism of moisture loss from fresh fruits and vegetables is by vapor-phase diffusion driven by a gradient of water vapor pressure at different locations. The thickness of the barrier and moisture permeability of coatings are important factors from the viewpoint of mass transfer rate. On the other hand, respiration causes a weight reduction because a carbon atom is lost from the fruit in each cycle (Labuza & Contrereas, 1981). Weight loss of cherries in cold storage was statistically lower than in those stored at ambient temperature due to temperature effects on vapor pressure difference and increased water retention. Sweet cherry (cv. Star King) when treated with 100% aloe vera gel and stored at 1°C, treated fruits showed lower weight loss (3.82%), compared to control (6.2%) after 16 days of storage (Martínez-Romero et al., 2006).

7.6.1.6   PAPAYA

Marpudi et al. (2011) reported that when papaya fruit treated with aloe vera gel (50%), it showed reduced weight loss (26%) compared to control (32%), during storage at 30 ± 3°C for 10 days.

7.6.1.7   AVOCADO

Feygenberg et al. (2005) applied carnauba wax based organic coating on avocado and mango which effectively reduced water loss, shrinkage, chlorophyll breakdown, and chilling injury symptoms.

7.6.1.8   APPLE

"Granny Smith" and "Red Chief" apples coated with aloe vera gel (1, 5, and 10% w/v) showed lower loss in weight compared to control, during storage at 2°C for 6 months (Ergun & Satici, 2012).

7.6.1.9   PEAR

Zhou et al. (2008) worked with Huanghua pears cv. Huanghua. Coatings used were shellac, Semperfresh, and carboxymethyl chitosan (CMC) during cold storage (4°C). Coatings were applied as Shellac@14.3 g/100 mL water, Semperfresh@1.0 g/100 mL water, CMC@ 2.0 g/100 mL water. The fruit coated with shellac and Semperfresh showed 5.82 and 6.94% weight loss, respectively, as compared to 8.47 and 7.84% weight loss in control fruit and CMC-coated fruits; this difference was significant. Previous studies indicated that approximately 3–5% weight loss leads to shriveling in apples and approximately 5% weight loss was the normal acceptable limit for grapes (Bai et al., 2002). However, in this study, all coatings reduced weight loss in pears during storage, and no shriveling was observed in any treatment, including in the control samples. The CMC and Semperfresh coatings induced greater weight loss because they are more hydrophilic than the shellac coating. Water loss can cause flesh softening, fruit ripening, and senescence by ethylene production and other metabolic reactions (Bai et al., 2002). Clearly, relatively lower weight loss in shellac-coated pears contributed to maintaining better quality of fruit during cold storage.

7.6.1.10   PLUM

Valero et al. (2013) reported that alginate treatments significantly decreased weight loss for all plum cultivars. Weight loss of fruit is due to the transpiration process which is determined by the gradient of water vapor pressure between the fruit and the surrounding air. Transpiration is usually reduced by both epidermal cell layer and cuticle. Thus, as fruit surface/volume ratio and epidermis and cuticle structure are different among plum culti-vars, differences in weight loss were observed in control fruit depending on cultivar. In addition, edible coatings act as an extra layer which also coats the stomata leading to a decrease in transpiration and in turn, to a reduction in weight loss, this being the primary beneficial effect of edible coatings, as has been demonstrated in a wide range of fruits including apricot, pepper, peach, sweet cherry, and litchi (Ayranci & Tunc, 2004; Díaz-Mula et al., 2012).

7.6.1.11   STRAWBERRIES

Velickova et al. (2013) worked with strawberries cv. Camarosa. Treatments given were Chitosan@ (0.8 g/100 g) with acetic acid (1 mL/100 mL) and 0.2 g of glycerol and 0.2 g of Tween 80, beeswax coatings, 0.5 g/100 g with 25% of Tween 80 and three-layer coating consisting of separate beeswax-chitosan-beeswax layer, composite coating (0.8 g/100 g chitosan and 10 g/100 g beeswax emulsion) and stored at 20°C and 35–40% RH. Chitosan-based coatings prolonged the storage period of strawberries for seven days at temperature of 20°C and relative humidity of 53% and slowed down their senescence process compared to uncoated strawberries, beeswax or as composite coating showed beneficial effect against fungal infection and weight loss reduction.

Aloe vera gel treatment (25, 50, 75, and 100% gel v/v) of strawberry fruit showed that, coating of fruit significantly reduced fruit weight loss, compared to control (Vahdat et al., 2009). Singh et al. (2011) also reported similar finding in strawberry fruit coated with aloe vera gel (1:3), during storage at 5°C for 16 days.

7.7   EDIBLE COATINGS AFFECT RESPIRATION AND ETHYLENE PRODUCTION

The fruits have a natural wax coating, which develops during the matu-ration and ripening processes. However, during rough handling of fruits, the natural shield gets destroyed and, therefore, bruising occurs during the packing and transport operations. Fruits and vegetables continue to respire even after harvest and use up all the oxygen within the produce, which is not replaced as quickly as by edible coating and produces carbon dioxide. This accumulates within the produce because it cannot escape as easily through coating. Eventually the fruit and vegetable will shift to partial anaerobic respiration that requires less oxygen (1–3%) (Park et al., 1994a, b; Guilbert et al., 1996; McHugh & Senesi, 2000). With less oxygen, the production of ethylene (which accelerates ripening process) is disrupted and physiological loss of water is minimized. Thus, the fruits and vegetables remain firm, fresh, and nutritious for longer period and their shelf life almost doubles. The natural barrier on fruit and vegetable and the type and amount of coating will influence the extent to which the internal atmosphere (oxygen and carbon dioxide) are modified and the level of reduction in weight loss.

7.7.1   PEAR

Respiration rate of stored fruits has been found increased with the advancement of storage period. Zhou et al. (2008) worked with Huanghua pears cv. Huanghua. Coatings used were shellac, Semperfresh, and CMC during cold storage (4°C). Coatings were applied as Shellac@14.3 g/100 mL water, Semperfresh@1.0 g/100 mL water, and CMC@ 2.0 g/100 mL water. The main characteristics of the respiration rates of the Huanghua pears treated with different kinds of coatings. According to the results, throughout the storage period, the respiration rates of coated pears significantly decreased. These values were only 59.9–79.1% of those of the control samples at the beginning of the cold storage period. By day 60, the respiration rates of the control samples were 1.25–1.36 times higher than those of the coated fruits. Previous studies indicated that the gas exchange between fruit and the atmosphere occurs partly by diffusion through open pores (stomates, lenticels, stem, and blossom scars) and partly by permeation through fruit skin (Bai et al., 2002), and that it occurs mainly through pores (Amarante et al., 2001).

The low respiration rate in carnauba wax treated fruit attributed due to reduced gas interchange and consequently low oxygen availability to the fruit tissues for respiration (Purvis, 1994). Further, reported by Barman et al. (2011) impact of combined application of Putrescine + carnauba wax proved better, this additive benefit arises because of antisenescence and barrier properties of Putrescine and carnauba wax, respectively. It was also reported that ethylene production of pomegranate fruits was suppressed when they have treated with Putrescine either alone or in combination with carnauba wax.

7.7.2 PLUM

Valero et al. (2013) coated four plum (*Prunus salicina Lindl.*) cv. Black Amber, Larry Ann, Golden Globe and Songold with alginate@ 1 and 3% w/v. Analytical observations were made after 7, 14, 21, 28, and 35 days at 2°C and after a 3 day period at 20°C (shelf-life). Ethylene production rate at harvest was $0.25-0.50/g kg^{-1} h^{-1}$ for all plum cultivars and remained at these low levels during storage at 2°C, without significant differences between control and treated plums. However, when plums were transferred to 20°C, ethylene production increased, and the four plum cultivars showed the typical climacteric ripening pattern of plum fruit, although some plum cultivars have a suppressed-climacteric phenotype, such as "Shiro", "Rubyred" (Abdi et al., 1997), "Angeleno" (Candan et al., 2008), and "TC Sun" (Díaz-Mula et al., 2009). However, edible coatings significantly inhibited ethylene production for all plum cultivars, especially in Alg-3 treated plums, in which the climacteric peak of ethylene production was highly inhibited. The barrier properties of the edible coatings also reduce the selective permeability to O_2 and CO_2 of the fruit surface leading to an increase in CO_2 concentration in the fruit tissues and a decrease in O_2 concentration (De Wild et al., 2005), which could be responsible for the reduced ethylene production rate in the alginate coated plums. Accordingly, ethylene production decreases due to the effect of elevated CO_2 concentration on inhibiting the conversion of S-adenosylmethionine (SAM) to 1-aminocyclopropane- 1-carboxylic acid (ACC) by ACC synthase (De Wild et al., 2005).

Eum et al. (2009) worked on plum cv. Sapphire, and applied carbohydrate based Versasheen@ 5% with sorbitol as plasticizer. The coated plums were stored at 25°C with 85% RH for 8 days. Ethylene production of uncoated and Versasheen-coated plums increased during room temperature storage. The coating of Versasheen without sorbitol showed a higher ethylene production than uncoated and Versasheen with sorbitol since 3 days room temperature storage. Using Versasheen without sorbitol, coating application did not affect ethylene efflux of the fruit, while Versasheen with sorbitol-coated plums delayed the ethylene production. Although, Versasheen without sorbitol coating showed higher ethylene levels than other treatment, ripening indicators such as color development and firmness changes were delayed. Similarly, in other cultivar of plums, the changes in acidity and color were at least ethylene-dependent while soluble solid content changes appeared to be ethylene independent. Some previous studies showed that the effect of coating application on ethylene production depended on the commodity and coating materials. Ethylene production of apple coated with polysaccharide-based compound was reduced, whereas those of plum coated with hydroxypropyl methylcellulose was not affected.

7.7.3 GRAPES

Valverde et al. (2005) studied the effect of aloe vera gel (1:3 diluted with distilled water) coating on ethylene evolution rate of table grape during storage at 1°C for 35 days. The result showed that it was significantly lower in treated fruits (0.45 mL g^{-1} h^{-1}), compared to control (0.90 mL g^{-1} h^{-1}).

"Crimson Seedless" grape when coated with aloe vera gel (1:3 diluted with distilled water) and stored at 1°C for 35 days, it significantly reduced respiration rate (13.14 mg kg^{-1} h^{-1}) compared to control (19.03 mg kg^{-1} h^{-1}) (Valverde et al., 2005). When the grapes were treated with aloe vera gel preharvest, similar result was found during postharvest storage at 2°C (Castillo et al., 2010).

7.7.4 CHERRY

Martınez-Romero et al. (2006) reported that sweet cherry (cv Star King) treated with aloe vera gel (1:3 diluted with distilled water) showed significantly lower respiration rate than control, during storage at 1°C for 16 days.

7.7.5 BREADFRUIT

Worrell et al. (2002) have investigated that low temperature storage of breadfruit at 13°C doubled the shelflife of this breadfruit to about 10 days by delaying the onset of the climacteric. All the coatings investigated delayed fruit softening slightly at both ambient temperature and 13°C, possibly by depressing internal O_2 and increasing internal CO_2 concentrations.

Breadfruit showed a comparatively high basal CO_2 production rate of 20–50 mL kg^{-1} h^{-1} as might be expected for such a perishable commodity. When fruit were stored at 13°C, peak CO_2 production (300 mL kg^{-1} h^{-1}) was reduced to one-fifth its value at ambient temperature and occurred between 10 and 16 days postharvest compared to 5–6 days postharvest at ambient temperature. Since the peaks for CO_2 and C_2H_4 production coincide, the same delay was seen in C_2H_4 production at 13°C as for CO_2 but the lower temperature depressed peak C_2H_4 production (1.5/L kg^{-1} h^{-1}) about eightfold.

7.7.6 STRAWBERRY, APPLE, AND MANGO

Velickova et al. (2013) studied that the coatings controlled the exchange of the gases between the fruit and the environment due to their different permeability to gases like CO_2 and O_2. Incorporation of the beeswax in the coating significantly reduced the respiration rate of the strawberries, which is in agreement with the reported effect of lipids on the gas barrier properties of polysaccharide films (Vargas et al., 2006). The values of 291.7 mL CO_2/ kg/h for the control samples (1) and 274.3 (2); 279.7 (3); 252.1 (4) and 227.9 (5) mL CO_2/kg/h for the coated samples obtained after 0.5 h are comparable with the results reported by Lima et al. (2010) when working with sliced and then coated apples and mangoes stored at 20°C. The modification of the gas balance in the microenvironment of the fruits led to their decreased respiration and hence better preservation.

Vargas et al. (2006) have investigated that Respiration rate in terms of both CO_2 generation and O_2 consumption, decreased throughout storage period.

Aloe vera, a tropical and subtropical plant has been used as an herbal remedy for regeneration and rejuvenation of human skin since ancient time in China, Japan, and India (Boudreau & Beland, 2006). Today, aloe vera gel derived from its leaf parenchyma tissue is commonly used for medical studies and cosmetic products. Although mostly used for medical studies (Shamim et al., 2004; Rosca-Casian et al., 2007), the gel has been tested for few fresh fruits by a postharvest research group from Spain since 2005 (Valverde et al., 2005). It has been reported to maintain postharvest quality of fruits during storage (Martínez-Romero et al., 2006; Dang et al., 2008; Ahmed et al., 2009; Castillo et al., 2010). The gel is applied to fruits as an edible coating which has a various favorable effect on fruits such as imparting a glossy appearance and better color, retarding weight loss, or prolonging storage/shelf-life by preventing microbial spoilage (Dang et al., 2008). The performance of aloe vera gel as edible coating is dependent on its composition (Dang et al., 2008). Its use offers an option to film packaging owing to their environment friendly characteristic (Rojas-Argudo et al., 2005). Further, aloe vera leaves are rich in bioactive compounds some of which are antioxidants those are broadly used in food engineering as preservatives such as mannans, antrachinon, c-glycoside, antron, antrakuinon, and lectine (King et al., 1995; Eshum & He, 2004). Carnauba wax is an edible coating material under the group of lipids, which is mainly imparted to reduce water loss and gloss (Baldwin et al., 1999). Carnauba wax is recovered from the underside of the leaves of a Brazilian palm tree (*Copernicia cerifera*). The thin wax film protects the leaves during rainy periods and in summer protects the leaves from dehydration. Carnauba wax listed by the Organic Material Review Institute (OMRI, 2003) as a regulated processing product. This wax is prepared without using organic solvent and does not include shellac or ammonia, it contains organic compounds such as oils, alcohol, salt, phospholipids, and natural vitamin.

7.8   EDIBLE COATING AFFECTING FRUIT DECAY

7.8.1   PLUMS AND NECTARINES

Goncalves et al. (2010) reported that the protective use of 4.5% carnauba wax significantly reduced brown rot incidence in plums and nectarines by 46 and 35%, respectively, when compared with the control. This treatment significantly reduced Rhizopus rot incidence by 39% in plums and 27% in nectarines. The possible mechanisms by which wax application reduces postharvest diseases in plums and nectarines include: (1) the formation of a physical barrier around the fruit that prevents pathogen entry; (2) modification of the atmosphere around the fruit; and (3) direct action of the wax to the pathogens (antifungal activity).

7.8.2   APPLE AND BANANA

El-Anany et al. (2009) studied the efficacy of Arabic gum (AG) as an edible coating was observed on apple fruit during cold storage. It was found that apple fruit coated with Arabic gum showed a significant delay in ripening and therefore, decayed slowly as compared to control. Another study was conducted by Maqbool et al. (2010) on banana fruit with use of chitosan. It was found that chitosan inhibited the growth of *Colletotrichum musae* as compared to the control. The highest fungicidal effects were found in those bananas treated with 1.5% chitosan concentration followed by 1.0% chitosan alone.

7.8.3   STRAWBERRIES

Wang and Gao (2013) observed that the Decay of strawberries coated with chitosan was reduced significantly. The percentages of decay were 21.5 and 11.7% for control and chitosan coated (0.5 g/100 mL) samples, respectively. The effect of chitosan in reducing microbial growth was more evident and clear for longer storage periods and at higher storage temperature. Han et al. (2004) studied the effect of Chitosan-based edible coatings, were used to extend the shelf-life and enhance the nutritional value of strawberries (*Fragaria × ananassa*) and red raspberries (*Rubus ideaus*) stored at either 2°C and 88% relative humidity (RH) for 3 weeks or −23°C up to 6 months. It was found that the decay incidence of fresh strawberries and red raspberries stored at 2°C and 88% RH were reduced significantly compared to uncoated (control) fruits. Vargas et al. (2006) worked on strawberries cv. Camarosa. Chitosan (1%, w/v) with oleic acid was coated on strawberries which were then cold stored at 4°C for 10 days. Coating protected strawberries against fungal infection (50% reduction).

7.9   EDIBLE COATINGS AFFECT PHYSICAL PARAMETERS

7.9.1   FRUIT FIRMNESS AND SOFTENING

7.9.1.1   MANGO AND PAPAYA

Fonseca et al. (2004) reported that mango (cv. Haden) coated with emulsion (containing a mixture of carnauba wax and acrylic resin) had less mass loss and remained firmer for a longer time, having more turgid aspect when stored at 26.5 + 3.5°C and 71.5 + 15.5% relative humidity up to 15 days after storage.

Papaya fruits treated with aloe vera gel (50%) and stored at 30 ± 3°C for 10 days maintained significantly higher firmness compared to control (Marpudi et al., 2011).

7.9.1.2   GRAPE

The effect of aloe vera gel (1:3 diluted with distilled water) coating on table grape firmness during storage at 1°C resulted in delayed softening process compared to control (Valverde et al., 2005). Preharvest application of aloe vera gel (1:3) exerted similar effect on grape, during postharvest storage at 2°C (Castillo et al., 2010).

7.9.1.3   SWEET CHERRY

In general, the fruit firmness followed a declining trend commensurate with advancement in the storage period. Yaman and Bayoundurlc (2002) worked on sweet cherry and applied coating and its concentration as Semper-fresh@10, 20 g/L, stored at ambient temperature $(30 \pm 3°C)$ and at humidity (40–50% RH), and stored at cold storage $(0°C)$ and at humidity (95–98% RH). Cherries stored at $0°C$ had higher firmness values than cherries stored at ambient temperature. So, cold temperature had a strong effect on the retention of firmness. At the same time, as the Semperfresh coating (SE) concentration increased, the firmness values increased. Retention of firmness can be explained by retarded degradation of insoluble protopectins to the more soluble pectic acid and pectin. During fruit ripening, depolymerization or shortening of chain length of pectin substances occurs with an increase in pectinesterase and polygalacturonase activities. Low oxygen and high carbon dioxide concentrations reduce the activities of these enzymes and allow retention of the firmness of fruits and vegetables during storage (Salunkhe et al., 1991).

Martınez-Romero et al. (2006) studied that aloe vera gel (1:3) coating of sweet cherry fruit was highly effective in maintaining firmness during cold storage at $1°C$ for 16 days. No significant change in firmness was observed throughout the experiment.

7.9.1.4   APPLE

Ergun and Satici (2012) reported that coating of apple fruit (cv. Granny Smith and Red Chief) with aloe vera gel (1, 5, and 10% w/v) significantly suppressed the loss in firmness compared to control, during storage at $2°C$ for 6 months.

7.9.1.5   PLUM

Valero et al. (2013) coated four plum (*Prunus salicina Lindl.*) cv. Black Amber, Larry Ann, Golden Globe and Songold with alginate@ 1 and 3% w/v. Analytical observations were made after 7, 14, 21, 28, and 35 days at $2°C$ and after a 3 day period at $20°C$ (shelf-life). They observed the softening process was faster when plums were transferred to $20°C$ after cold storage, since after 7 days of cold storage + shelf-life firmness levels in control plums were close to those found after 35 days of cold storage. However, alginate edible coatings slowed down the softening process for all plum cultivars, either during cold storage or subsequent shelf-life, the effect being significantly higher for the Alg-3 than the Alg-1 treatment. The effect of alginate edible coating on delaying the softening process was also evident after the shelf-life period for all plum cultivars, which showed firmness levels after 35 days at 2°C + shelf-life similar to those found in control fruit after just 7 days at 2°C + shelf-life. Changes in cell wall composition, especially cell wall mechanical strength and cell-to-cell adhesion are the most important factors contributing to firmness loss during fruit on-tree ripening or after harvesting, the activity of cell wall hydrolyzing enzymes being enhanced by ethylene in climacteric fruit (Valero & Serrano, 2010). In plums, the main cell wall-degrading enzymes are polygalacturonase, pectin methylesterase, 1,4-d-glucanase/glycosidase and galactosidase (Manganari et al., 2008) Thus, the inhibition of ethylene production observed in alginate-coated plums could be responsible for their lower softening process with respect to control fruit.

Hardness during ripening in climacteric fruit, such as Huanghua pears, is generally attributed to degradation of the cell wall and loss of turgor pressure in the cells reduced by water loss (Lohani et al., 2004; Khin et al., 2007). The coating treatments may maintain hardness by inhibiting water loss. Coating may also inhibit the activities of pectin degrading enzymes closely related to fruit softening by reducing the rate of metabolic processes during senescence (Conforti & Zinck, 2002; Zhou et al., 2008), which also contributed to fruit hardness maintenance. Previous studies have reported a similar performance of delaying softening by shellac coating in apples (Bai et al., 2002), by Semperfresh coating in quinces (Yurdugul, 2005), and by chitosan coating in citrus fruit (Chien et al., 2007). Compared with the Semperfresh coating, shellac coating and CMC coating were more efficient in reducing changes in the texture profile analysis values of pear flesh during storage.

7.9.1.6 STRAWBERRIES

Velickova et al. (2013), reported that the Strawberries soften considerably during ripening due to degradation of the middle lamella of the cell wall of cortical parenchyma cells (Perkins-Veazie, 1995).The three-layer coating of the strawberry with beeswax-crosslinked chitosan beeswax led to significant softening of the strawberry surface, while the other coating formulations did not cause significant softening of the surface when compared to the fresh strawberries.

Malmiri et al. (2011) studied the effect of chitosan and glycerol concentrations on the firmness of coated banana is clearly observed as; an increase in chitosan concentration at glycerol concentrations less than 1.3% (w/v), there was a beneficial effect on firmness retention.

Martinez-Romero et al. (2006) observed the aloe vera treatment significantly reduced the firmness losses (more than 50%) during cold storage + shelf life compared with control fruit. The softening process in sweet cherries has been reported to be dependent on the increase in polygalacturonase, ß-galactosidase, and pectin methylesterase activities.

Vahdat et al. (2009) reported that treatment of strawberry fruit with aloe vera gel (25, 50, 75, and 100% v/v) maintained significantly higher firmness, compared to control. In a different experiment, Singh et al. (2011) also observed similar result in strawberry fruit, after treating with aloe vera gel (1:3), during storage at 5°C for 16 days.

7.9.2 EDIBLE COATINGS AFFECTING FRUIT PEEL COLOR

7.9.2.1 SAPOTA

Peel color is most important quality parameter of fruits which influence consumer appeal as well as internal condition. Baldwin et al. (1999) have reported that beneficial effects of carnaba wax include improvement of appearance, imparted an attractive natural-looking sheen to the fruit. Ergun et al. (2005) have reported that Mamey sapote fruit treated with wax+1-MCP had a richer, salmon color in the pulp than that of control fruit.

7.9.2.2 POMEGRANATE

Barman et al. (2011) also reported that the changes of peel color during maturation process in pomegranate fruit are associated with synthesis of anthocyanin pigments. It was investigated that control fruits showed marked reduction of hue angle and increase in chroma value as compared to treated fruits. The lower hue and higher chroma values under Putrescine + carnauba wax treated fruits attributed to delayed maturation process in comparison to control fruits. Yaman and Bayoundurlc (2002) have reported that Sucrose polyester coatings were also effective in increasing lightness.

7.9.2.3 PLUM

Eum et al. (2009) investigated that the Surface color of "Sapphire" plums changed from green to red and dark black during room temperature storage. The main color changes of plum were identified by hue angle and correlated to the anthocyanin as well as chlorophyll alterations. L* value was not different in all treatments. However, throughout the storage period, the value was diminished as shown in previous results. Also, a* value and hue angle were decreased. The decrease of a* value and hue angle indicates that plums became redder and darker during ripening and senescence. Valero et al. (2013) coated four plum (*Prunus salicina Lindl.*) cv. Black Amber, Larry Ann, Golden Globe and Songold with alginate@ 1% and 3% w/v. Analytical observations were made after 7, 14, 21, 28, and 35 days at 2°C and after a 3 day period at 20°C (shelf-life). It was found that Skin color also changed during storage in all plum cultivars, to dark purple in "Blackamber" and "Larry Ann" and to deep yellow in "Golden Globe" and "Songold." Color changes were delayed by Alg-1 and Alg-3 edible coatings, without significant differences between them, except for "Golden Globe" plum, where Alg-3 was the most effective. During cold storage, color changes were lower than after the shelf life periods.

7.9.2.4   STRAWBERRIES AND RASPBERRIES

Vargas et al. (2006) studied that no significant differences, in terms of hue or chroma, were found due to coatings. Coatings led to a decrease in the luminosity of samples, which became significant, Chroma and hue coated samples did not change significantly during storage, although uncoated strawberries were slightly lower in hue and chroma, which can also be attributed to the surface drying. Han et al. (2004) reported similar color changes in chitosan-coated strawberries during cold storage. Han et al. (2004) studied in red raspberries and found that chitosan coating alone showed the best control of fruit color during storage.

7.9.2.5   GRAPES

Valverde et al. (2005) studied that, aloe vera gel treatment of table grapes significantly delayed changes in color during storage at 1°C for 35 days. In another experiment, Castillo et al. (2010) also found that preharvest treatment of grapes with aloe vera gel (1:3) significantly delayed diminution of hue angle and color changes compared to control, during storage at 2°C.

7.9.2.6 CHERRY

Martınez-Romero et al. (2006) reported that aloe vera gel treatment of sweet cherry fruit maintained significantly higher hue angle compared to control during cold storage at 1°C for 16 days.

7.9.2.7 PAPAYA

Aloe vera gel (50%) treatment of papaya fruit showed that it delayed the development of peel color compared to control, during storage at room temperature (Marpudi et al., 2011).

7.9.2.8 APPLE

The effect of aloe vera gel (1, 5, and 10% w/v) treatment on color development of "Granny Smith" and "Red Chief" apples showed that, L*, a*, and b* values increased continuously with the advancement of storage period. However, the increase in a* value was slower, compared to control fruit, during storage at 2°C for 6 months (Ergun & Satici, 2012).

7.9.2.9 PINEAPPLE

Nimitkeatkai et al. (2006) reported that pineapple fruit when treated with edible coating and stored at cold storage (10°C, 90–95% RH) showed lower hue angle reduction in treated fruits in comparison to control.

7.10 EDIBLE COATINGS AFFECTING BIOCHEMICAL PARAMETERS

7.10.1 TOTAL SOLUBLE SOLIDS, TITRATABLE ACIDITY, AND ASCORBIC ACID

7.10.1.1 PEAR

Zhou et al. (2008) worked with Huanghua pears cv. Huanghua. Coatings used were shellac, Semperfresh and CMC during cold storage (4°C). Coatings were applied as Shellac@14.3 g/100 mL water, Semperfresh@1.0 g/100 mL water, CMC@ 2.0 g/100 mL water. The total soluble solids (SSCs), titratable acidity (TA) and ascorbic acid levels in Huanghua pears decreased in all treatments after 60 days of storage. Compared with the control samples, the shellac coated pears had higher SSCs, TA and ascorbic acid levels, and the difference was significant. Soluble solids and organic acids of fruits are substrates that are consumed by respiration during storage (Yaman and Bayoundurlc, 2002). Ascorbic acid is primarily regulated by ascorbic acid oxidase and phenoloxidase, whose activities are influenced by the oxygen contents in the storage condition (Yaman and Bayoundurlc, 2002). In this study, shellac coating was more effective in the retention of SSCs and the TA and ascorbic acid levels because of the lower gas permeability of shellac coating that inhibited the respiratory rates and retarded the overall metabolic activities of pears during storage.

7.10.1.2   CHERRIES

It was observed by Yaman and Bayoundurlc (2002) that acidity increased with increasing Semperfresh concentrations. Semperfresh coatings were effective in reducing the ascorbic acid loss for both ambient and cold temperatures storage conditions. The reduction of ascorbic acid loss in coated cherries was due to the low oxygen permeability of sucrose polyester coating which lowered the activity of the enzymes and prevented oxidation of ascorbic acid. The effect of low temperature significantly reduced the ascorbic acid loss. This shows the effect of temperature on the activities of the related enzymes. Ascorbic acid is lost due to the activities of phenoloxidase and ascorbic acid oxidase enzymes during storage (Salunkhe et al., 1991). Semperfresh showed the largest rise in total sugars and the largest fall in starch content. Martınez-Romero et al. (2006) observed that TA content of sweet cherry fruit was significantly higher in aloe vera gel (1:3) coated fruit compared to control, during storage at 1°C up to 16 days.

7.10.1.3   STRAWBERRY

Strawberry fruit when treated with aloe vera gel (1:3), and stored at 5°C, treated fruits maintained significantly higher TSS (8.4°B), compared to control (7.0°B) (Singh et al., 2011).

7.10.1.4   APPLE AND PLUM

Ergun and Satici (2012) observed that apple fruit (cv. Granny Smith and Red Chief) when coated with aloe vera gel (1, 5, and 10% w/v) and stored at 2°C for 6 months, TSS content of fruit declined slightly. However, this decline was much slower compared to control.

Plums (*Prunus salicina Lindl.* cv. Sapphire) when treated with edible coating materials and stored at 20°C and 85% RH, reduced ethylene production (Eum et al., 2009). Valero et al. (2013) have investigated that the alginate edible coating delayed acidity losses in all plum cultivars. Decrease in total acidity is also typical during postharvest storage of fleshy fruit, including plums, has been attributed to the use of organic acids as substrates for the respiratory metabolism in detached fruit (Díaz-Mula et al., 2009; Valero & Serrano, 2010).

7.10.1.5   PAPAYA

The effect of aloe vera gel (50%) treatment on TSS content of papaya fruit showed that, it delayed the increase in TSS content compared to control (Marpudi et al., 2011). Martınez-Romero et al. (2006) reported that sweet cherry fruit when treated with aloe vera gel (1: 3), it significantly delayed the increase in TSS content during storage at 1°C for 16 days.

7.10.1.6   GRAPES

Table grapes (cv. Crimson Seedless) when coated with aloe vera gel (1:3 diluted with distilled water), treated fruits showed significantly delayed increase in TSS compared to control, during storage at 1°C for 35 days (Valverde et al., 2005). On the contrary, when aloe vera gel was applied preharvest, it did not exert any effect on TSS on grapes (Castillo et al., 2010).

7.10.2   PHENOL CONTENT AND ANTIOXIDANT ACTIVITY

7.10.2.1   GRAPES

Sánchez-González et al. (2011) have investigated that the antioxidant capacity of the samples sharply increase during the first 3 storage days (from 51 to 68%), but afterwards hardly increased at all, regardless of the treatment. Enzymatic browning could also contribute to the formation of antioxidant compounds, since an increase in polyphenol oxidase and peroxidase activities has been observed in grapes during postharvest storage. Phenolic acids (cynnamic and benzoic, esterified or not with tartaric acid) are mainly present in white grapes. These compounds are highly oxidative, producing brown compounds that also show antioxidant activity. Previous work dealing with the oxidative process in salad tomatoes during ripening also revealed changes in oxidative and antioxidative parameters (Jimenez et al., 2002). The levels of the aqueous-phase antioxidants increased during the ripening process and this increase was associated with significant changes in their redox status, becoming more reduced as ripening progressed. Observations of the phenol content of the samples showed no significant differences. The phenol content sharply decreased significantly from 121 to 82 mg/100 g during the first 3 cold storage days for all the treatments, regardless of the coating treatment, and a progressive, slow decay occurred afterwards (reaching a value of 68 mg/100 g in the last control) as has been previously observed in grapes (Valero et al., 2006) and in other non-climateric fruit, such as strawberry, concomitant with natural phenol content decay that occurs in grape maturation and postharvest stages. The activity of phenylalanine ammonia-lyase (PAL) is key to the phenolic accumulation in grapes and this activity decreases in the maturation and postharvest stages

7.10.2.2 STRAWBERRY

Wang and Gao (2013) studied with strawberries stored at temperatures greater than 0°C showed an increase in antioxidant capacity. The decline in antioxidant activity in untreated fruit at the end of storage might be due to senescence and decay. This indicated that chitosan treatment not only can extend shelf-life, but also can retain higher antioxidant activity in strawberries after prolonged storage. Strawberries stored at 5 and 10°C showed increases in antioxidant activities after 6 days (5°C) and 3 days (10°C) then declined for untreated fruit. P-Coumaroyl glucose was the predominant phenolic compound with an initial concentration of 25.8 mg/g fw. In chitosan treated (0.5, 1.0, and 1.5 g/100 mL) fruit.

7.10.3   ANTHOCYANINS AND PIGMENTS

7.10.3.1   PLUMS

Valero et al. (2013) have investigated that the skin color in purple plums is due to anthocyanins, the main anthocyanin quantified in "Blackamber" and "Larry Ann" plums being cyanidin 3-glucoside followed by cyanidin 3-rutinoside, as previously reported for these and other purple plum cultivars (Tomás-Barberán et al., 2001; Wu & Prior, 2005; Díaz-Mula et al., 2009). Both anthocyanins increased with the progress of cold storage, these increases being lower in alginate-coated plums than in control ones. Thus, alginate treatment delayed color change in purple plum cultivars by retarding the anthocyanin synthesis associated to the postharvest ripening process (Serrano et al., 2009; Díaz-Mula et al., 2012). Similarly, strawberry treated with alginate at 2% showed lower increases in total anthocyanin than controls (Fan et al., 2009).

7.10.3.2   STRAWBERRIES &&

Wang and Gao (2013) have reported that Pelargonidin 3-glucoside was the predominant anthocyanin with an initial content of 424.5 mg/g fw. strawberries. In fruit treated with chitosan coating, total anthocyanins also increased but at a slower pace. They also did not show the decline at the later part of the storage as the control samples. Therefore, chitosan-treated fruit maintained higher anthocyanins than the control samples at the one explanation for this finding could be related to the fact that total phenolic and anthocyanin contents generally increase with temperature. Cordenunsi et al. (2005) found an increase in anthocyanin during storage and the rate of anthocyanin accumulation increased with increasing temperature. These results indicated that there was still anthocyanin biosynthesis after harvesting and end of storage.

7.10.4   LIPID PEROXIDATION AND ENZYMATIC ACTIVITIES

7.10.4.1   LOQUAT AND PEAR

Singh and Singh (2012) have observed that it is widely accepted that the chilling injury involves membrane damage or dysfunction which is primarily caused by the lipid peroxidation. Membrane lipid peroxidation is initiated either enzymatically or by reactive oxygen species. Lipid peroxidation (LOX) activity increased in response to chilling stress and may be involved in occurrence of chilling injury through enhanced lipid peroxidation as reported in loquat. Lipid peroxidation also contributes to the production of superoxide radicals which are among the reactive oxygen species contributing to accumulation of thiobarbituric acid (TBARMs) and increased oxidative stress. The data showed that lipid peroxidation activity was greatly reduced under controlled atmosphere (CA) storage compared to modified atmosphere storage and normal air, but increased significantly during ripening at $21 \pm 1°C$ after cold storage. In "Blanquilla" pears, lipid peroxidation activity in fruit under optimum CA ($2\% O_2 + 0.7\% CO_2$) was reported to be significantly lower than under stressful atmospheres high in CO_2 ($2\% O_2 + 5\% CO_2$) (Larrigaudière et al., 2001). The lower lipid peroxidation activity under CA was coincident with the reduced incidence and severity of chilling injury in "Blackamber" plums. Similarly, the lower lipid peroxidation activity was also associated with the lower core browning in "Blanquilla" pears held under optimum CA (Larrigaudière et al., 2001). These observations support the role of lipid peroxidation in occurrence of physiological disorders in fruit. Thiobarbituric acid is a marker for the degree of lipid peroxidation and oxidative stress (Hodges & Forney, 2000). The accumulation of thiobarbituric acid was also lower in CA stored fruit compared to modified atmosphere storage and normal air which indicated the lower level of lipid peroxidation in CA stored fruit. CA storage under optimal conditions has been known to inhibit the increase in thiobarbituric acid concentration in spinach (Hodges & Forney, 2000) and pear (Larrigaudière et al., 2001).

7.10.4.2   PLUM

Surface coating of "Sapphire" plums with a carbohydrate-based formulation has been reported to delay the accumulation of thiobarbituric acid (Eum et al., 2009). The degree of lipid peroxidation could explain the incidence and severity of chilling injury in Japanese plums. Eum et al. (2009) worked on plum cv. Sapphire, and applied carbohydrate based Versasheen@ 5% with sorbitol as plasticizer. The coated plums were stored at 25°C with 85% RH for 8 days. After 4-day room temperature storage, differences in malondialdehyde (MDA) value of plums were detected between uncoated and Versasheen with and without sorbitol. After 8-day room temperature storage, the production of MDA in Versasheen-coated plums reached the same level of uncoated plum after 4-day storage. The treatment with and without sorbitol did not influence the MDA value. Lipid degradation and peroxidation can provide early responses in many tissues of horticultural crops undergoing ripening and senescence. Hydrogen ion can easily be removed from double bond of unsaturated fatty acid and lipid radical formed through this reaction. Many lipid hydroxyl radical and active oxygen species might be proliferated as chain reaction and resulted in oxidative stress leading to peroxidative damage in the chloroplast and thylakoid membrane. Finally it could enhance chlorophyll degradation. The increment of MDA, which is a decomposition product of the oxidation of polyunsaturated fatty acids, is a measure for large production of active oxygen species. The coatings also significantly reduced the MDA production pointing to diminished lipid peroxidation at results in oxidative stress leading to peroxidative damage in the membranes.

Yu et al. (2012), have investigated that the MDA contents of all the samples continuously increased during the entire storage period. No significant difference was observed in the MDA content between the jujubes coated with chitosan + nano-silicon dioxide and chitosan alone. The application of chitosan + nano-silicon dioxide coating also delayed the MDA increase in jujube. After 32 days, the MDA content of the jujube coated with chitosan + nano-silicon dioxide was 0.38 L/mol g^{-1}, which was 15.6% lower than that of the jujube coated with chitosan alone. MDA is originated of cytoplasmic membrane oxidation, and it may indicate the degree of cell senescence. MDA content of the jujube coated with the compound film was the lowest, and the reason was probably that higher activities of superoxide dismutase, peroxidase, and chloramphenicol acetyltransferase could quickly eliminate the free radical. Thus, the harm to the cytoplasmic membrane by the free radical was minimized to the least degree.

Thus, the classification of edible coating effect on fruit crop can be underlined as physical, physiological, and biochemical. The activities carried out by edible coating to extend the shelf life retaining fruit sensory characteristics are actually controlled by these physical, physiological, and biochemical effects, which in turn is affected by the type of edible coating used. The changes in fruit morphology by the application of edible coatings seems simple, but the changes carried out by these coatings are needed to be understood well according to the consumer demand, so that the better selection of fruit coating can be made for application. The knowledge of effects of edible coating on fruits will lead to development of new options and evolution of edible coating in field of fresh fruits postharvest handling.

KEYWORDS

- **edible coatings**
- **coating formulations**
- **commercial edible coatings**
- **green technology**

REFERENCES

Abdi, N.; Holford, P.; McGlasson, W. B.; Mizrahi, Y. Ripening Behaviour and Responses to Propylene in Four Cultivars of Japanese Type Plums. *Postharv. Biol. Technol.* **1997**, *12*, 21–34.

Ahmed, M. J.; Singh, Z.; Khan, A. S. Postharvest Aloe Vera Jel-coating Modulates Fruit Ripening and Quality of 'Arctic Snow' Nectarine Kept in Ambient and Cold Storage. *Int. J. Food. Sci. Tech.* **2009**, *44*, 1024–1033.

Amarante, C.; Banks, N. H.; Ganesh, S. Relationship Between Character of Skin Cover of Coated Pears and Permeance to Water Vapour and Gases. *Postharvest Biol. Technol.* **2001**, *21*, 291–301.

Ayranci, E.; Tunc, S. The Effect of Edible Coatings on Water and Vitamin C Loss of Apricots (*Armeniaca vulgaris* Lam.) and Green Peppers (*Capsicum annum* L.). *Food Chem.* **2004**, *87*, 339–342.

Bai, J. H.; Baldwin, E. A.; Hagenmaier, R. H. Alternative to Shellac Coatings Provides Comparable Gloss, Internal Gas Modification and Quality for 'Delicious' Apple Fruit. *Hort Sci.* **2002**, *37*, 559–563.

Baldwin, E. A.; Burns, J. K.; Kazokas, W. Effect of Two Edible Coatings with Permeability Characteristics on Mango (*Mangifera indica* L.) Ripening during Storage. *Postharvest Biol. Technol.* **1999**, *17*, 215–226.

Barman, K.; Asrey, R.; Pal, R. K. Putrescine and Carnauba Wax Pretreatments Alleviate Chilling Injury, Enhance Shelf Life and Preserve Pomegranate Fruit Quality during Cold Storage. *Sci. Hortic.* **2011**, *130*, 795–800.

Boudreau, M. D.; Beland, F. A. An Evaluation of the Biological and Toxicological Properties of Aloe Barbadensis (Miller), Aloe Vera – Review. *J. Environ. Carcinog. Ecotoxicol.* **2006**, *24*, 103–54.

Bourtoom, T. Edible Films and Coatings: Characteristics and Properties. *Int. Food Res. J.* **2008**, *15* (3), 237–248.

Candan, A. P.; Graell, J.; Larrigaudiere, C. Roles of Climacteric Ethylene in the Development of Chilling Injury in Plums. *Postharvest Biol. Technol.* **2008**, *47*, 107–112.

Castillo, S.; Navarro, D.; Zapataa, P. J.; Guillena, F.; Valeroa, D.; Serrano, M.; Martínez-Romero, D. Antifungal Efficacy of Aloe Vera In Vitro and Its Use as a Pre Harvest Treatment to Maintain Postharvest Table Grape Quality. *Postharvest. Biol. Technol.* **2010**, *57*, 183–188.

Chien, P. J.; Sheu, F.; Lin, H. R. Coating Citrus (Murcott tangor) Fruit with Low Molecular Weight Chitosan Increases Postharvest Quality and Shelf Life. *Food Chem.* **2007,** *100,* 1160–1164.

Conforti, F. D.; Zinck, J. B. Hydrocolloid-Lipid Coating Affect on Weight Loss, Pectin Content, and Textural Quality of Green Bell Peppers. *Food Chem. Toxicol.* **2002,** *67,* 1360–1363.

Cordenunsi, B. R.; Genovese, M. I.; Nascimento, J. O.; Aymoto, H. N. M.; dos Santos, R. J.; Lajolo, F. M. Effects of Temperature on the Chemical Composition and Antioxidant Activity of Three Strawberry Cultivars. *Food Chem.* **2005,** *45,* 4589–4594.

Dang, K. T.; Singh, Z.; Swinny, E. E. Edible Coatings Influence Fruit Ripening, Quality and Aroma Biosynthesis in Mango Fruit. *J. Agric. Food Chem.* **2008,** *56,* 1361–1370.

De Wild, H. P. J.; Balk, P. A.; Fernandes, E. C. A.; Peppelenbos, H. W. The Action Site of Carbon Dioxide in Relation to Inhibition of Ethylene Production in Tomato Fruit. *Postharvest. Biol. Technol.* **2005,** *36,* 273–280.

Dhall, R. K. Advances in Edible Coatings for Fresh Fruits and Vegetables: A Review, *Crit. Rev. Food Sci. Nutr.* **2013,** *53* (5), 435–450.

Díaz-Mula, H. D.; Serrano, M.; Valero, D. Alginate Coatings Preserve Fruit Quality and Bioactive Compounds during Storage of Sweet Cherry Fruit. *Food Bioprocess Technol.* **2012,** *5,* 2990–2997.

Díaz-Mula, H. M.; Zapata, P. J.; Guillén, F.; Martínez-Romero, D.; Castillo, S.; Serrano, M.; Valero, D. Changes in Hydrophilic and Lipophilic Antioxidant Activity and Related Bioactive Compounds during Postharvest Storage of Yellow and Purple Plum Cultivars. *Postharvest Biol. Technol.* **2009,** *51,* 354–363.

Duan, J.; Wu, R.; Strick, B. C.; Zhao, Y. Effect of Edible Coatings on the Quality of Fresh Blueberries (Duke and Elliott) Under Commercial Storage Conditions. *Postharvest Biol. Technol.* **2011,** *59,* 71–79.

El-Anany, A. M.; Hassan, G. F. A.; Rehab Ali, F. M. Effects of Edible Coatings on the Shelf Life and Quality of Anna Apple (Malus Domestica Borkh) during Cold Storage. *J. Food Technol.* **2009,** *7* (1), 5–11.

Ergun, M.; Satici, F. Use of Aloe Vera Gel as Biopreservative for 'Granny Smith' and 'Red Chief' Apples, *J. Anim. Plant Sci.* **2012,** *22* (2), 363–368.

Ergun, M.; Sargent, S. A.; Fox, A. J.; Crane, J. H.; Huber, D. J. Ripening and Quality Responses of Mamey Sapote Fruit to Postharvest Wax and 1-Methylcyclopropene Treatments. *Postharvest Biol. Technol,* **2005,** *36,* 127–134.

Eshum, K.; He, Q. Aloe Vera: A Valuable Ingredient for the Food Pharmaceutical and Cosmetic Industries–A Review. *Crit. Rev. Food. Sci. Nutr.* **2004,** *44* (2), 91–96.

Eum, H. L.; Hwang, D. K.; Linke, M.; Lee, S. K. Influence of Edible Coating on Quality of Plum (Prunus Salicina Lindl. cv. 'Sapphire'). *Eur. Food Res. Technol.* **2009,** *29,* 427–434.

Fan, Y.; Xu, Y.; Wang, D.; Zhang, L.; Sun, J.; Sun, L.; Zhang, B. Effect of Alginate Coating Combined with Yeast Antagonist on Strawberry (Fragaria × Ananassa) Preservation Quality. *Postharvest Biol. Technol.* **2009,** *53,* 84–90.

Feygenberg, O.; Hershkovitz, V.; Ben-Arie, R.; Jacob, S.; Pesis, E.; Nikitenko, T. Postharvest Use of Organic Coating for Maintaining Bio-Organic Avocado and Mango Quality. *Acta Hort.* **2005,** *682,* 507–512.

Fonseca, M. J. O.; Salomao, L. C. C.; Cecon, P. R.; Puschmann, R. Fungicides and Wax in Postharvest Preservation of Mango 'Haden'. *Acta Hort.* **2004,** *645,* 557–563.

Goncalves, F. P.; Martins, M. C.; Silva, G. J.; Lourenc, S. A.; Amorim, L. Postharvest Control of Brown Rot and Rhizopus Rot in Plums and Nectarines Using Carnauba Wax. *Postharvest Biol. Technol.* **2010**, *58,* 211–217.

Guilbert, S.; Gontard, N.; Gorris, L. G. M. Prolongation of the Shelf Life of Perishable Food Products Using Biodegradable Films and Coatings. *LWT Food Sci. Technol.* **1996**, *29,* 10–17.

Han, C.; Zhao, Y.; Leonard, S. W.; Traber, M. G. Edible Coatings to Improve Storability and Enhance Nutritional Value of Fresh and Frozen Strawberries (Fragaria × Ananassa) and Raspberries (*Rubus ideaus*). *Postharvest Biol.Technol.* **2004**, *33,* 67–78.

Hernandez, E. Edible Coatings from Lipids and Resins. In *Edible Coatings and Films to Improve Food Quality;* Krochta, J. M., Baldwin, E. A., Nisperos-Carriedo, M., Eds.; Technomic Publishing Company: Lancaster, PA, 1991; pp 279–303.

Hernandez, E. Edible Coating from Lipids and Resins. In *Edible Coatings and Films to Improve Food Quality*; Krochta, J. M., Balwin, E. A., Niperos-Carriedo, M. O., Eds.; Technomic Publishing Company: Basel, Switzerland, 1994; pp 279–303.

Hodges, D. M.; Forney, C. F. The Effects of Ethylene, Depressed Oxygen, and Elevated Carbon Dioxide on Antioxidant Profiles of Senescing Spinach Leaves. *J. Exp. Bot.* **2000**, *51,* 645–655.

Jimenez, A.; Creissen, G.; Kular, B.; Firmin, J.; Robinson, S.; Verhoeyen, M.; Mullineaux, P. Changes in Oxidative Processes and Components of the Antioxidant System during Tomato Fruit Ripening. *Planta.* **2002**, *214,* 751–758.

Kester, J. J.; Fennema, O. R. Edible Films and Coatings: A Review. *Food Technol.* **1986**, *40,* 47–59.

Khin, M. M.; Zhou, W.; Yeo, S. Y. Mass Transfer in the Osmotic Dehydration of Coated Apple Cubes by Using Maltodextrin as the Coating Material and Their Textural Properties. *J. Food Eng.* **2007**, *81,* 514–522.

King, G. K.; Yates, K. M.; Greenlee, P. G.; Pierce, K. R.; Ford, C. R.; Mcanalley, B. H.; Tizard, I. R. The Effect of Acemannan Immunostimulant in Combination with Surgery and Radiation Therapy on Spontaneous Canine and Feline Fibrosarcomas. *J. Am. Anim. Hosp. Assoc.* **1995**, *31* (5), 439–447.

Kinsella, J. E.; Phillips, L. G. Film Properties of Modified Proteins. In *Food Protein*; Kinsella, J. E., Soucie, W. G., Eds.; The American Oil Chemist's Society: Champaign, IL, 1979; pp 78–99.

Krochta, J. M.; Mulder-Johnston, C. D. Edible and Biodegradable Polymer Films: Challenges and Opportunities. *Food Technol.* **1997**, *51,* 61–74.

Labuza, T.; Contrereas-Medellin, R. Prediction of Moisture Protection Requirements for Foods. *Cereal Food World.* **1981**, *26,* 335.

Larrigaudière, C.; Pintó, E.; Lentheric, I.; Vendrell, M. Involvement of Oxidative Processes in the Development of Core Browning in Controlled-Atmosphere Stored Pears. *J. Hort. Sci. Biotech.* **2001**, *76,* 157–162.

Lohani, S. Trivedi, P. K.; Nath, P. Changes in Activities of Cell Wall Hydrolases during Ethylene-Induced Ripening in Banana: Effect of 1-MCP, ABA and IAA. *Postharvest Biol. Technol.* **2004**, *31,* 119–126.

Malmiri, J.; Osman, A.; Tan, C. P. Development of an Edible Coating Based on Chitosan-Glycerol to Delay 'Berangan' Banana (Musa Sapientum cv. Berangan) Ripening Process. *Int. Food Res. J.* **2011**, *18* (3), 989–997.

Manganari, G. A.; Vicente, A. R.; Crisosto, C. H. Effect of Pre-Harvest and Post-Harvest Conditions and Treatments on Plum Fruit Quality. CAB Reviews: Perspectives in Agriculture, Veterinary Science, Nutrition and Natural Resources. 3, No. 009, 2008.

Maqbool, M.; Ali, A.; Ramachandran, S.; Smith, D. R.; Alderson, P. G. Control of Posthar-vest Anthracnose of Banana Using a New Edible Composite Coating. *Crop Prot.* **2010,** *29,* 1136–1141.

Marpudi, S. L.; Abirami, L. S. S.; Pushkala, R.; Srividya, N. Enhancement of Storage Life and Quality Maintenance of Papaya Fruit Using Aloe Vera Based Antimicrobial Coating. *Ind. J. Biotechnol,* **2011,** *10,* 83–89.

Martinez-Romero, D.; Alburquerque, N.; Valverde, J. M.; Guillen, F.; Castillo, S.; Valero, D.; Serrano, M. Postharvest Sweet Cherry Quality and Safety Maintenance by Aloe Vera Treat-ment: A New Edible Coating. *Postharvest Biol. Technol.* **2006,** *39,* 93–100.

McHugh, T. H.; Senesi, E. Apple Wraps: A Novel Method to Improve the Quality and Extend the Shelf Life of Fresh-Cut Apples. *J. Food Sci.* **2000,** *65,* 480–485.

Nimitkeatkai, H.; Srilaong, V.; Kanlyanarat, S. Effect of Semi-Active Modified Atmosphere on Internal Browning of Cold Stored Pineapple. *Acta Hort.* **2006,** *712,* 649–653.

Olivas, G. I.; Davila-Avina, J. E.; Salas-Salazar, N. A.; Molina, F. J. Use of Edible Coatings to Preserve the Quality of Fruits and Vegetables during Storage. *Stewart Postharvest Rev.* **2008,** *4* (3), 1–10.

OMRI. OMRI Brand Name Products List – Processing and Handling Materials by Supplier. Organic Materials Review Institute: Eugene, OR (www.omri.org), 2003.

Park, H. J.; Chinnan, M. S.; Shewfelt, R. L. Edible Corn-Zein Film Coatings to Extend Storage Life of Tomatoes. *J. Food Process. Preserv.* **1994a,** *18,* 317–331.

Park, J. W.; Testin, R. F.; Rank, H. J.; Vergano, P. I.; Wlter, C. I. Fatty Acid Concentration Effect on Textile Strength, Elongation and Water Vapour Permeability of Laminated Edible Films. *J. Food Sci.* **1994b,** *59,* 916–919.

Perkins-Veazie, P.; Collins, J. K. Culitvar and Maturity Effect Postharvest Quality of Fruit from Erect Blackberries. *HortSci.* **1996,** *31*(2), 258–261.

PFA. Prevention of Food Adulteration Act, 19th ed.; International Law Book Co.: Delhi, India, 2008.

Prasad, K.; Sharma, R. R. Screening of Mango Genotypes for the Incidence of Lenticel Browning, a New Postharvest Problem. *Ind. J. Agri. Sci.* **2016,** *86* (9), 1169–71.

Prasad, K.; Sharma, R. R.; Srivastav, M. J. Postharvest Treatment of Antioxidant Reduces Lenticel Browning and Improves Cosmetic Appeal of Mango (*Mangifera Indica* L.) Fruits without Impairing Quality. *Food Sci. Technol.* **2016,** *53* (7), 2995–3001. DOI 10.1007/ s13197-016-2267-z

Purvis, A. C. Interaction of Waxes and Temperature in Retarding Moisture Loss from and Chilling Injury of Cucumber Fruits during Storage. *Proc. Fla. State Hort. Soc.* **1994,** *107,* 257–260.

Raghav, P. K.; Agarwal, N.; Saini, M. Edible Coating of Fruits and Vegetables: A Review. *Int. J. Sci. Res. Mod. Educ.* **2016,** *1* (1), 188–204.

Rojas-Argudo, C.; Perez-Gago, M. B.; delRio, M. A. Postharvest Quality of Coated Cherries 'Burlat' as Affected by Coating Composition and Solids Content. *Food Sci. Tech. Int.* **2005,** *11* (6), 417–424.

Rosca-Casian, O.; Parvu, M.; Vlase, L.; Tamas, M. Antifungal Activity of Aloe Vera Leaves. *Fitoteropia.* **2007,** *78,* 219–222.

Rupak, K.; Suman, K. Morpholine: A Glazing Agent for Fruits and Vegetables Coating/ Waxing. *Int. J. Sci. Technol. Eng.* **2016,** *2* (11), 694–697.

Salunkhe, D. K.; Boun, H. R.; Reddy, N. R. *Storage Processing and Nutritional Quality of Fruits and Vegetables;* CRC Press Inc.: Boston, MA, 1991, pp 156–161.

Sánchez-González, L.; Pastor, C.; Vargas, M.; Chiralt, A.; González-Martínez, C. Effect of Hydroxypropylmethylcellulose and Chitosan Coatings with and Without Bergamot Essential Oil on Quality and Safety of Cold-Stored Grapes. *Postharvest Biol. Technol.* **2011,** *60,* 57–63.

Serrano, M.; Díaz-Mula, H. M.; Zapata, P. J.; Castillo, S.; Guillén, F.; Martínez-Romero, D.; Valverde, J. M.; Valero, D. Maturity Stage at Harvest Determines the Fruit Quality and Antioxidant Potential Alter Storage of Sweet Cherry Cultivars. *J. Agric. Food Chem.* **2009,** *57,* 3240–3246.

Shamim, S.; Ahmed, S. W.; Azhar, I. Antifungal Activity of Allium, Aloe, and Solanum Species. *Pharm. Biol. J.* **2004,** *42* (7), 491–498.

Singh, D. B.; Singh, R.; Kingsly, A. R. P.; Sharma, R. R. Effect of Aloe Vera Coatings on Fruit Quality and Storability of Strawberry (Fragaria × Ananassa). *Indian J. Agric. Sci.* **2011,** *81* (4), 407–412.

Singh, S. P.; Singh, Z. Postharvest Oxidative Behaviour of 1-methyl cyclopropene Treated Japanese Plums (*Prunus salicina* Lindell) during Storage under Controlled and Modified Atmospheres. *Postharvest Biol. Technol.* **2012,** 74, 26–35.

Smith, S.; Geeson, J.; Stow, J. Production of Modified Atmospheres in Deciduous Fruits by the Use of Films and Coatings. *Hortic. Sci.* **1987,** *22,* 772–776.

Vahdat, S.; Ghazvini, R. F.; Ghasemnezad, M. Effect of Aloe Vera Gel on Maintenance of Strawberry Fruits Quality. *Acta Hortic.* **2009,** *877,* 875–884.

Valero, D.; Serrano, M. Postharvest Biology and Technology for Preserving Fruit Quality. CRC-Taylor & Francis: Boca Raton, FL, 2010.

Valero, D.; Mula-diaz, H. M.; Zapata, P. J. Effect of Alginate Edible Coating on Preserving Fruit Quality in Four Plum Cultivars during Postharvest Storage. *Postharvest Biol. Technol.* *2013,* 77, 1–6.

Valero, D.; Valverde, J. M.; Martínez-Romero, D.; Guillén, F.; Castillo, A.; Serrano, M. The Combination of Modified Atmosphere Packaging with Eugenol or Thymol to Maintain Quality, Safety and Functional Properties of Table Grapes. *Postharvest Biol. Technol.* **2006,** *41,* 317–327.

Valverde, J. M.; Valero, D.; Martinez-Romero, D.; Guillen, F.; Castillo, S.; Serrano, M. Novel Edible Coating Based on Aloe Vera Gel to Maintain Table Grape Quality and Safety. *J. Agric. Food. Chem.* **2005,** *53,* 7807–7813.

Vargas, M.; Albors, A.; Chiralt, A.; Gonz'alez-Mart'ınez, C. Quality of Cold-stored Strawberries as Affected by Chitosan–oleic Acid Edible Coatings. *Postharvest Biol. Technol.* **2006,** *41,* 164–171.

Velickova, E.; Winkelhausen, E.; Kuzmanova, S.; Alves, B. D. Impact of Chitosan-beeswax Edible Coatings on the Quality of Fresh Strawberries (Fragaria ananassa cv Camarosa) under Commercial Storage Conditions. *Food Sci. Technol.* **2013,** *52,* 80e92.

Wang, S. W.; Gao, H. Effect of Chitosan-based Edible Coating on Antioxidants, Antioxidant Enzyme System, and Postharvest Fruit Quality of Strawberries (Fragaria × Aranassa Duch). *Food Sci. Technol.* **2013,** *52,* 71–79.

Worrell, D. B.; Carrington, C. M. S.; Huber D. J, The Use of Low Temperature and Coatings to Maintain Storage Quality of Breadfruit, *Artocarpus Altilis* (Parks.) Fosb. *Postharvest Biol. Technol.* **2002,** *25,* 33–40.

Yaman, O.; Bayoundurlc, L. Effect of an Edible Coating and Cold Storage on Shelf Life and Quality of Cherries. *LWT Food Sci. Technol.* **2002,** *35,* 146–150.

Yu, Y.; Zhang, S.; Ren, Y.; Li, H.; Zhang, X.; Di, J. Jujube Preservation Using Chitosan Film with Nano-silicon Dioxide. *J. Food Eng.* **2012,** *113,* 408–414.

Yurdugul, S. Preservation of Quinces by the Combination of an Edible Coating Material, Semperfresh, Ascorbic Acid and Cold Storage. *Eur. Food Res. Technol.* **2005,** *220,* 579–586.

Zhou, R.; Mo, Y.; Li, Y. Quality and Internal Characteristics of Huanghua Pears (*Pyrus pyrifolia* Nakai, cv. Huanhhua) Treated with Different Kinds of Coating during Storage. *Postharvest Biol. Technol.* **2008,** *49,* 171–179.

CHAPTER 8

PACKAGING OF FRESH-CUT FRUITS AND VEGETABLES

OLUWAFEMI J. CALEB, MARTIN GEYER, and
PRAMOD V. MAHAJAN*

Department of Horticultural Engineering, Leibniz Institute for Agricultural Engineering (ATB), Max-Ethy-Allee 100, 14469 Potsdam, Germany

Corresponding author. E-mail: pmahajan@atb-potsdam.de

CONTENTS

ABSTRACT

Consumer demand for freshness and convenience food has led to the evolution and increased production of fresh-cut fruits and vegetables. Moreover, this may represent a way to increase the consumption of fresh fruits and vegetables (FFV) and therefore, a benefit for the crops sector economy. Since the increase in convenience for the consumer has a detrimental effect on product quality, attention must be focused on extending shelf-life while maintaining quality. Fresh-cut fruits and vegetables (FFC) are very important to retailers. Quality and range of FFC on offer have a major influence on where customers shop. Packaged salads and pre-cut fruits are occupying more shelf space as they continue to gain acceptance by customers. The latest trend is to combine a range of fruits and/or vegetables into meal packs which are ready for consumption, with varying weight portions to meet the needs of families and individual consumers

8.1   INTRODUCTION

The segment of ready-to-eat fresh-cut consumer products is one of the few that has shown consistent growth in the last few years (Lange, 2000). Consumers are looking for convenience high quality fresh-cut ready-to-eat mixed fruits in order to overcome some of the pressures of modern lifestyles and this segment of the market has seen unprecedented growth in recent times. Consumer demand for freshness and convenience food has led to the evolution and increased production of fresh-cut fruits and vegetables. Moreover, this may represent a way to increase the consumption of fresh fruits and vegetables (FFV) and therefore, a benefit for the crops sector economy. Since the increase in convenience for the consumer has a detrimental effect on product quality, attention must be focused on extending shelf-life while maintaining quality. Fresh-cut produce offers many advantages, including cost control, waste reduction, variety and selection, consistent quality, and less in-store labor. Shipping costs are also reduced because most of the waste is eliminated at the processor/packer level.

Fresh-cut fruits and vegetables (FFC) are very important to retailers. Quality and range of FFC on offer have a major influence on where customers shop. Retailers are willing to reward entities which can assemble large supplies of consistent quality products with greater shelf-life on an all-year-round basis, achieved through the introduction of modern post-harvest handling techniques and cold-chain management. With a view to meeting consumer demand, new and more varied products have been introduced to the market. Packaged salads and pre-cut fruits are occupying more shelf space as they continue to gain acceptance by customers. The latest trend is to combine a range of fruits and/or vegetables into meal packs which are ready for consumption, with varying weight portions to meet the needs of families and individual consumers (Axtman, 2007). It is important to give priority to the flavor than appearance in determining postharvest life or post-cutting life of fresh produce (Kader, 2010).

8.2 FRESH-CUT PRODUCE PHYSIOLOGY

Fresh-cut produce is different from intact fruits and vegetables in terms of its physiology, handling and storage requirements, since processing results in disruption of tissue and cell integrity, with a concomitant increase in enzymatic, respiratory, ethylene, and microbiological activity. Further, cutting induces elevated ethylene production rates that may stimulate respiration rate and promote ripening of climacteric fruits and consequently accelerate microbial growth (Brecht, 1995). Type of cut, for example, sliced, finely shredded, coarsely shredded, cubed, grated etc., may also affect the respiration rate and senescence of fresh-cut fruits and vegetables (Iqbal et al., 2009). Minimal processing of fresh-cut produce involves sorting, cleaning, washing, trimming/peeling/deseeding/coring, and cutting (such as chopping, slicing, shredding, chunking, and dicing). Moreover, fresh-cut processing leads to major tissue disruption as enzymes and substrates, normally sequestered within the vacuole, become mixed with other cytoplasmic and nucleic substrates and enzymes leading to browning of cut surfaces and microbial contamination. Fresh-cut fruits are still physiologically active and respond to wounding. The first responses to wounding relate to increase in both respiration rate and ethylene production (Brecht, 1995). This has been largely studied for cut-vegetables that showed that the respiration rate of fresh-cut carrots can increase up to three-fold directly after cutting (Iqbal et al., 2009). However, the increase is momentary, mainly due to initial stress response, as respiratory, activity decreases to an equilibrium value within 24 h of processing for fresh-cut pineapple (Finnegan et al., 2013). The Weibull model has been successfully used to predict the influence of time on respiration rate with further combination of Arrhenius equation accounting for both the effect of time and temperature (Caleb et al., 2012a, 2012b; Waghmare et al., 2013). Emerging trend is to mix fresh-cut fruits together in a ready-to-eat form; however, there is still a lack of basic scientific information on the effect of changing the proportions of fresh-cut fruits in a mixture on respiration rate and ethylene production rate. From the recent study by Mahajan et al. (2014), there was a synergetic effect of ethylene produced by one fruit on to the ethylene production rates of other fruits in a given mixture.

8.3   MODIFIED ATMOSPHERE PACKAGING

During storage and distribution, the packaging of a fresh-cut produce plays a critical role in quality preservation and shelf-life extension. Packaging is essential in protecting the product against the outside environment and preventing mechanical damage and chemical or biological contamination. It allows the control of water loss, which is an important symptom of loss of quality, and controls the flux of O_2 and CO_2, building up and/or maintaining an optimal atmosphere within the packaging headspace. Packaged salads and pre-cut fruits are occupying more shelf space, as products continue to gain acceptance by customers. Restaurants, fast-food outlets, and institutional food-service operators are seeking to reduce labor costs by buying processed, ready-to-use or ready-to-eat fresh-cut fruits. The latest trend is to mix a range of fresh-cut fruits into small packs which are ready for consumption, with varying weight portions to meet the needs of families and individual consumers (Fig. 8.1).

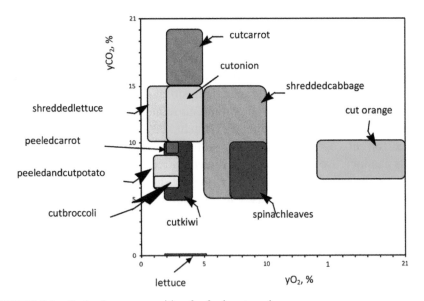

FIGURE 8.1   Optimal gas composition for fresh-cut produce.

Modified atmosphere packaging (MAP) of fresh produce relies on modifying of the atmosphere inside the package, achieved by the natural interplay between two processes, the respiration of the product and the transfer of gases through the packaging, which leads to an atmosphere richer in CO_2 and poorer in O_2. This atmosphere can potentially reduce the respiration rate, ethylene sensitivity, and physiological changes. MAP generally involves the packaging of a whole or fresh-cut product in plastic film bags, and can be either passive or active. Their main differences are shown in Table 8.1.

TABLE 8.1 Type of Modified Atmosphere Packaging (MAP) for Fresh-cut Produce.

	Passive	Active
Definition	Modification of the gas composition inside the package due to interplay between the product respiration and the package permeability	Modification of the gas composition inside the package by flushing the package headspace with desired gas mixture
Equilibrium time	1–2 days to 10–12 days	1–2 h
Products suitable for	Mushrooms, carrots, strawberry, spinach	Cut apples, minimally processed leafy green vegetables
Cost	Modification occurs naturally, so no extra cost involved if the package is properly designed	Extra investment is required for special machinery, that is, gas mixer, gas flushing, and packaging machine
Labeling requirements	No	Yes

In passive MAP, the equilibrium concentrations of O_2 and CO_2 are a function of the product weight and its respiration rate, which is affected by temperature, surface area, perforations, thickness, and permeability to gases of the films used in packaging. In active MAP, the desired atmosphere is introduced in the package headspace before heat sealing, but the final atmosphere will eventually be a function of the same factors that affect passive MAP. Correct equilibrium atmosphere can delay respiration and senescence, and slow down rate of deterioration, thereby extending product storage life. One of the main disadvantages of a passive MAP is the long time necessary to reach a condition of dynamic steady state, close to the optimal gas composition. To reach the optimal storage condition in such a long time can be detrimental to the stability of the package. In contrast, by packing a product by means of an active MAP the equilibrium time is very short. Nevertheless, the improvement in the equilibrium time is paid with an increment in the cost. FFC are prewashed and packaged in bags flushed with either normal air or recommended modified atmosphere (Kader, 2010).

8.4   SELECTION OF PACKAGING MATERIALS

Nowadays, minimizing the time required to achieve equilibrium coupled with the creation of the atmosphere best suited for the extended storage of a given product is the main objective of MAP design. Since the 1980s, MAP has evolved by a pack and pray procedure that may have economic and safety hazard consequences to a "trial and error" approach which is an extremely time-consuming procedure. The design of a package depends on a number of variables: the characteristic of the product, its mass, the recommended atmospheric composition, the permeability of the packaging material to gases and its dependence on temperature, and the respiration rate of the product as affected by different gas compositions and temperature. Therefore, to ensure an optimal gas composition during product shelf-life, an engineering approach has to be followed to model all the variables that play a critical role. Simulation of a MAP system is the most appropriate method to allow proper MAP design and thereby obtain a successful commercial product.

Sound use of MAP entails consolidated knowledge of the food-package environment interaction: when a product is packed, it is surrounded by a gaseous mixture, whose composition depends on the interactions between the food product, the package material, and the environment. If the packed food is a fresh-cut product, such interactions concern the respiration metabolism of the product: The product exchanges gas with the surrounding atmosphere, consuming O_2, and CO_2. Due to the respiration process, O_2, and CO_2 concentration gradient between the head space and the environment is generated. Thus, a gas flow is activated through the packaging material due to film permeability to O_2 and CO_2. It later became clear that an efficient use of MAP technology should be based on knowledge of the effect of gas on product quality in order to find the optimal conditions for each product. An extensive number of models have been developed to predict respiration rates of FFV under MAP conditions (Kader, 2010). An example of prediction models is that of Mahajan et al. (2007) where the Pack-in-MAP® software for optimum packaging solutions for FFV was developed. Behind the software are extensive database on product respiration rate, optimum temperature, and optimum ranges of O_2 and CO_2 and gas permeability of packaging materials commonly used in MAP (Fig. 8.2). The software is based on a series of mathematical algorithms to simulate the evolution of internal gas composition in the packaging as a result of food respiration and mass transfer through the packaging material and, when used in a reverse manner, to identify the window of gas permeability that satisfies food requirements, size, and the number of micro-perforations, if needed.

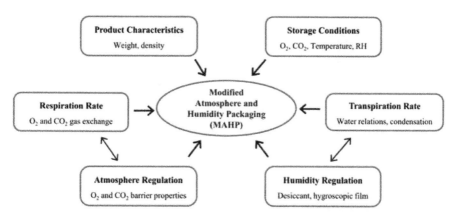

FIGURE 8.2   Factors involved in designing the modified atmosphere packaging for FFV.

8.5   COMMERCIALLY AVAILABLE PACKAGED FRESH-CUT PRODUCE

A market survey was conducted on fresh-cut fruit packages commercially available in the major supermarket chains in Ireland (Mahajan et al., 2011). It was evident from the market survey that there is an increase in the range of materials, the material structures, and the type of forming and filling equipment that is now used while designing flexible and rigid packages for the fruit and vegetable sector. Analysis showed that packaging styles have evolved over time showing major interest in fresh-cut-fruits, specially mixed, and packed together in ready-to-eat format. The number of fruit combinations varied from two to six different fruits packed in a single package. Fruits commonly used were blends of melon and pineapple chunks, whole grapes, apple slices, orange slices, whole and cut strawberry, mango chunks/sticks, whole blueberry, and kiwi slices. Figure 8.3 shows the different packaging formats being used by four different retailers. These include tray, cup, pouch, flow-wrap, and stand-up pouch with total package volume ranging from 250 to 800 cm^3. The trays were manufactured using rPET and PET material. The lidding material appeared to be orientated polypropylene except one retailer which used clamshell trays (PET) for packaging of cut-fruits. While it is necessary to consider the O_2 permeability and beta ratio for designing MAP, it is also equally important to consider other properties such as sealability, printability, coefficient of friction, anti-fog, film clarify, source, and price.

| Tray | Cup | Pouch | Flow wrap | Stand-up pouch |

FIGURE 8.3   Typical packaging formats used in fresh produce industry.

Further on, most of the packages studied deviated from the optimal gas composition recommended for maintaining quality and extending shelf-life of fresh and fresh-cut fruits. More than 60% of the packages studied were found to have O_2 concentration between 9 and 21% and CO_2 concentration between 0 and 12%. Although high O_2 concentration is not harmful, it does not help in gaining benefits of MAP technology. The extreme gas composition (1.3% O_2 and 14.8% CO_2, both crossing the critical limit) was found to be in a package containing melon–pineapple–mango sticks. This might be due to a very small net breathable film packaging area of 23.6 cm^2 available for gas exchange. The ratio of net product weight per unit volume of package was found to be between 0.4 and 0.6 g/cm^3 for all the packages tested. The package with pineapple chunks contained the highest amount of liquid (6–10% of net product weight) which might have oozed out through the cut tissues of pineapple chunks.

Fruit and vegetable processors currently rely on human judgment, custom, and practice when deciding how to pack their produce. The current practice is usually haphazard and is often referred to ironically as "pack and pray" (use the industry norm when packing and pray nothing goes wrong). The standard approach is to use a packaging material that is commonly used, or to ask packaging suppliers to recommend a film that is "typically used." If the end product looks satisfactory then the industry is largely unconcerned, in the belief that other options are both limited and expensive. Some processors will perhaps attempt more than one material, but usually packaging price is the main driver when selecting a particular alternative. Quality problems which are particularly obvious (e.g., excessive condensation on the inside of the package) are usually solved by randomly piercing a few holes. It would be possible to calculate exactly the size and number of holes that would result in the ideal packaging solution (permeability of the package) from respiration rate data, packaging material properties and first principle mathematical equations, but very few people working in the fruits and vegetable industry have the knowledge, or the time, to do so. Such approach was taken by Pack-in-MAP software for packaging design of fresh produce (Mahajan et al., 2007).

The oversimplified and inaccurate way of designing packaging systems at present are not only an issue of unnecessary losses due to shortened or insufficiently extended shelf-life, but more important than that, a real public health hazard. When inadequate packaging results in shortening shelf-life, the consequence is loss and waste, but when it results in inability to deal with high respiration rates leading to total O_2 depletion (anoxia), the package becomes an ideal ground for the growth of anaerobic bacteria, some of which are extremely dangerous pathogens. Potential contamination associated with unsuitable packaging design is a disaster waiting to happen: food poisoning due to consumption of contaminated salads has occurred in other countries in the past (the USA, UK) which is already well known. The challenge faced by the industry is to effectively and efficiently select a packaging solution for fresh produce.

8.6 FUTURE NEEDS

Current MAP design considers the respiration rate of product as the only important parameter for deciding target gas barrier properties required to achieve an equilibrium-modified atmosphere. However, besides in-package gas composition it is also important to take into consideration the in-package level of humidity, in order to avoid condensation and/or mold and bacterial development in MAP systems (Sousa-Gallagher et al., 2013). It is well known that the in-package humidity is influenced by respiration and transpiration of the fresh produce as well as the water vapor permeability of the packaging material (Fig. 8.2). However, most polymeric materials (polyethylene, polypropylene, or polyvinyl chloride) used in MAP have lower water vapor permeability relative to transpiration rates of fresh produce; therefore, most water molecules evaporated from the produce do not escape through the film and remain within the package, enhancing the water vapor pressure in the package microenvironment. Under these near-saturation conditions, even minor temperature fluctuation may result in condensation inside the package resulting in produce sliminess and enhancement of microbial growth and decay of produce (Linke & Geyer, 2013). Therefore, major challenge of modified atmosphere and humidity packaging (MAHP) is finding a solution to create optimal atmosphere and reducing the risk of water condensation in the package while still maintaining produce weight loss as low as possible. Recent developments on high water vapor permeable films such as Xtend, Nature Flex, or bio-degradable films resulted in the use of hygroscopic additives located within the package headspace or directly integrated into the packaging material. However, the hygroscopic additives should not be used for fresh produce, which have high water activity, in order to avoid excessive weight loss of the packed food. Recently, Singh et al. (2010) and Rux et al. (2015) reported that the humidity absorption of the trays with NaCl improved the quality and shelf-life of fresh mushrooms. In this study, different percentage of NaCl was incorporated in the polymer matrix of a film from which trays were produced. The results indicate that the amount of water vapor absorbed by the tray is directly proportional to the percentage of salt incorporated in the trays, which enhanced the total appearance of the package. The maximal capacity and the rate of moisture absorption by the humidity regulating trays need to be studied further in order to confirm if the trays have enough moisture absorption capacity to prevent condensation for the selected horticultural crops. Additionally, questions about sustainability and consumer expectations of such humidity regulating trays have to be taken into consideration. Other active MAP technologies have been emerging that adsorb substances such as O_2, ethylene, moisture, CO_2, flavors/odors or release substances such as CO_2, antimicrobial agents, antioxidants, and flavors (Mahajan et al., 2014). Such technologies are important for shelf-life extension of fresh and fresh-cut fruits and vegetables.

KEYWORDS

- **fresh-cut fruits and vegetables**
- **packaged salads**
- **pre-cut fruits**
- **ready-to-eat products**
- **packaging**

REFERENCES

Axtman, B. Make the Cut: New Insights into the Consumer can Help Grocers Keep Pole Position in the Race to Sell More Fresh-cut Produce. Progressive Grocer: Chicago, IL, Oct, pp 60–64, 2007.

Brecht, J. K. Physiology of Lightly Processed Fruits and Vegetables. *Hort. Sci.* **1995,** *30* (1), 8–22.

Caleb, O. J.; Mahajan, P. V.; Opara, U. L.; Witthuhn, C. R. Modelling the Effect of Time and Temperature on Respiration Rate of Pomegranate Arils (cv. 'Acco' and 'Herskawitz'). *J. Food Sci.* **2012b,** *77* (4), E80–E87.

Caleb, O. J.; Mahajan, P. V.; Opara, U. L.; Witthuhn, C. R. Modelling the Respiration Rates of Pomegranate Fruit and Arils. *Postharvest Biol. Technol.* **2012a,** *64,* 49–54.

Finnegan, E.; Mahajan, P. V.; O'Connell, M.; Francis, G. A.; O'Beirne, D. Modelling Respiration in Fresh-cut Pineapple and Prediction of Gas Permeability Needs for Optimal Modified Atmosphere Packaging. *Postharvest Biol. Technol.* **2013,** *79,* 47–53.

Iqbal, T.; Rodrigues, F. A. S.; Mahajan, P. V.; Kerry, J. P. Mathematical Modeling of the Influence of Temperature and Gas Composition on the Respiration Rate of Shredded Carrots. *J. Food Eng.* **2009,** *91* (2), 325–332.

Kader, A. A. Future of Modified Atmosphere Research. *Acta Hort.* **2010,** *857,* 213–217.

Lange, D. L. New Film Technologies for Horticultural Commodities. *Hort. Technol.* **2000,** *10,* 487–490.

Linke, M.; Geyer, M. Condensation Dynamics in Plastic Film Packaging for Fruit and Vegetables. *J. Food Eng.* **2013,** *116,* 144–154.

Mahajan, P. V.; Lucas, A.; Edelenbos, M. Impact of Mixtures of Different Fresh-cut Fruits on Respiration and Ethylene Production Rates. *J. Food Sci.* **2014,** *79* (7), 1366–1371.

Mahajan, P. V.; Caleb, O. J.; Zora, S.; Watkins, C.; Geyer, M. Postharvest Treatments of Fresh Produce. Philosophical Transactions of the Royal Society (A): *Math. Phys. Eng. Sci.* **2014,** *372,* 20130309.

Mahajan, P. V.; Oliveira, F. A. R.; Montanez, J. C.; Frias, J. Development of User-friendly Software for Design of Modified Atmosphere Packaging for Fresh and Fresh-cut Produce. *Innov. Food Sci. Emerg. Technol.* **2007,** *8,* 84–92.

Mahajan, P. V.; Sousa-Gallagher, M. J. *Analysis of Commercially Available Packages of Fresh-cut Fruits,* Proceedings of II International Conference on Quality Management of Fresh Cut Produce, Torino, Italy, 17–21 July 2011,

Rux, G.; Mahajan, P. V.; Geyer, M.; Linke, M.; Pant, A.; Sängerlaub, S.; Caleb, O. J.; Application of Humidity-regulating Tray for Packaging of Mushrooms. *Postharvest Biol. Technol.* **2015,** *108,* 102–110.

Singh, P.; Saengerlaub, S.; Stramm, C.; Langowski, H. C. Humidity Regulating Packages Containing Sodium Chloride as Active Substance for Packing of Fresh Raw Agaricus Mushrooms; Kreyenschmidt, J. ed.; Proceedings of the 4th International Workshop Cold Chain Management: Bonn, Germany, 2010.

Sousa-Gallagher, M. J.; Mahajan, P. V.; Mezdad, T. Engineering Packaging Design Accounting for Transpiration Rate: Model Development and Validation with Strawberries. *J. Food Eng.* **2013,** *119,* 370–376.

Waghmare, R. B.; Mahajan, P. V.; Annapure, U. S. Modelling the Effect of Time and Temperature on Respiration Rate of Selected Fresh-cut Produce. *Postharvest Biol. Technol.* **2013,** *80,* 25–30.

CHAPTER 9

ENGINEERING PROPERTIES OF PACKAGING FILMS

STEFANO FARRIS*

Department of Food, Environmental and Nutritional Sciences, DeFENS, Packaging Division, University of Milan, Via Celoria 2 – 20133, Milano, Italy

*E-mail: stefano.farris@unimi.it

CONTENTS

ABSTRACT

The design and development of food packages requires a deep knowledge of the inherent properties of packaging materials together with the understanding of the kinetics possibly leading to the quality decay of food matrices. The food-packaging interactions thus rise as an aspect of primary relevance to convey to the consumers safe, convenient, high quality, and minimally processed foods. The packaging materials have therefore, to be properly selected with the goal of assisting the preservation of food products, for example, by reducing the respiration rates, controlling the moisture uptake/ loss, or preventing mechanical damages. At the same time, the presentation of the packaged food product must be kept in mind as it can influence the final choice made by consumers. The goal of this chapter is to provide an introduction to the basic physicochemical properties of packaging materials intended for food products. Mechanical barrier, optical, and surface properties will be first discussed. The last part of the chapter is devoted to a special property of plastic films specifically intended for minimally processed fruits and vegetables, tha t is, the anti-fog property, which is emerging as a sought-after feature especially from an aesthetic point of view. Both common strategies and recent advances will be reviewed.

9.1   INTRODUCTION

The selection of packaging materials intended for food products must fulfill specific requirements linked to the main functions of packaging, such as containment, protection, convenience, and communication. Depending on the food, each of the above functions can weigh differently on the final packaging configuration. Nevertheless, from an engineering perspective, whatever packaging system is used should aim at extending the shelf life of the food product while acting as a "silent seller." Toward these goals, physicochemical properties of packaging materials must be carefully evaluated and properly set. Among others, mechanical, barrier, optical, and surface properties play a role in dictating the shelf life of the foods and the final consumer choice. This applies even more for packaging materials for fruits and vegetables, where accelerated metabolism and fast aging restricts their consumption to short time spans. Because plastic films (including bioplastics) are the most widely used packaging material for fruits and vegetables, our focus will be on this class of polymeric materials. In addition, despite the large number of films tested for packaging fruits and vegetables, practical operations make use of only a few materials, namely: polyolefins, including polypropylene (PP) and polyethylene (PE); vinyl compound polymers such as polystyrene (PS) and polyvinyl chloride (PVC); and polyethylene terephthalate (PET) belonging to the polyester family. More recently, niche products are packaged using polylactic acid (PLA), a biodegradable thermoplastic aliphatic polyester derived from renewable resources, such as cornstarch.

9.2 MECHANICAL PROPERTIES

Mechanical properties are generally referred to as the reaction of a material in response to an applied load. More conveniently, for example, in food packaging, it is relevant to investigate the capability of a given material to withstand mechanical stresses, especially tensile and compressive stress. In a typical tensile test, the specimen (a simple rectangular strip or a dogbone-shaped piece) is placed between a fixed jaw and a moveable head. The specimen is then stretched by moving the movable head at a constant speed. Given the thickness and the width of the specimen, the cross-sectional area of the specimen can be easily calculated and the force is converted to stress (force per unit surface). Using "stress" (e.g., in MPa) rather than "force" (e.g., in N) is most useful for comparison purposes. From the stress-strain plot, three main parameters are determined usually by a software-assisted procedure: elastic (Young's) modulus; tensile strength; and elongation at break (Table 9.1). The elastic modulus, usually expressed as MPa, is the slope of the first linear part of the stress-strain curve. It represents the elastic (reversible) component of the material, that is, the part of the stress that can be absorbed by the material without leading to irreversible deformations. In practice, the higher the elastic modulus, the stiffer is the material. However, because the elastic modulus is an intensive property while stiffness is an extensive property, the previous relationship applies for specimens of the same shape tested under the same conditions. Tensile strength is the maximum stress that the material is able to withstand before rupture. Depending on where the stress value is taken on the stress-strain plot, three main tensile strengths can be defined, namely: (1) yield strength, that is, the maximum tensile strength in the elastic region; (2) breaking strength, that is, the tensile stress at the breaking point; and (3) ultimate strength, that is, the highest stress point along the stress–strain curve. The same parameters are gathered from compressive tests. Tensile tests evaluate the behavior of a material under elongation forces that can apply, for example, during unwinding operations of the plastic webs in most converting and packaging

TABLE 9.1 Mechanical Properties of Typical Polymer Films Used for Food Packaging Application.

Material				Mechanical Properties				
	Elastic modulus (MPa)	Tensile strength, break (MPa)	Elongation at break (%)	Coefficient of friction (static)			Coefficient of friction (dynamic)	
				mat/met[a]	mat/mat[b]	mat/met	mat/mat	
OPP	1800	40	400	0.45	1.27	0.35	0.75
LDPE	400	12	650	0.46	0.18	0.86	0.64
PET	3000	200	100	0.34	0.54	0.27	0.42
PS	3350	55	3	0.35	0.50	–	–
Nylon (PA 6,6)	2000	250	100	0.6	0.25	0.5	–
PLA	3500	55	6	–	0.55	–	–

The values reported are average values, which depend on finishing and presence of additives in the final film.
[a]Material surface on itself.
[b]Material surface against steel surface.

plants or during stretching of plastic films intended to wrap foods. On the contrary, compressive tests aim to describe the material's behavior under a compressive force. One of the most common test is the so-called "puncture test," in which the material is subjected to a punctual compression. This test is particularly suitable to simulate packaging failures possibly occurring during packaging, transport, and storage operations due, for example, to protruding parts of packaging machinery and contact with the retail shelves and display racks.

Another relevant mechanical test is the quantification of the coefficient of friction (COF), which reflects the resistance of a surface in order to slide on another surface. Although the frictional properties of a material strongly rely on physicochemical properties of the surface (e.g., topography, atomic composition, and polarity), they are included in the "mechanical properties" group because COF measurements are run in a universal testing machine similarly to those used for tensile and compression tests. Most tests are carried out according to two main setups. In a typical "polymer-to-polymer" setup, one piece of the specimen is made to slide against the surface of the same material. This kind of test aims to get information about the frictional properties of a material against itself. In a typical "polymer-to-metal" setup, one piece of the specimen is made to slide against a metallic surface, simulating when the material unrolls from a reel and runs on forming, printing, or laminating equipment. In practice, the film to be tested is wrapped around a metallic sled of a given weight, which is connected to the cell load of the machine by yarn. The sled is placed on top of a metallic plate, which is covered or not with the same material. Pulling the cross member of the equipment by the motion system will make the slide move on the plane for a defined path length at a specific speed. The software will continuously record the force opposing to the movement. The COF value (f) is eventually calculated as the ratio of the recorded frictional force (F) to the gravitational force (F_g, given by the weight of the slide); thus it is expressed by a non-dimensional quantity. Two different COF values are usually calculated. The static COF is the resistance opposing the force necessary for the initial motion of the slide on the surface underneath. The kinetic (or dynamic) COF is the resisting force opposing the force necessary to keep the slide moving at a given speed (the test speed). Depending on the application, packaging materials are sought with high or low COF values. For example, to speed up the packaging operations by avoiding abrupt blocking of the plastic webs running on the packaging machinery, low COF values (especially kinetic COF) are sought. Conversely, when vending units have to be stacked one on top of the other (during both transportation and storage at the marketplaces);

it is of utmost important to avoid collapsing the pile. Therefore, high COF values would be preferred.

9.3   TRANSPORT PROPERTIES

The capability of a packaging material to act as a barrier against the transfer of gas and vapors plays a key role in dictating the quality decay of the food product. While in most circumstances high barrier performance is sought to limit the detrimental effect of, for example, oxygen (oxidations) and water vapor (sogginess, microbial growth), the exchange of gases and/or vapors from the external environment to the internal atmosphere of the package is necessary for some applications, such as freshly roasted coffee and fresh fruits and vegetables. Therefore, the selection of the most suitable packaging material must arise from a careful consideration of the food matrix and the different performance of packaging materials (Table 9.2).

TABLE 9.2 Permeability Parameters (O_2TR and WVTR) and Permselectivity Values of Typical Polymer Films Used for Food Packaging Applications.

Material	Permeability properties				
	O_2TR^a	$WVTR^b$	O_2/N_2	CO_2/N_2	CO_2/O_2
OPP (20 μm)	1700	6	4.6	12	2.7
LDPE (40 μm)	2600	10	2.9	19	6.5
PET (12 μm)	110	40	4.4	31	7.0
PS (25 μm)	3500	140	3.8	30	7.9
Nylon (PA 6,6) (12 μm)	44	410	3.8	16	14.2
PLA (40 μm)	510	150	–	–	–

The values reported are average values, which depend on finishing and presence of additives in the final film.
[a]cm^3 (STP) m^{-2} 24 h^{-1} at 23°C, 0% RH, and 1 atm pressure difference.
[b]g (STP) m^{-2} 24 h^{-1} at 38°C and 90% RH.

The transfer of gases and vapors through the package occurs according to two main mechanisms. Permeation is a mechanism that involves three sequential steps: (1) solubilization (adsorption) of the permeant molecule on the material's surface; (2) diffusion of the molecule across the material's thickness; and (3) desorption, that is, the release of the molecule inside the package atmosphere. These steps are gradient-driven, namely, they occur from the high concentration side to the low concentration side of the package through the molecular voids (free volume) and interstices of the material. Permeation is affected by the inherent characteristics of the package, such as crystallinity degree, surface, and bulk chemical composition, and presence of additives. Other factors that influence permeation—such as temperature, especially for hydrophilic polymers (e.g., Nylon, EVOH, and PVOH), relative humidity—pertain to the external environment. A second mechanism of gas and vapor transfer is named leak, which takes place across the material's defects such as pinholes (e.g., those found in aluminum foil) and channel leaks (e.g., due to deficient seals). Differently than permeation, leak only consists of a diffusion step of the permeant from the high-concentration side to the low-concentration side of the material. Pinholes and leaks have a great impact on the barrier properties of a package because the transfer of gases and vapors is 5–7 orders of magnitude greater than permeation phenomena. It is estimated that the gas transfer through a pinhole of 1 mm² area equals the gas transfer that would occur through 1 m² of the same undamaged material. Therefore, the potential occurrence of both pinholes and leaks should be prevented.

At steady state, that is, when the diffusion flux of the permeant is constant across the film, the permeation rate equation is:

$$Q = P'A\frac{p_A - p_B}{L} = \frac{P'A}{L}\Delta p$$

where Q is the steady-state permeation rate; P' is the permeability coefficient (or simply permeability), namely the product of diffusion coefficient (D) and solubility coefficient (S); A is the surface area of the film; L is the film thickness; and Δp is the partial pressure gradient of the permeant, that is, the partial pressure difference across the film. The usefulness of the above equation relies on the simplicity of taking into account the parameters influencing the permeation phenomenon, namely the package parameters (P', A, and L), and environmental variables (p_A and p_B). Temperature and relative humidity, though not explicitly indicated, are included in the equation, both influencing the permeation rate as stated before.

Despite the importance of this parameter for several applications, confusion still exists as far as the use of terms and units for permeation are concerned. In this chapter, we have decided to adhere to the terms and units adopted by the ASTM standards. Accordingly, permeability coefficient (or simply permeability), P', is defined as the amount of permeant passing through a material per unit area, per unit time, per unit thickness, and per unit partial pressure difference on the two sides of the film at steady state conditions. In practice, the use of P' is discouraged because the linear relationship between permeability and thickness (i.e., the permeation decreases linearly with thickness) will only apply for highly homogeneous materials. The SI unit of P' is mol m^{-1} s^{-1} Pa^{-1}. The temperature and relative humidity conditions set during the test must be explicitly stated.

Permeance (P) is the amount of permeant passing through a material per unit area, per unit time, and per unit partial pressure difference on the two sides of the film at steady state conditions, for a given thickness. Because P is not normalized by the thickness, it can be used for both homogeneous and heterogeneous materials. The SI unit of P' is mol m^{-2} s^{-1} Pa^{-1}.

The most widely adopted parameter to quantify the barrier property of a material is the transmission rate (TR), which becomes gas transmission rate (GTR) in the case of permeant gases (e.g., oxygen and carbon dioxide) and water vapor transmission rate (WVTR) in the case of water vapor as a permeant. TR is the amount of permeant passing through a film per unit area and per unit time at steady state. Therefore, TR values must always be accompanied by the test conditions (temperature and relative humidity), the thickness of the film, and the partial pressure difference between the two sides of the sample. The SI unit of TR is mol m^{-2} s^{-1}. In practice, a common unit is cm^3 (STP) m^{-2} 24 h^{-1} for gases, where 1 cm^3 at standard temperature and pressure (STP) is 44.62 mmol and 1 atm is 0.1013 MPa. The same unit for water vapor is g (STP) m^{-2} 24 h^{-1}.

There are two main methods that allow for measuring the permeability of gases and vapors across a material. The isostatic method is based on the establishment of the same total pressure on both sides of the specimen (i.e., the film), which is placed between two semi-chambers where an inert gas (usually nitrogen) is continuously flushed. After a first conditioning step of the sample, the test gas is then flushed. The test gas will thus permeate across the sample and be carried to the detector by the inert gas. The partial pressure of the test gas is kept constant throughout the analysis to ensure the same driving force at any time. A typical permeation profile is displayed in Figure 9.1a. The plot considers the amount of gas passing through the sample (e.g., O$_2$TR) over time. The final plot can be split in three main regions. In the first region, the "lag phase" is due to the molecules of the permeant dissolving into the polymer matrix. The duration of this step depends strongly on the chemical affinity between permeant and polymer. A steep rising tract then follows, which is due to the desorption of the molecules of the permeant from the film thickness to the detector. The third part of the plot is a plateau, in correspondence with when a steady state condition is reached, that is, the rates of molecular transfer on both sides of the film counterbalance each other. This is the final permeability value of the material tested. Interestingly, the final plot obtained from an isostatic approach allows us to estimate the diffusion coefficient (D) using the half-time method:

$$D = L^2 / (7.205\ t_{1/2})$$

where $t_{1/2}$ is the time needed to reach one half of steady-state permeation rate and L is the thickness of the specimen. However, it must be kept in mind that this method is suitable only for polymers showing a Fickian behavior for molecular diffusion, that is, D does not depend on the initial concentration of the permeant. For example, the diffusion coefficient of oxygen through polyethylene is the same if the oxygen partial pressure is 20.9% (air) or 100%.

In the quasi-isostatic method, there is no continuous flushing of carrier and test gas. Rather, the test gas is allowed to accumulate from one semi-chamber to the other that previously was flushed with the carrier gas. A small amount of gas is thus taken at fixed time intervals and analyzed by a gas chromatograph. The profile of the "amount of permeant vs time" plot in a typical quasi-isostatic method shows a rising tract immediately following the first lag time (Fig. 9.1b). The slope of the linear part of the plot is assumed as the steady state permeation. The estimation of the diffusion coefficient (D) can be made according to the following equation:

$$D = L^2 / 6\tau$$

where τ is the lag time given by the x-intercept of the steady-state straight line (Fig. 9.1b).

The apparatus involving two semi-chambers is used to test films and sheets, that is, 2D packaging materials. However, there could be the need for determining the barrier properties of finished packages, for example, pouches, bags, and trays. The setup used for 3D samples involves a hermetically sealed chamber (usually a polycarbonate cylinder), inside of which the 3D object is fixed to one inlet and one outlet for the carrier gas (Fig. 9.2). The headspace outside the sample is filled with the test gas until a partial pressure of one is reached. When the measurement starts, the carrier gas will convey continuously the gas permeated across the package walls to the detector, according to the isostatic configuration. The final permeation is usually expressed as mL pack^{-1}24 h^{-1}.

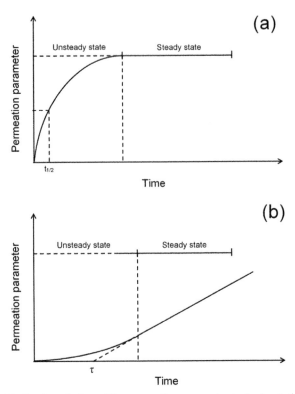

FIGURE 9.1   Idealized evolution of the permeation parameter in the isostatic (a) and quasi-isostatic (b) methods.

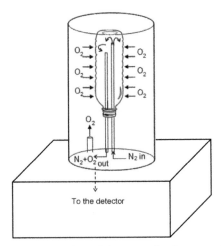

FIGURE 9.2   Schematic representation of the setup for the permeability measurement of 3D packages.

9.4 OPTICAL PROPERTIES

The optical properties of materials rely on the interaction between the material and the electromagnetic radiation, which spans from the cosmic rays (shortest wavelength) to the radio waves (longest wavelength). However, the most relevant range for food packaging deals with ultraviolet (UV) and visible (Vis) light, namely the spectral range within ~200–800 nm wavelength (the boundary between the two being centered at ~320 nm). In particular, it is of interest to know how and to what extent the energy associated to UV and Vis light interacts with the packaging material in terms of absorbance, transmittance, and reflectance. From a practical point of view, the interest in UV–Vis light concerns two main aspects, that is, the shelf life of food and their display. As far as the former is concerned, it is important to note that both UV and Vis light represent potential factors of degradation for many food matrices. This specifically applies for foods sensitive to light (e.g., fatty and meat products) where photosensitizers such as chlorophyll, pheophytins, porphyrins, riboflavin, myoglobin, and synthetic colorants can absorb energy from light and transfer it to triplet (atmospheric) oxygen to form singlet oxygen, thus triggering detrimental reactions such as oxidation and discoloration of the food matrix. The acquisition of the UV–Vis spectra of materials provides adequate information on the capability of a material to act as a barrier to light. In general, pristine plastics are poor barriers to both UV and Vis light, with the only exception being PET, which exhibits a cut-off in the transmittance spectrum at around 310 nm. For this reason, the use of additives acting as UV/Vis filters is widespread for many different applications, especially when the display of the product has to be preserved. Alternatively, the use of aluminum foil or metalized films allows achieving a total barrier against the UV–Vis light, though the display of the product is precluded.

Other optical properties are more specifically related to aesthetic attributes of the product. Among others, transparency, haze, and gloss greatly influence the consumers' perception of the package and their final choice. According to ASTM D1746, transparency is defined in terms of specular transmittance, that is, that part of the incident light transmitted by the material in the same direction of the incident radiant flux:

$$\text{Transparency} = (T_{540-560}/T_t) \times 100$$

This property is quantified simply by recording the transmittance value between 540 and 560 nm (i.e., at the highest sensitivity of the human eye) using a spectrophotometer, placing the sample between the source and the detector. Transparency is intimately linked to the hedonistic property known as "see-through," namely, the ability of a material to transmit image-forming light. Typical transparency values for commonly used packaging materials are reported in Table 9.3.

Haze represents the amount of transmitted light diffused (scattered) by the sample, namely the total luminous transmittance voided of the specular luminous transmittance. According to the ASTM D1003, haze is defined as the amount of incident light transmitted by the material deviating more than 2.5° from the direction of the incident radiant flux.

TABLE 9.3 Optical Properties of Typical Polymer Films Used for Food Packaging Application.

Material	Optical properties		
	Transparency[a] (%)	Haze (%)	Gloss
OPP (20 µm)	91	1.5	95 (45°)
LDPE (40 µm)	85	5.5	70 (45°)
PET (12 µm)	84	3.1	80 (45°)
PS (25 µm)	45	1.0	92 (60°)
Nylon (PA 6,6) (12 µm)	88	1.6	180 (60°)
PLA (40 µm)	93	1.2	125 (60°)

The values reported are average values, which depend on finishing and presence of additives in the final film.
[a]Specular transmittance at 550 nm.

$$Haze = (T_d/T_t) \times 100$$

where T_d is the diffused transmittance and T_t is the total transmittance. Haze may be due to high surface roughness as well as to the presence of scattering centers, such as the presence of fillers and/or additives added to the main polymer matrix. In practice, haze is responsible for the opacity of materials, that is., the loss of contrast. The setup used for the determination of haze consists of a spectrophotometer mounting an integrative sphere that is able to trap all the light diffused by the sample that otherwise would be lost, eventually leading to an underestimation of the total transmitted light. It should be here noted that the part of the incident light transmitted by an angle smaller than 2.5° is defined as clarity. The difference between transparency and clarity is subtle and depends on the instrument configuration adopted. While transparency only considers the light transmitted in the same direction of the incident light (also known as "direct" or "regular" transmittance), clarity accounts for a "narrow angle scattering" of the transmitted light, the quantification of which must be assessed with a spectrophotometer coupled with an integrative sphere. However, clarity also is often associated with the see-through attribute of packaging materials.

Gloss is defined as the amount of the incident radiant flux reflected by the material surface at a specific angle, which is the same as the angle of incidence. In practice, gloss provides an indication of the shiny appearance of a material surface. As a rule, the higher the shiny attribute, the more appreciated the final display of the packaging by consumers. The instrumental apparatus used to quantify the gloss is the glossmeter, which operates at angles of 20, 45, and 60°. The gloss attribute of materials surface can be improved by some finishing operations, such as metallization, coating, and lacquering.

9.5 CONCLUDING REMARKS ON PACKAGING MATERIALS FOR FRESH FRUITS AND VEGETABLES

The packaging materials intended for fresh fruits and vegetables must fulfill specific requirements linked to both the physiology and the shelf life of the packaged food. Fresh fruits and vegetables are "living" organisms, that is, they breathe by consuming oxygen and producing carbon dioxide, water, and energy (heat). The respiration rate can be different depending on both the type of fruit/vegetable and on the manipulation before packaging, namely washing, peeling, cutting, shredding, etc. However, the respiration process will always lead to the accumulation of CO_2 and consumption of O_2 inside the package, which eventually yields an anaerobic atmosphere, with subsequent deterioration and senescence of the fruit/vegetable and rapid spoilage. In addition, condensed water inside the package will promote microbial spoilage, enzymatic browning, and tissue rotting. Therefore, one of the goals to extend the shelf life of fresh fruits and vegetables (especially fresh-cut/minimally processed products) is the control of the internal atmosphere, e.g., by acting on the respiration rate of the product. Storing the products at a temperature between 0 and 6°C is the first step to slow down the respiration rate. However, altering the atmosphere inside the package (modified atmosphere packaging, MAP) has been suggested as another valid strategy, which foresees the use of CO_2, O_2, and sometimes N_2 as the most commonly used gases in MAP. High and low concentrations of CO_2 and O_2, respectively, have proven beneficial for most fruits and vegetables to control the respiratory metabolism. The use of MAP has to be related to the permeability properties of packaging materials. Barrier materials are not suitable because a high CO_2-rich atmosphere will be quickly established. Theoretically, a packaging solution that allows for keeping the selected CO_2/O_2 concentration stable over time would be the best choice. Assuming a respiratory quotient (RQ) equal to 1.0, the above condition can be achieved by selecting a packaging material with a permselectivity (i.e., β, which is the ratio of CO_2/O_2 permeation coefficients) equal to 1.0. However, β value for most plastic packaging materials ranges between 4 and 8. While different approaches have been proposed to achieve this target, such as coating deposition and micro-perforation, none of them seems to have succeeded satisfactorily, as demonstrated by the limited commercial applications.

At the same time, a package of fresh fruits and vegetables should meet other important criteria as far as, for example, its mechanical performance. In particular, resistance to pinholes and machinability are of utmost importance, especially when bags and trays have to be designed. Therefore, excellent puncture strength should be measured in order to ensure packaging integrity during both transportation and storage. At the same time, low COF of the packaging material running on the metallic parts of the manufacturing equipment would ensure high-speed production lines, that is, high throughput. As important are the optical properties of the packaging materials intended for fresh fruits and vegetables. Because of the great impact on the consumers' choice, an adequate display of the packaged product can be firstly achieved by properly selecting plastics with low haze values and high transparency. For this reason, one of the most suitable materials is oriented polypropylene, which finds wide use for freshly cut and minimally processed vegetables such as salads. However, initial excellent optical properties are not a sufficient condition to keep the display of the product unchanged during the shelf life of the product. One of the problems encountered during the storage of fresh fruits and vegetables concerns the fog formation on the inner side of the package. Fog formation is a consequence of environmental changes in temperature and humidity: a decrease of film surface temperature (due to both the heat released during respiration and to the low temperature of the retail refrigerator aisles) below the dew point causes the condensation of water vapor present inside the package. The thermodynamic difference between solid and liquid phase (the free surface energy of a polyolefin is ~28–30 mJ m^{-2}, whereas the surface tension of water is 72 mJ m^{-2}) is the reason for the water droplet standing on the plastic surface, that is, water tries to minimize the contact interface with the solid. The final effect is the appearance of a "foggy" layer that modifies the optical properties of the material, hiding the contents of the package by scattering of the incident light in all directions, due to the newly appeared droplets. In particular, both shape and size of the droplet have great influence on the see-through property of the material. Most of the antifog packaging materials available on the market include additives (e.g., nonionic surfactants such as sorbitan esters, polyoxyethylene esters, glycerol esters, and polyglycerol esters) that migrate from the bulk to the surface of the plastic films with the goal of increasing the hydrophilic features of the hydrophobic surfaces of polyolefins such as PP and PE. However, this approach involves two main technical disadvantages: (1) the migration kinetics of the additives is not easily quantified, which means that the antifog attribute may appear beyond the in-service life of the packaging material; (2) the antifog additives can be washed away from the film surface by the condensed vapor, which otherwise means that the claimed antifog properties tend to decrease over time. To overcome these drawbacks, recent developments have suggested the use of thin antifog coatings made of biopolymers (e.g., polysaccharides) able to decrease the free surface energy of the plastic polymer, thus improving the water wettability and, eventually, avoiding the occurrence of small water droplets. This means the water on the internal side of the package will be still present; however, in another form (a thin continuous layer instead of discrete particles). This different form will indeed prevent the increase in haze due to moisture condensation.

KEYWORDS

- **packaging**
- **consumer**
- **plastic films**
- **tensile strength**
- **specimen**

REFERENCES

Al-Ati, T.; Hotchkiss J. H. The Role of Packaging Film Permselectivity in Modified Atmosphere Packaging. *J. Agric. Food Chem.* **2003,** *51,* 4133–4138.

Brody, A. A.; Zhuang, H.; Han, J. H. *Modified Atmosphere Packaging for Fresh-Cut Fruits and Vegetables;* Wiley-Blackwell: Hoboken, NJ, 2011.

Fonseca, S. C.; Oliveira, F. A. R.; Brecht, J. K. Modelling Respiration Rate of Fresh Fruits and Vegetables for Modified Atmosphere Packages: A Review. *J. Food Eng.* **2002,** *52,* 99–119.

Introzzi, L.; Fuentes-Alventosa, J. M.; Cozzolino, C. A.; Trabattoni, S.; Tavazzi, S.; Bianchi, C. L.; Schiraldi, A.; Piergiovanni, L.; Farris, S. 'Wetting Enhancer' Pullulan Coating for Anti-Fog Packaging Applications. *ACS Appl. Mater. Interfaces.* **2012,** *4,* 3692–3700.

Larsen, H.; Liland, K. H. Determination of O_2 and CO_2 Transmission Rate of Whole Packages and Single Perforations in Micro-Perforated Packages for Fruit and Vegetables. *J. Food Eng.* **2013,** *119,* 271–276.

Lee, D. S.; Yam, K. L.; Piergiovanni, L. *Food Science Packaging and Technology;* CRC Press: Boca Raton, FL, 2008.

Mistriotis, A.; Briassoulis, D.; Giannoulis, A.; D'Aquino, S. Design of Biodegradable Bio-based Equilibrium Modified Atmosphere Packaging (EMAP) for Fresh Fruits and Vegetables by Using Micro-Perforated Poly-Lactic Acid (PLA) Films. *Postharvest Biol. Technol.* **2016,** *111,* 380–389.

Oliveira, M.; Abadias, M.; Usall, J.; Torres, R.; Teixidó, N.; Viñas, I. Application of Modified Atmosphere Packaging as a Safety Approach to Fresh-Cut Fruits and Vegetables – A Review. *Trends Food Sci. Technol.* **2015,** *46,* 13–26.

Piergiovanni, L.; Limbo, S. *Food Packaging;* Springer-Verlag: Milano, Italy, 2010.

Yam, K. *The Wiley Encyclopedia of Packaging Technology,* 3rd ed.; John Wiley & Sons Inc.: New York, NY, 2010.

CHAPTER 10

PREDICTIVE MODELING FOR PACKAGED FRUITS AND VEGETABLES

AMIT KUMAR, SHUBHRA SHEKHAR, and KAMLESH PRASAD*

Department of Food Engineering and Technology, Sant Longowal Institute of Engineering and Technology, Longowal 148106, Punjab, India

Corresponding author. E-mail: profkprasad@gmail.com

CONTENTS

ABSTRACT

The shelf-life of packaged fruits and vegetables is extensively affected by temperature and atmospheric condition. Food packaging has a vital role in maintaining the secure food supply chain. After processing, agricultural produce are packaged in different packaging materials based on the biochemical constituents of food commodities. Glass or polyethylene terephthalate (PET) bottles are basically utilized for aseptic filling of juices and beverages. Polyvinyl chloride (PVC) and PET are considered as an excellent packaging material for the frozen food items, whereas paper, paperboards, and laminated papers are usually used to package dry fruits. Testing the shelf-life of packaged fruits and vegetables stored at different temperatures and humidity is not only expensive but time consuming and requires skill. Therefore, predictive model can be developed for different conditions and these can be used for effective estimation of shelf-life of packaged fruits and vegetables.

10.1   INTRODUCTION

Fruits and vegetables are indispensable portion of our diet. The benefits of fresh-cut fruits or veg-fruits are their high nutrients, low amount of additives and ease of use. However, these products are highly susceptible to microbial invasion and quality deterioration (Caleb et al., 2012). The shelf-life of packaged fruits and vegetables is extensively affected by temperature and atmospheric condition. Food packaging has vital role in maintaining the secure food supply chain. After processing, agricultural produce are packaged in different packaging materials based on the biochemical constituents of food commodities. Glass or polyethylene terephthalate (PET) bottles are basically utilized for aseptic filling of juices and beverages. Polyvinyl chloride (PVC) and PET are considered as an excellent packaging material for the frozen food items, whereas paper, paperboards, and laminated papers are usually used to package dry fruits. Testing the shelf-life of packaged fruits and vegetables stored at different temperatures and humidity is not only expensive but time consuming and requires skill. Therefore, predictive model can be developed for different conditions and these can be used for effective estimation of shelf-life of packaged fruits and vegetables (Siripatrawan & Jantawat, 2008). Continuous studies and research work are progressing to explore and to develop a rapid, accurate, precise, and cost-effective shelf-life simulation model that help in the prediction of quality and shelf-life of packaged fruits and vegetables.

10.2 FRUITS AND VEGETABLES

Fruits are edible and delectable portion of plants that are derived from one or more ovaries after fertilization and some case it may develop from the accessory tissues. Fruits which are evolved without fertilization of ovary are known as parthenocarpic fruit. Seed is the characteristic feature of gymnosperm and angiosperm plants that are generally enclosed in fruits. Fruits are classified into drupe (mango and cherry), pome or accessory fruit (apple and pear), berry or soft fruits (strawberry and blueberries), synconium (fig), nut or dried seeds (walnut and cashew nut), hesperidium (oranges), sorosis (mulberry), and coenocarp (jack fruit). Similarly, on the basis of edible parts, vegetables are classified as root (carrot, turnip, and radish), stem (asparagus, taro, and kohlrabi), leaves (lettuce, cabbage, and spinach) flowers (broccoli and cauliflower), and fruits (sweet corn, lima bean, cucumber, and eggplant). Based on its use, vegetables are classified as salad crops (lettuce and chicory), bulb crops (onion and garlic), pulses (peas and beans), potherbs (mustard and spinach), and cucurbits (pumpkin and cucumber). Further, on the basis of shelf-life of fruits, veg-fruits, or vegetables may be classified as:

- Perishable,
- Semi-perishable,
- Non-perishable.

Perishable commodities deteriorate rapidly and become unreliable to consume, if it is stored in inappropriate condition. Shelf-life of perishable foods depends upon processing, packaging, and environmental conditions at every stage of storage and distribution. Refrigeration may reduce the deterioration rate of perishable foods, as microbial growth and chemical changes could be minimized at low temperature. Fruits like ripe banana, mango, strawberries, figs, and watermelons; some vegetables like red tomatoes, broccoli, and beans are good example of perishable food category.

Semi-perishable foods have low water contents and generally do not require ideal refrigeration, but have a limited shelf-life (green pineapples and unripe banana). Some fruits and vegetables (potatoes, onions, pumpkins, carrot, etc.), flour, grain products, dried fruits are excellent example of semi-perishable food. Non-perishable foods are those which have prolonged shelf-life as they do not get easily deteriorated by bacteria and other microorganism and also do not need refrigeration. Food commodities like dehydrated fruits and vegetables, canned soups and vegetables, breakfast cereals, pasta, tea, and nuts are good example of non-perishable food. These products can also be used during emergency and catastrophe (Table 10.1).

TABLE 10.1 Moisture Content, a_w, and pH of Some Fruits and Vegetables at 20°C.

Food	a_w	pH	Moisture content
Banana	0.979 ± 0.001	4.84 ± 0.05	79.9 ± 0.6
Apple	0.988 ± 0.002	4.58 ± 0.02	88.5 ± 0.3
Orange	0.995 ± 0.002	3.96 ± 0.19	90.1 ± 0.4
Grapes	0.977 ± 0.001	3.94 ± 0.22	85.7 ± 1.2
Water melon	0.991	5.60	90.9
Lemon	0.989 ± 0.001	2.72 ± 0.07	94.2 ± 0.3
Tomato	0.996 ± 0.001	4.27 ± 0.10	95.9 ± 0.7
Radish	0.997	6.40	87.4
Carrot	0.992 ± 0.001	5.85	91.5
Fig	0.971	5.34	77.5
Potato	0.997 ± 0.001	6.12 ± 0.07	79.9 ± 1.1
Pineapple marmalade	0.912	3.37	54.6

Source: Salguero et al. (1993).

10.3   MATURATION AND RIPENING

Fruits can be considered as a fleshy portion of plant that may be sweet or sour in taste and can be consumed in raw state while vegetables are generally consumed in cooked form. Some vegetables like carrots, tomatoes, and radish can also be considered as fruits since it may be consumed without cooking. Unripe fruits and vegetables have generally high phenolic contents which impart astringency, bitterness, and act as repellent for damaging pest and insects. Maturity, chemical composition, handling, storage, and packaging have a great impact on the quality of fruits and vegetables. Starch content of fruits significantly decreases during ripening, as a consequence of enzymatic hydrolysis of starch into sugar. Total soluble solid (TSS) content increases during ripening of banana, plantain, and many other fruits due to increase in sugar content. The extent of soluble solids increase depends on the variety of fruits. Quality parameters of most fruits were significantly affected by its varieties, growing environment and storage condition (Belayneh et al., 2014). Due to enhancement in pectin methylesterase (PME) activity, breakdown of pectin and softening of fruits take place during ripening. Respiration rate (RCO_2) of agricultural produce is also increased during the ripening of fruits.

10.4 PACKAGING OF FRUITS AND VEGETABLES

Primary purpose of packaging is to maintain the quality and freshness of fruits and vegetables until it reaches the consumers. Role of suitable packaging material in maintaining the quality of food product and its influence on microbial growth cannot be ignored. Appropriate packaging system for fresh fruits and vegetables plays a critical role in prolonging the shelf-life of horticultural produce. In any horticultural sector of the world, precise packaging system is required, as fresh produce has limited shelf-life even stored at ambient conditions. Apart from protection from the outside environment and ruinous microbial contaminants, packaging also communicates the nutritional and ingredient information about food products. Most underdeveloped and developing countries utilizes low-cost trays, baskets, and sacks as packaging materials. However, while carrying these fresh produce to a long distance, one cannot be depended on these traditional packaging materials. For commercial and safe transportation, plastic crates are generally used, which provide higher level protection of agricultural produce and is also easier in hygienic handling.

Before processing or packaging, the agricultural products are checked for climacteric or non-climacteric nature. Climacteric fruits like banana, plantain, mango, apple, papaya, and many others have high RCO_2 and also have high carbon dioxide and ethylene emission rate. Excess release of plant hormone (ethylene) is an excellent initiator of ripening, where remarkable changes in flavor, color, aroma, firmness, and other physiological modification takes place. Non-climacteric fruits usually do not require ethylene for ripening. Fruits like orange, kinnow, and watermelon are some examples of non-climacteric fruits. Knowledge of ethylene signal during ripening and storage can help to regulate the fruit ripening and senescence. Charcoal with palladium chloride would act as an excellent absorber of endogenously produced ethylene gas and hence reduce the softening rate of banana and kiwifruits. It also minimizes the chlorophyll depletion in spinach leaves (Abe & Watada, 1991). During storage of harvested fruits, 1-methylcyclopropene is employed to retain its quality as it opposes ethylene activity and plays a significant role in ripening-related processes of both climacteric and non-climacteric fruits. Li et al. (2016) studied the effect of 1-methylcyclopropene on various non-climacteric fruits and observed that, it can inhibit the rachis browning in grapes, pericarp browning in litchi, and in pomegranate it facilitates to inhibit scald development. It also inhibits the de-greening and any color alteration in strawberry, olive, and various citrus fruits. The consequences of 1-methylcyclopropene on valuable agricultural produce have also been reviewed by Bower et al. (2003), Aguayo et al. (2006), Sharma et al. (2010), and Li et al. (2015). A successful model can be developed for predicting the shelf-life and quality of food by observing the changes in perishable food as the function of temperature and relative humidity (Kwolek & Bookwalter, 1971).

Optimal oxygen and carbon dioxide levels during storage and packaging enhance the shelf-life of fruits and vegetables. Blending of ethylene vinyl acetate with oriented polypropylene and low-density polyethylene can generate an efficient packaging material which will raise the shelf-life of veg-fruits like carrots and tomatoes, as it has good gas permeability (Ahvenainen, 1996). Different packaging materials are used for different agricultural produce with aim to sustain the quality parameter of fruits and vegetables during transportation. These packaging materials are aluminum foil, glass, polyethylene, polypropylene, polyurethane, polyvinylidene chloride, and other (Mangaraj et al., 2009). These materials are durable, good barrier property, resistant to water as well as chemicals and are thermally stable but extensive use of these polymeric films, raises a severe environmental concerns (Mistriotis et al., 2016). Biodegradable films are an excellent alternative of these polymeric materials and can be used for the packaging of agricultural produce, as it has good mechanical integrity and barrier property. Mostafa et al. (2015) developed eco-friendly biodegradable plastic from agricultural wastes, which not only helps in agricultural residues management but also in resource conservation and sustains the eminence of the environment. Great advantage of the biopolymer is its resistance toward acid and salt, hence can be utilized in both the medicine and food industries.

10.5 QUALITY AND SHELF-LIFE OF PACKAGED FRUITS AND VEGETABLES

Quality parameters such as weight loss, pH, firmness, color are examined to ensure the associated safety of food. Changes, which specify the deterioration and spoilage of food materials, include textural changes, rancidity, loss of flavor, color change, increase in microbial count, loss of pigments, and nutrition. The characteristics of fruits and vegetables like aroma, peel color, texture are the deciding factors of its acceptance or rejection. Some important parameters which have significant impact on the quality and shelf-life of fruits and vegetables are discussed in the following sections.

10.5.1 WATER ACTIVITY (a_w)

All fruits and vegetables have an optimal a_w range and deviation from these range may results in its quality loss and rejection. Water activity is the ratio between partial water vapor pressure of a product to the partial water vapor pressure of pure water at constant temperature and it ranged from 0 to 1.00. For Hazard analysis and critical control points (HACCP) program, water activity of fresh fruits and vegetables are taken as critical control points considering its role in chemical and enzymatic reactions. Growth of microorganism can be effectively inhibited by reducing the water activity of products. Moisture loss in fruits and vegetables is mainly due to difference in water activity between the surface of agricultural produce and the surrounding. Different salts like $MgCl_2$, KNO_3, $KC_2H_3O_2$, NaCl, LiCl, $(NH_4)_2SO_4$, and K_2CO_3 are used to maintain constant water activities in food products to avoid any unexpected spoilage and wastage.

10.5.2 RESPIRATION AND TRANSPIRATION RATE

In any fresh fruits and vegetables, respiration process involves conversion of sugar and oxygen into carbon dioxide and water vapor during its metabolic activity (Mannapperuma et al., 1991). The ratio of CO_2 evolved to O_2 consumed is known as respiratory quotient, which helps in *designing* the package system. If RCO_2 of fresh produce is higher than its shelf-life is shorter. Quality as well as shelf-life of fresh fruits and vegetables decline as a consequence of high RCO_2. At 0–10°C storage temperature range, it was observed that Q_{10} of RCO_2 varies between 2.0 and 8.6 in several fresh-cut fruits and vegetables (Watada et al., 1996). Using respire meter the effects of intrinsic factor and extrinsic factors on the RCO_2 of freshly cut pineapple chunks were determined at 4°C (Finnegan et al., 2013). Mathematical model to predict RCO_2 was developed using the sample weights, level of CO_2 at the inlet and outlet and flow rates. RCO_2 (mL CO_2/kg h) can be calculated as:

$$R_{co2} = \frac{V_f \times 60}{W \times 100} \cdot (CO_2^{out} - CO_2^{in}) \qquad (10.1)$$

where V_f (mL/min) is the gas flow rate, W (kg) is the weight of pineapple sample and CO_2^{out} and CO_2^{in} is the percent of CO_2 concentration at the outlet and inlet of the respire meter. Equations (10.2) and (10.3) are used for the estimation of target barrier properties required for packaging of fresh-cut pineapple fruit.

$$R_{O_2} \cdot W = OTR \cdot A \left(O_2^{out} - O_2^{in} \right) \qquad (10.2)$$

$$R_{CO_2} \cdot W = CTR \cdot A \left(CO_2^{in} - CO_2^{out} \right) \qquad (10.3)$$

where oxygen transmission rate (OTR) (mL/m^2 day atm) is the required transmission rate for O_2 and carbon dioxide transmission, CTR (mL/m^2 day atm) is required transmission rate for CO_2. Considering exponential decay, a mathematical model to predict the RCO$_2$ of packaged fresh-cut pineapple fruit is presented in eq (10.4).

$$R_{CO_2} = R_{CO_2}^{eq} + \left(R_{CO_2}^{i} - R_{CO_2}^{eq} \right) e^{-(\alpha t)} \qquad (10.4)$$

where $R_{CO_2}^{eq}$ and $R_{CO_2}^{i}$ are equilibrium and initial RCO$_2$ (mL/kg h), t is the storage time (h) and α is the constant coefficient varied from 0.11 to 0.27. Equation (10.5) was developed by combining eqs (10.1) and (10.4), which is used to fit the experimental data acquired during storage of fresh-cut pineapples. It was reported that RCO$_2$ gradually decline and average initial and equilibrium RCO$_2$ of pineapple chunks were found to be 8.52 and 2.64 mL/kg h.

$$CO_2^{out} = \frac{\left[R_{CO_2}^{eq} + \left(R_{CO_2}^{i} - R_{CO_2}^{eq} \right) e^{-\alpha t} \right] \cdot W \times 100}{V_f \times 60} + CO_2^{in} \qquad (10.5)$$

By predicting the O_2 and CO_2 concentrations inside bag headspace, efficient modified atmosphere packaging (MAP) technique can be developed to maintain quality of fresh fruits and vegetables (Fig. 10.1). MAP also extends the shelf-life of products by reducing the metabolic activities, delaying enzymatic degradation as well as reducing microbial growth (Kader, 1986).

Application of MAP on sugar snap peas was investigated by Elwan et al. (2015). They utilized highly perforated polypropylene packages, non-perforated polypropylene packages, and micro-perforated polypropylene packages to examine the storability, shelf-life, and quality of sugar snap peas. Experimental results showed that the increase in CO_2 level and decrease in O_2 level is slow in the case of micro-perforated polypropylene packages, while non-perforated polypropylene packages show high concentration of CO_2 and significant level of off odors. It is also reported that of micro-perforated polypropylene packages with 12 holes maintain the highest values of total sugar content, chlorophyll, vitamin C and have a higher score in visual quality, crispness, firmness, and taste.

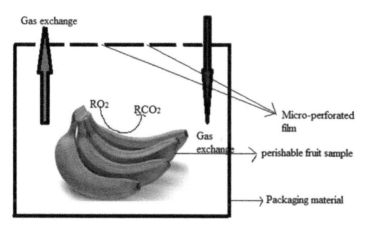

FIGURE 10.1 Modified atmosphere packaging containing fruit sample.

It is very essential to recognize the impact of time and temperature on the RCO_2 of fresh-cut fruits and vegetables to develop a MAP. RCO_2 and respiratory quotient of various fresh-cut fruits and vegetables stored in controlled atmosphere was demonstrated (Watada et al., 1996). Mathematical model was developed by Waghmare et al. (2013) to predict the RCO_2 of fresh-cut produce. Oxygen and carbon dioxide concentration (%) can be calculated using eqs (10.5) and (10.6):

$$y_{O_2} = y_{O_2}^i - \frac{R_{O_2}.W}{V_f}\cdot(t-t_i)\times 100 \tag{10.5}$$

$$y_{CO_2} = y_{CO_2}^i + \frac{R_{CO_2}.W}{V_f}\cdot(t-t_i)\times 100 \tag{10.6}$$

where $y_{O_2}^i$ and $y_{CO_2}^i$ are O_2 and CO_2 concentration (%) at the initial time t_i (h) and R_{O_2} and R_{CO_2} are RCO_2 (mg kg^{-1} h^{-1}), W is the total weight of the product (kg) and V_f is the free volume inside the glass jar (mL). A simple Arrhenius-type equation was used in model fitting as presented in eqs (10.7) and (10.8) which illustrates the temperature as a function of RCO_2.

$$R_{O_2} = R_{O_2,ref} \times e^{\left[[-E_a,_{O_2}/R]\left(\frac{1}{T}-\frac{1}{T_{ref}}\right)\right]} \tag{10.7}$$

$$R_{CO_2} = R_{CO_2,ref} \times e^{\left[[-E_a,_{CO_2}/R]\left(\frac{1}{T}-\frac{1}{T_{ref}}\right)\right]} \tag{10.8}$$

where $R_{O_2,ref}$ and $R_{CO_2,ref}$ are initial RCO_2 (mg kg^{-1} h^{-1}) at reference temperature (T_{ref}, K), R is the universal gas constant and E_a,O_2 and E_a,CO_2 are activation energy (kJ mol^{-1}). Substituting R_{O_2} and R_{CO_2} in eqs (10.5) and (10.6) secondary model was reported to estimate oxygen and carbon dioxide concentration (%), which is presented in eqs (10.9) and (10.10).

$$y_{O_2} = y_{O_2}^i - \left[R_{O_2,ref} \times e^{\left[[-E_a,_{O_2}/R]\left(\frac{1}{T}-\frac{1}{T_{ref}}\right)\right]}\right]\frac{W}{V_f}(t-t_i) \times 100 \tag{10.9}$$

$$y_{CO_2} = y_{CO_2}^i - \left[R_{CO_2,ref} \times e^{\left[[-E_a,_{CO_2}/R]\left(\frac{1}{T}-\frac{1}{T_{ref}}\right)\right]}\right]\frac{W}{V_f}(t-t_i) \times 100 \tag{10.10}$$

Bhande et al. (2008) and Torrieri et al. (2010) studied the RCO_2 of banana fruit and broccoli (*Brassica rapa* var. Sylvestris) to design suitable packaging system. Appearance, texture, weight, and other quality parameters of fresh fruits and vegetables are significantly affected by transpiration rate. When fruits and vegetables are developing on mother plant then water and other important micronutrient are continuously supplied from the soil. While in harvested agricultural produce, if water loss is not controlled then fresh agricultural produce may appear wilted and lose its freshness. A systematic experimental setup was developed by Mahajan et al. (2008) to predict the transpiration rate and also to monitor the mass loss of defect-free mushrooms at different relative humidity and surrounding temperatures. Mathematical model to compute the transpiration rate is shown in the following eq (eq. 10.11).

$$TR = -\frac{1}{A_s}\frac{dM}{dt} \tag{10.11}$$

where TR is the transpiration rate (dM/dt) is the change in mass of mushroom per unit time and surface area (A_s) can be calculated eq (10.12) as:

$$A_s = d \times (M)^b \tag{10.12}$$

where d (0.738) and b (0.029) are the constants, obtained from the experimental data. After integrating with the limits of initial mass of mushroom (M_i) to mass (M) at time t, the eq (10.13) to estimate water loss is:

$$M = \left[M_i^{(1-b)} + d \times (b-1) \times \text{TR} \times t \right]^{\frac{1}{1-b}} \tag{10.13}$$

It was observed that mass loss in mushroom was higher at 76% relative humidity and 16°C as compared to 96% relative humidity and 4°C. According to Newton's law, change in volume of water (cm³) is expressed as:

$$\frac{dV}{dt} = -K_i \times A_s \times (a_{wi} - a_w) \tag{10.14}$$

In terms of change in mass of mushroom with respect to time, eq (10.14) can be expressed as:

$$\frac{dM}{dt} = -\rho \times K_i \times A_s \times (a_{wi} - a_w) \tag{10.15}$$

where ρ is the density of water, K_i (cm h⁻¹) is the mass transfer coefficient, and a_w and a_{wi} are the water activity of container and mushroom. Rearranging the eq (10.15) and combining with eq (10.11) new eq (10.16) was obtained which describes the transpiration rate in term of water activity:

$$\frac{dM}{dt \times A_s} \text{ or TR} = \rho \times K_i \times (a_{wi} - a_w) \tag{10.16}$$

Once again integrating eq (10.16) with the limits M_i and M of mushroom and storage time 0 to t, another equation to predict the mass of mushroom was obtained as shown in eq (10.17):

$$M = \left[M_i^{1-b} + (b-1) \times \rho \times K_i \times d \times (a_{wi} - a_w) \times t \right]^{\frac{1}{1-b}} \tag{10.17}$$

Equation (10.17) was modified to integrate the overall effect of temperature on mass transfer coefficient (K_i) as shown in eq (10.18):

$$M = \left[M_i^{1-b} + (b-1) \times \rho \times K_i \times d \times (a_{wi} - a_w) \times (1 - e^{-aT}) \times t \right]^{\frac{1}{1-b}} \tag{10.18}$$

Equation (10.16) was thus integrated with the temperature term to predict transpiration rate at all combinations of temperature and RH, shown in eq (10.19):

$$TR = \rho \times K_i \times \left(a_{wi} - a_w\right) \times \left(1 - e^{-at}\right) \tag{10.19}$$

Gallagher et al. (2013) investigated the impact of temperature and RH on strawberries, which is extremely valuable and perishable in nature. They also developed a predictive model for quantifying transpiration rate in packaged strawberries. It was observed that the transpiration rate of strawberries was in the range of 0.24–1.16 g/kg h and by increasing relative humidity of the container from 76 to 96% there was considerable decrease in TR at 5°C. Whereas, when temperature is reduced from 15 to 5° C at 96% RH then transpiration rate is decreased by 47%. Other valuable predictive model to quantify the transpiration rate of beneficial agricultural produce was developed by Sastry and Buffington (1982), Kang and Lee (1998), and Song et al. (2002).

10.5.3 TEMPERATURE

It is very important to maintain proper temperature conditions as any fluctuation in temperature will affect the quality and may result in food deterioration. Time–temperature indicators (TTI), which act as a predictive temperature sensor, are used for packaged food commodities. TTI are attached to the packaged food and it signifies the temperature abuse or subjection above or below a reference temperature in the form of visual color alteration which is generally irreversible in nature (Taoukis & Labuza, 2003).

Oliveira et al. (2012) evaluated the effect of temperature and number of film perforations on quality of mushrooms and developed shelf-life kinetic model for a MAP for sliced mushrooms, one of the highly perishable food items due to high moisture content and transpiration rate. MAP as well as storage at low temperature can effectively slow down the quality deterioration of fresh-cut mushrooms and applied Weibull model to reflect the time-dependent degradation of quality eq (10.20).

$$\frac{X - X_e}{X_0 - X_e} = exp\left[-\left[\frac{t}{\alpha}\right]\right]^{\beta} \tag{10.20}$$

where X_0 and X_e are initial and equilibrium quality parameter, X is the quality parameter at time t, A is the scale parameter and β is the shape parameter. The influence of temperature on α or β parameter was described using Arrhenius equation (10.21).

$$k = k_{ref} \exp\left[-\frac{E_a}{R_c}\left[\frac{1}{T} - \frac{1}{T_{ref}}\right]\right] \qquad (10.21)$$

where k is α or β parameter, k_{ref} is the frequency factor for α or β parameter at the reference temperature, E_a (kJ mol^{-1}) is the activation energy, R_c is the gas constant, T (K) is the temperature and T_{ref} (283.15 K) is the reference temperature. Corbo et al. (2004) kinetically modeled the shelf-life of minimally processed cactus pear fruit to examine the effects of storage temperature and microbial indices on fruit quality to produce ready-to-eat fruit. Microbial colony formed during storage was modeled using Gompertz equation modified by Zwietering et al. (1990) presented in eq (10.22):

$$y = k + A \times \exp\left\{-\exp\left[\left(\frac{\mu_{max} \times e}{A}\right) \times (\lambda - t) + 1\right]\right\} \qquad (10.22)$$

where y is cell load (log CFU g^{-1}), k the initial level of the dependent variable, A is the maximum cell load at stationary phase, μ_{max} (Δlog CFU g^{-1} per day) is the maximal growth rate, t is the time and λ the lag time. The correlation between acceptability times (At) for bacteria and temperature was represented by linear equation (10.23).

$$At = a + bT \qquad (10.23)$$

where a signifies the theoretical shelf-life at 0°C, T (°C) is the temperature and b (the slope) signifies the increase of degradative reaction rates with increase in 1°C temperature (Singh, 1994).

Tchango et al. (1998) studied the combined effects of temperature and pH to predicting the contamination level of guava nectar by *Candida pelliculosa* using a turbidity method. Using an exponential regression, relation between optical density and population density was also established and it was found that maximal growth occurs at 37° C and pH 6.25. Considering the temperature and relative humidity of moisture sensitive foods, Cardoso and Labuza (1983) and Pieglovanni et al. (1995) had developed mathematical model for estimating the shelf-life of valuable food products.

10.5.4 MICROBIAL SPOILAGE

The growth of pathogenic microorganism on fresh fruits and vegetables results in spoilage and wastage, which decreases its market value. Several microorganisms activate undesirable reactions that deteriorate quality and cause food-borne infections. Normally, pathogenic microorganisms present on the peel will not survive for longer time but if it enters inside the juicy part of fruits and vegetable it will grow and multiply rapidly due to presence of abundant nutrient at ambient conditions. It is observed that carbon dioxide is produced in the packaged food in case of microbial contamination. Different indicators or devices are available, which indicate the increase of CO_2 during microbial growth. Numerous predictive models signifying the concentration of carbon dioxide during bacterial growth exist, which help to design a suitable packaging system (Powell et al., 2015). Predictive mathematical model was implemented using MATLAB by Silva et al. (2014) to study the *Byssochlamys fulva* spores growth on papaya pulp. They reported that increase in microorganism in pulp was best described using a modified Gompertz model (eq 10.24).

$$y = y_0 + y_{max} \exp\left[-\exp\left\{\frac{\mu \exp(1)}{y_{max}}(\lambda - t) + 1\right\}\right] \tag{10.24}$$

Differential form eq (10.25) of the modified Gompertz model proposed by Van Impe et al. (1992) was employed as microbes grow and multiply under dynamic temperature conditions.

$$\frac{dy}{dt} = \frac{\mu \exp(1)}{y_{max}} y \ln\left(\frac{y_{max}}{y}\right) \tag{10.25}$$

where μ is the radial growth rate (mm day^{-1}) and λ is the lag time (h), estimated using eq (10.26).

$$\sqrt{\mu} = b(T - T_{min})\left[1 - \exp\left\{c(T - T_{max})\right\}\right] \text{ and } \ln(\lambda) = \frac{p}{T - q} \tag{10.26}$$

where T signifies temperature (K), T_{min} and T_{max} signify minimum and maximum temperature for growth, p and q are parameters of hyperbolic model, b and c are parameters of the extended square-root model, y, y_0 and y_{max} are radius of colony at time t (cm), initial radius of colony (cm) and maximum radius of colony (cm). Fernandez et al. (2011) designed automatic classifiers to predict growth/no-growth of food-borne microorganisms like *Staphylococcus aureus*, *Listeria monocytogenes*, and *Shigella flexneri* based on artificial neural network (ANN). It is a valuable alternative for mathematical modeling. The accuracy and significant classification rate was observed from area under the receiver operating characteristic (ROC) curve and root mean squared error (RMSE). Multi-classification model based on Smote Memetic Radial Basis Function (SMRBF) developed by Navarro et al. (2010) using ANN to determine the microbial growth/no growth interface. The Baranyi's DMFit Excel add-in software was employed by Walter et al. (2009) to examine and predict the response of *L. monocytogenes* in the fresh coconut water kept at different temperature.

$$y(t) = y_0 + \mu_{max} A(t) - \frac{1}{m}\left[1 + \frac{e^{m\mu_{max} A(t)} - 1}{e^{m(y_{max} - y_0)}}\right] \qquad (10.27)$$

where $A(t)$ can be calculated as

$$A(t) = t + \frac{1}{\mu_{max}} \ln\left(\frac{e^{(-\mu_{max} t)} + q_0}{1 + q_0}\right) \qquad (10.28)$$

Lag time (h) or λ is calculated as

$$\lambda = \frac{\ln\left(1 + \frac{1}{q_0}\right)}{\mu_{max}} \qquad (10.29)$$

where $y(t)$ is the $\log_{10}$ of the cell concentration at time t(h), y_0 and y_{max} (k) are the initial and maximum cell concentration, μ_{max} is the maximal specific growth rate, q_0 compute the physiological state of the cell at $t = t_0$, m parameter is associated with the curvature after exponential phase. Experimental work supports the higher k value at elevated storage temperature.

10.5.5 QUALITY ASSESSMENT OF FRUITS AND VEGETABLES

Peel color and texture of any fruit is an important quality parameter which decides its acceptance or rejection. Consumer opinion about the food quality depends on visual appearance (Nisha et al., 2011). Shelf-life as well as physio-chemical characteristics or changes in fruits and vegetables can be correlated well from peel color or visual appearance (Prasad et al., 2010; Prasad, 2015a, 2015b). When achromatic light is reflected from any object it is perceived by our eye and the valuable information of electromagnetic light is converted as sensation of color by our brain (Dobrzanski & Rybczynski, 2002). Food color depends on the chemical, biochemical and microbial changes during ripening and postharvest handling and storage. Quality attributes like flavor and pigments contents can be predicted through the use of electric nose and the color measurement (Pathare et al., 2013). Color is an important part of the memory depiction and bears the descriptive property of an object (Hanna & Remington, 1996). To describe the color of an object there is various color models available with some advantage and limitation. Some of these are RGB, *CMYK,* HSV, L*a*b* (CIELAB), and L*u*v*. RGB color space (red (R), green (G), and blue (B) chromaticity's are most commonly used. RGB color space is an additive color space (Susstrunk et al., 1999), which is convenient for the computer graphics. Color value of any object varies from individual to individual depending on its age, brightness, and intensity. In 1931, *Commission International de l'Eclairage* (CIE), specify a new color space XYZ that represent color on numerical form, but it is not uniform so in 1976 a new system is developed by CIE that is L*a*b* which is more uniform and better related to the human perception (Radzevicius et al., 2014). The mechanism of postharvest changes in green color due to chlorophyll degradation and appearance of yellow color due to presence of β-carotene has been elaborated (Ankita & Prasad, 2015). Our work to classify and predict the banana maturity based on color value in order to predict the correlation between the changes in physico-chemical characteristics during ripening. Good quality, defect free and mature green bananas were used. After removing the latex from the surface, cleaned with 0.5% magnesium sulfate solution. Fruits were rinsed in running tap water for 5 min and apparently surface dried. After treating with 500 ppm ethephon, it was stored at $20 \pm 1°C$ under packaged condition of relative humidity (85–90%) until over ripened.

The changes in peel value during ripening of banana in two days interval are shown in Figure 10.2. Computer vision system (CVS) was employed to inspect the associated changes on surface color. The image of fruit was acquired using image capturing chamber (Fig. 10.6). Color value of fruits and vegetables are generally evaluated using different instruments. Visual color inspection is carried out in a standard illumination condition with reference color chart to compare the color of sample. A standard peel color scale (Chiquita Brands Inc.) was employed as a reference color chart to inspect the color of banana (Pathare et al., 2013). Colorimeters and spectrophotometers are two major instruments used for the color measurement in food industries. In CVS after acquiring the image, any distortions or noise present in the image was removed during the pre-processing stage. The aim of pre-processing is to improve and highlight the region of interest of the image. After pre-processing step, feature either in the form of size, shape, or color is extracted from the image.

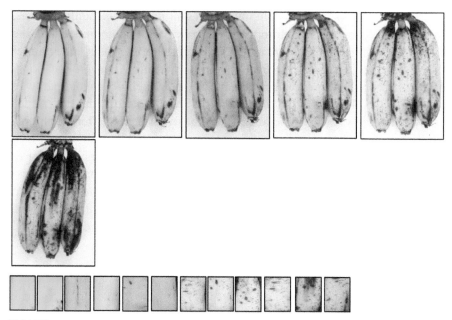

FIGURE 10.2 Image of ripening banana fingers with region of interest.

Each pixel of selected region was acquired from the image by color sensors, so as to obtain the RGB using CVS. While L*a*b* color space is an international standard for color measurement, is non-linear and color value can be easily interpreted (Hashim et al., 2012). RGB color values were transformed to L*a*b* color values using standard color conversion developed algorithm, shown in Figure 10.3. The color value obtained was employed as input for classification and prediction using ANN.

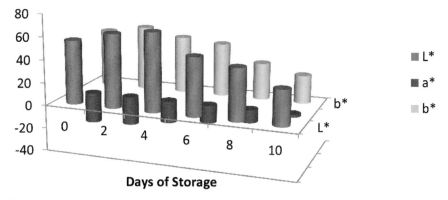

FIGURE 10.3 Time-dependent variations in color values.

Pigments or biochemical compounds present in fruits like anthocyanins, carotenoids, lycopene, or chlorophyll has to be measured to understand the associated changes in the color value of any fruits. Hortensteiner and Krautler (2011) reported that, during fruit ripening, chlorophyll is broken down through Pheophorbide A oxygenase (PAO) mediated biochemical pathway to remove the efficient phototoxic pigments from the senescing cell. Chlorophyll breakdown is an imperative signal of fruit ripening. During chlorophyll degradation, large variety of fluorescent chlorophyll catabolites (FCC) were formed and accumulated in the peels of fresh ripe bananas (Muller & Krautler, 2010). De Ell et al. (2003) reported that the chlorophyll fluorescence changes can possibly help the researchers to gain valuable information on early responses to postharvest stress in chloroplast containing fruits and vegetables. Decrease in total chlorophyll, chlorophyll "a" and chlorophyll "b," during ripening of banana is shown in Figure 10.4.

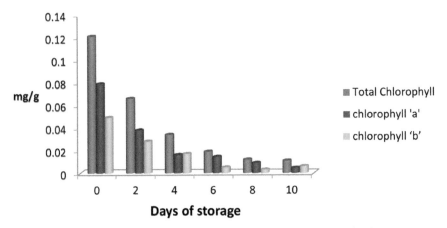

FIGURE 10.4   Time-dependent changes in chlorophyll content of ripening banana.

Predictive model based on color value and texture to quantify shelf-life help to reduce wastage of fruits and vegetables. CVS is a promising tool, which is used nowadays to inspect the texture and color value of the fruits, as it is precise, cost effective, remarkably fast and serves consistent performance. Texture of fruits and vegetables are attributable to cell wall material, water content, and type of tissue. Changes in pectin, cellulose, and hemicelluloses during storage of fresh produce causes softening of fruits and hence texture of fruits changes. Quality of valuable agricultural commodities can be judged from the surface texture, which can be computed from digital images. Application of texture analysis technique using computer-mediated vision has great importance in food industries nowadays. Overripening process or spotting in bananas was characterized by a rise in the fractal value. The image was recorded using developed CVS. The acquired image was segmented from the background and its surface intensity was derived. Surface intensity obtained from segmented image was quantified using the fractal theory. It was observed that surface intensity obtained from images of over ripen banana having "'senescent spotting" exhibits a fractal behavior as its Fourier power spectrum follows the power-law scaling and fractal dimensions increase with the ripening process. Firmness of banana fruit decreases during ripening (Fig. 10.5).

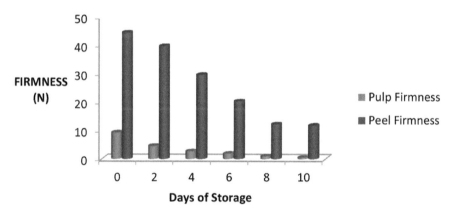

FIGURE 10.5 Time-dependent changes in firmness of ripening banana.

Four different types of texture are statistical texture, transform based texture, structural texture, and model-based texture. Statistical texture can be acquired by employing statistical approach from the higher-order of pixel grey values of target images. It is widely used in food industries due to its high accuracy and short evaluation time. Structural texture can be acquired using some structural element or structural primitives, created from grey values of pixels. Transform based texture can also be obtained by using statistical measurements of transform image and this method is apt for both micro-texture and macro-texture patterns. Images are transformed by convolution mask, Fourier transform and wavelet transform. Model-based texture can be obtained by calculating the coefficients of model developed by simulating the dependence of pixels and its neighboring pixels (Zheng, 2006). Quevedo et al. (2008) employed the fractal texture Fourier image to determine senescent spotting in banana (*Musa cavendish*). Fractal texture shows a change in pixel intensity which carries information about structure of target object.

10.5.6 COMPUTER-MEDIATED VISION FOR QUALITY ASSESSMENT

Nowaday, decrease in price of electronic commodity like computer, camera, and other aids, computer-mediated dimensional characterization and inspections are gaining popularity (Prasad et al., 2012; Singh & Prasad, 2013). It is a type of non-destructive approach, which is also suitable for online monitoring (Prasad, 2015a). Various scientists work on CVS to grade and sort the agricultural commodities. Experimental setup for CVS consists of various units for the acquisition and further interpretation.

10.5.6.1 IMAGE CAPTURING CAMBER (ICC)

A small wooden frame was used to construct an Image Capturing Camber where the image of sample was obtained (Fig. 10.6).

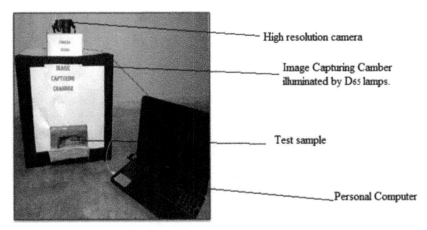

FIGURE 10.6   Setup for computer vision system.

10.5.6.2 ILLUMINATION SYSTEM

Illumination is an essential prerequisite of image acquisition, as the quality of image depends on the lighting condition. Image obtained in uniform illumination has low noise, which reduces the time and cost of the operation as well as increase the accuracy and efficiency of system. For the illumination, four standard lamps of color temperature 6500 K was arranged at an angle of 45° above the sample as the diffuse reflection responsible for the color occurs at 45° from the incident light. Four lamps were covered with light diffusers to assure uniform lighting inside the chamber.

10.5.6.3 DIGITAL CAMERA

A color digital camera, Canon Power Shot SX510 HS model with 13 megapixels of resolution (Fig. 10.6) was placed vertically above the sample at a distance of nearly 30 cm. Images of the bananas were taken with the following camera settings: image of sample was taken in manual mode with no flash, focal length 4 mm, exposure time 1/80, ISO speed- ISO 100, F-stop (aperture stop)-f/4 and stored in JPEG format. The camera was connected to the personal computer (PC) through cable as shown in Figure 10.6.

In computer-mediated vision, entire surface can be considered for the color measurement and valuable information can be obtained from the image of the objects (Sadegaonkar & Wagh, 2015). Postharvest losses of fruits and vegetables as well as demand of high quality and safe product, need fast, automatic, and reliable system, which will assist to meet the demand of consumers and suppliers. Manual sorting of fruits include visual inspection and handpicking which is time-consuming, labor intensive aspect and there is possibility of human error which can be reduced by using computer-mediated vision (Mahendran et al., 2012). Perishable fruits like mango, apple, and banana, if not adequately transported then its quality deteriorate with declined consumer acceptance (Yousef, 2011). Computer-based system for sorting and inspecting fruits and other agricultural products can help to overcome this problem. Mendoza and Aguilera (2004) classified the banana using color value. Jatropha fruit was classified by Syal et al. (2103) through its feature value extraction using fuzzy logics, Toylan and Kuscu (2014) designed a real-time machine vision system for classifying apple on multicolor space, while three ripeness phases: pre-climacteric, climacteric, and senescence, was taken as reference by Rivera et al. (2014) to classify mango using CVS. Set up for the computer vision required high-resolution camera, image capturing chamber, proper illumination, personal computer, and software for image analysis or feature extraction (Prasad, 2015a). Different illuminants are available like gas-filled tungsten lamp, Fluorescent lamps, D_{50}, D_{55}, or D_{65}. These illuminants operate at different color temperature.

Ohali et al. (2011) obtained the shape and the size of dates from the color image taken with a uniform color background and classified using ANN. Prabha and Kumar (2015) used the image processing technique to identify the maturity level of banana fruit using its color and size value. Banana from different stage was used for that purpose to classify it as under-mature, mature, and over-mature. The mean color and area were extracted from the image and used to develop mean color intensity algorithm and area algorithm. The study is beneficial and is employed during post-harvest processing. Each pixel of selected region was read from the image by color sensors, so RGB color space is generally obtained using CVS. RGB color space is device dependent (Yam & Papadakis, 2004) and cannot signify, color differences in a consistent scale and also similarity of two colors are not easily recognize using this color space. While L*a*b* color space being an international standard for color measurement, is non-linear and color value can be easily interpreted (Hashim et al., 2012).

10.5.6.4 SOFT COMPUTING TECHNIQUE TO CLASSIFY AND PREDICT SHELF-LIFE

Soft computing relies on advanced algorithms like genetic algorithm, genetic programming, fuzzy logic, neural network, and other upcoming algorithms to solve real-world complex problems which cannot be solved using conventional or earlier algorithms. Many complex biological events like folding of protein, gene expression, classification or prediction of fruits and vegetables maturity are solved using soft computing techniques.

ANN is popular and accepted to a great extent over past few years especially due to the arrival of modern high-speed computer. ANN is inspired by our complex human central nervous system, which consists of a huge number of interconnected neurons. Neurons are the building block of our nervous system that can transmit or receive electro-chemical signal and communicate with other cells of our body. Every process of our brain relies upon the neurons network, which arouses the interest of the researcher to develop a program or software which will act as our central nervous system and solve our practical problem. But, it is not easy for computer system to mimic our biological brain as it consists of 100 billion of neurons, so ANN is developed by considering a network of few neurons. An ANN is designed by establishing connection between different processing elements that work in parallel (Riad & Mania, 2004). Processing elements act as neurons and are interconnected to each other. As our brain acquires information every day and use it where required, same thing is employed for ANN, which requires training before its implementation (Singh et al., 2015). A trained neural network acts as an "expert" and categorizes the information that has been given to analyze. It can be used to resolve the pattern recognition, classification, and prediction task, which is difficult to solve using physical equations. Various parameters can be incorporated in single model, when ANN algorithm was employed. In the field of agriculture, the role of ANN cannot be neglected as it is actively used to classify the fruits and separate the bruised one from the valuable food items (Fig. 10.7).

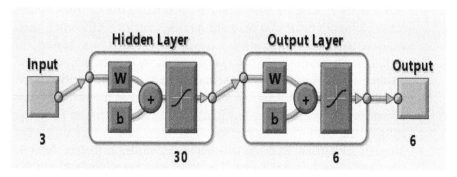

FIGURE 10.7   Neural network.

The workflow for the neural network design for classification purpose includes:

1. *Defining a problem:* Input vectors and target vectors are arranged as columns in a matrix.
2. *Create the network:* Network is created using input layer, hidden layer, and output layer with feed forward back propagation neural network and tan-sigmoid transfer functions in both the output layer and the hidden layer. Generally, 20–30 neurons are used in one hidden layer. More neurons and layer in the network requires more computation time, but we can use the network to solve further complicated problems.
3. *Train the network:* The network is generally trained by scale conjugate gradient back propagation neural network model if pattern recognition or classification task has to achieve as they are faster and their memory requirement was small. Input vectors and target vectors are divided into three sets before training that is training set, validation set, and test set. Training would stop if anyone condition

occurred that is when the maximum number of epochs was reached or maximum time had exceeded or mean square error (MSE) of the validation sample increases.

Training can be initiated after entering the command:

net = train (net, inputs data, and targets data).

Training carries out a loop of calculation and proceeds through the specified sequence of inputs, calculating the output, error, and adjusting the network. But the network training does not guarantee that the resulting network performs efficiently. We have to check the new values of weights and biases by computing the network output for each input vector. If a network does not perform efficiently then we have to train the network again with new weights and biases.

4. *To find the validation error and performance of trained network:* The training errors, validation errors, and test errors can be verified from the performance plot. Confusion matrix also shows the various types of errors that occurred in the final trained network. The diagonal cells of the confusion matrix show the number of cases that were properly classified, and the off-diagonal confusion matrix demonstrates the misclassified cases. The *ROC curve* is a plot between the true positive rate (sensitivity) and the false positive rate (1—specificity). In a perfect test that is 100% sensitivity and 100% specificity, colored lines of ROC curve lies in the in the upper-left corner.

5. *Use the network:* After validation of the trained network, it is used for the classification of many other unknown samples.

Ripening banana was classified with the help of Neural Pattern Recognition Tool (nprtool) of MATLAB software. In nprtool, the Inputs were classified into a set of target categories with the help of ANN. Banana was classified in different classes based on L*a*b* value obtained from image of sample. Each class represents the days of maturity, that is, class-1 represent 3rd day after harvest, class-2 represent 5th day after harvest and so on. A two layer feed forward, with sigmoid hidden and output neuron was used for the classification purpose. The ANN was trained using scale conjugate gradient back-propagation (trainscg). Network training function would train the network until its weight, net input, and transfer functions had derivative functions. With respect to the weight and bias, back propagation was used to calculate the performance. Training would stops if any one condition occurred that is when the maximum number of epochs was reached or maximum time had exceeded or MSE error of the validation sample increases.

To perform the ANN the dataset was divided into three parts:

- Training set (80%)
- Validation set (10%)
- Test set (10%).

Training set was used to build the model. The network was trained with the help of training set to rectify the weight. Validation data set was applied to check the accuracy and efficiency of the ANN algorithm. It was also used to confirm that network was not over fit (that is free of noise) and the error was within some range. It decides a stopping point for the algorithm. If both the training and validation performance was similar then test data set is used to measure the accuracy of the network with unknown data.

The network's performance was measured with the help of MSE and percent error (% E). MSE is the average square difference between the outputs and targets. Lower the value of MSE lesser will be the error. Similarly, percent error indicates the proportion of samples, which were classified incorrectly or assign to the wrong category. Lower % E indicates that there was very less misclassification. Confusion matrix and ROC were used to check the performance of a classification model. Confusion matrix is in tabular form which is relatively simple to understand and illustrate the performance of an algorithm. Whereas, ROC is a normally used graph that notify the performance of a classifier.

Confusion matrix (Fig.10.8) shows the errors occurred in the final trained network used for validating pattern recognition applications. The diagonal cells (green colored) in each table demonstrate the number of cases that were properly classified. The off-diagonal cells (red colored) display the misclassified cases. At the bottom right, the blue cell displays the total percent of correctly classified cases (Beale et al., 2015). High number of correct responses in the green squares and the low numbers of incorrect responses in the red squares indicate higher accuracy in classification. The overall accuracy was 99% as illustrated in blue cell.

The ROC is also used to check the excellence of classifiers. Colored lines in each axis stand for the ROC curves for each of the four categories of this simple test problem. The ROC curve is a plot between the true positive rate (sensitivity) and the false positive rate (1—specificity). A good test would confirm points in the upper-left corner, with 100% sensitivity and 100% specificity (Beale et al., 2015). Figure 10.9 shows that the network performance which was almost perfect as the colored lines lie in the upper left corner. For predicting the shelf-life or the days of maturity based on color value obtained after image processing, neural fitting tool (nftool) was employed.

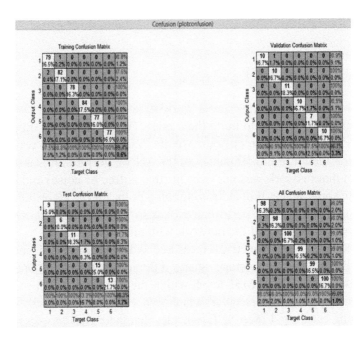

FIGURE 10.8   Confusion matrix.

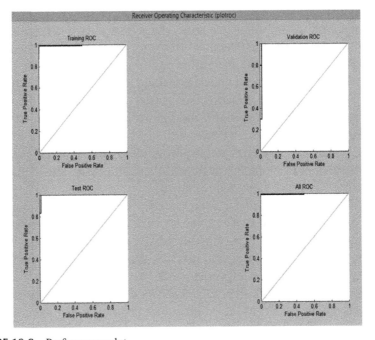

FIGURE 10.9   Performance plot.

A large number of fruits and vegetables are classified using ANN based on physical parameters like size, color, and external defects. Bhatt and Pant (2015) described an apple classification system based on machine vision and back-propagation ANN. Scaled conjugate gradient was used as training algorithm and it was observed that the best validation performance was 0.025042 at epoch 111 and overall 96% accuracy was confirmed in the confusion matrix.

10.6 CONCLUSION

After proper harvesting and storage of valuable agricultural produce, adequate packaging is required for safe transportation of fresh horticultural produce. Different packaging materials are available in the market, which may be used according to internationally accepted norms and information. Bio-chemical constituent of food commodities is also taken under consideration while developing a precise *packaging design*. Some films like polyethylene, polypropylene, polyurethane do not degrade easily and also have an adverse effect on our ecosystem. So, while selecting a packaging material its impact on environment should not be overlooked. Predictive model based on the microbial growth, temperature, respiration, and transpiration rate can help to estimate the shelf-life as well as the quality of packaged fruits and vegetables and also help to develop a MAP technology. ANN approach is effective and proficient for banana classification and in predicting the ripening process and shelf-life of banana. Further, this approach can be used to classify most of the fruits and vegetables. Systematic investigation and continuous research works were carried out by several scientists, which assist to develop a valu-able model that not only help to predict but also enhance the shelf-life and maintain the quality attributes of packaged fruits and vegetables.

KEYWORDS

- **fruits–vegetables**
- **packaging**
- **shelf-life**
- **temperature**
- **perishable**

REFERENCES

Abe, K.; Watada, A. E. Ethylene Absorbent to Maintain Quality of Lightly Processed Fruits and Vegetables. *J. Food Sci.* **1991,** *56,* 1589–1592.

Aguayo, E.; Jansasithorn, R.; Kader, A. A. Combined Effects of 1-Methylcyclopropene, Calcium Chloride Dip, and/or Atmospheric Modification on Quality Changes in Fresh-Cut Strawberries. *Postharvest Biol. Technol.* **2006,** *40,* 269–278.

Ahvenainen, R. New Approaches in Improving the Shelf Life of Minimally Processed Fruits and Vegetables. *Trends Food Sci. Technol.* **1996,** *7* (6), 179–187.

Ankita. K. P.; Prasad, K. Characterization of Dehydrated Functional Fractional Radish Leaf Powder. *Der Pharm. Lett.* **2015,** *7* (1), 269–279.

Beale, M. H.; Hagan, M. T.; Demuth, H. B. Neural Network Toolbox™ User's Guide, *the Math Works, Inc.*

Belayneh, M.; Workneh, T. S.; Belew, D. Physicochemical and Sensory Evaluation of Some Cooking Banana (Musa spp.) for Boiling and Frying Process. *J. Food Sci. Technol.* **2014,** *51* (12), 3635–3646.

Bhande, S. D.; Ravindra, M. R.; Goswami, T. K. Respiration Rate of Banana Fruit under Aerobic Conditions at Different Storage Temperatures. *J. Food Eng.* **2008,** *87* (1), 116–123.

Bhatt, A. K.; Pant, D. Automatic Apple Grading Model Development Based on Back Propagation Neural Network and Machine Vision, and its Performance Evaluation. *AI Soc.* **2015,** *30* (1), 45–56.

Bower, J. H.; Biasi, W. V.; Mitcham, E. J. Effects of Ethylene and 1-MCP on the Quality and Storage Life of Strawberries. *Postharvest Biol. Technol.* **2003,** *28,* 417–423.

Caleb, O. J.; Opara, U. L.; Witthuhn, C. R. Modified Atmosphere Packaging of Pomegranate Fruit and Arils: A Review. *Food Bioprocess Technol.* **2012,** *5,* 15–30.

Cardoso, G.; Labuza, T. P. Prediction of Moisture Gain and Loss of Packaged Pasta Subjected to a Sine Wave Temperature/Humidity Environment. *J. Food Technol.* **1983,** *18,* 587–606.

Corbo, M. R.; Altieri, C.; Amato, D. D.; Campaniello, D.; Del Nobile, M. A.; Sinigaglia, M. Effect of Temperature on Shelf Life and Microbial Population of Lightly Processed Cactus Pear Fruit. *Postharvest Biol. Technol.* **2004,** *31,* 93–104.

De Ell, J. R.; Toivonen, P. M. A. Use of Chlorophyll Fluorescence in Postharvest Quality Assessments of Fruits and Vegetables. In *Practical Applications of Chlorophyll Fluorescence in Plant Biology;* Springer: New York, NY, 2003; pp 203–242.

Dobrzanski, Bohdan. Jr.; Rybczynski, Rafał. Color as a Quality Factor of Fruits and Vegetables. In *Physical Methods in Agriculture;* Springer: Berlin, Germany, 2002; pp 375–397.

Elwan, M. W. M.; Nasef, I. N.; Seifi, S. K. E.; Hassan, M. A.; Ibrahim, R. E. Storability, Shelf-Life and Quality Assurance of Sugar Snap Peas (Cv. Super Sugar Snap) Using Modified Atmosphere Packaging. *Postharvest Biol. Technol.* **2015,** *100,* 205–211.

Fernandez, J. C.; Hervas, C.; Estudillo, F. J. M.; Gutierrez, P. A. Memetic Pareto Evolutionary Artificial Neural Networks to Determine Growth/No-Growth in Predictive Microbiology. *Appl. Soft Comput.* **2011,** *11,* 534–550.

Finnegan, E.; Mahajan, P. V.; Connell, M. O.; Francis, G. A.; Beirne, D. O. Modelling Respiration in Fresh-cut Pineapple and Prediction of Gas Permeability Needs for Optimal Modified Atmosphere Packaging. *Postharvest Biol. Technol.* **2013,** *79,* 47–53.

Gallagher, M. J. S.; Mahajan, P. V.; Mezdad, T. Engineering Packaging Design Accounting for Transpiration Rate: Model Development and Validation with Strawberries. *J. Food Eng.* **2013,** *119,* 370–376.

Hanna, A.; Remington, R. The Representation of Color and Form in Long Term Memory. *Mem. Cognit.* **1996**, *24* (3), 322–330.

Hashim, N.; Janius, R. B.; Baranyai, L.; Rahman, R. A.; Osman, A.; Zude, M. Kinetic Model for Colour Changes in Bananas during the Appearance of Chilling Injury Symptoms. *Food Bioprocess Technol.* **2012**, *5,* 2952–2963.

Hortensteiner, S.; Krautler, B. Chlorophyll Breakdown in Higher Palnts. *Biochim. Biophys. Acta – Bioener.* **2011**, *1807* (8), 977–988.

Kader, A. A. Biochemical and Physiological Basis for Effects of Controlled and Modified Atmospheres on Fruit and Vegetables. *Food Technol.* **1986**, *40,* 99–104.

Kang, J. S.; Lee, D. S. A Kinetic Model for Transpiration of Fresh Produce in a Controlled Atmosphere. *J. Food Eng.* **1998**, *35* (1), 65–73.

Kwolek, W. F.; Bookwalter, G. N. Prediction Storage Stability from Time Temperature Data. *Food Technol.* **1971**, *25,* 1025–1031.

Li, L.; Kaplunov, T.; Zutahy, Y.; Daus, A.; Porat, R.; Lichter, A. The Effects of 1- Methyl-cyclopropene and Ethylene on Postharvest Rachis Browning in Table Grapes. *Postharvest Biol. Technol.* **2015**, *107,* 16–22.

Li, L.; Lichter, L.; Chalupowicz, D.; Gamrasni, D.; Goldberg, T.; Nerya, O.; Ben-Arie, R. Porat, R. Effects of the Ethylene-action Inhibitor 1-Methylcyclopropene on Postharvest Quality of Non-climacteric Fruit Crops. *Postharvest Biol. Technol.* **2016**, *111,* 322–329.

Mahajan, P. V.; Oliveira, F. A. R.; Macedo, I. Effect of Temperature and Humidity on the Transpiration Rate of the Whole Mushrooms. *J. Food Eng.* **2008**, *84,* 281–288.

Mahendran, R.; Jayashree, G. C.; Alagusundaram, K. Application of Computer Vision Technique on Sorting and Grading of Fruits and Vegetables. *J. Food Process. Technol.* **2012**, *5,* 1–7.

Mangaraj, S.; Goswami, T. K.; Mahajan, P. V. Application of Plastic Films for Modified Atmosphere Packaging of Fruits and Vegetables: A Review. *Food Eng. Rev.* **2009**, *1,* 133–158.

Mannapperuma, J. D.; Singh, R. P.; Montero, M. E. Simultaneous Gas Diffusion and Chemical Reaction in Foods Stored in Modified Atmospheres. *J. Food Eng.* **1991**, *14* (3), 167–183.

Mendoza, F.; Aguilera, J. M. Application of Image Analysis for Classification of Ripening Bananas. *J. Food Sci.* **2004**, *69,* 471–477.

Mistriotis, A.; Briassoulis, D.; Giannoulis, A. Design of Biodegradable Bio-based Equilibrium Modified Atmosphere Packaging (EMAP) for Fresh Fruits and Vegetables by Using Micro-perforated Poly-lactic Acid (PLA) Films. *Postharvest Biol. Technol.* **2016**, *111,* 380–389.

Mostafa, N. A.; Farag, A. A.; Abo-dief, H. M.; Tayeb, A. M. Production of Biodegradable Plastic from Agricultural Wastes. *Arab. J. Chem.* **2015**, doi:10.1016/j.arabjc.2015.04.008

Muller, T.; Krautler, B. Chlorophyll Breakdown as Seen in Bananas: Sign of Aging and Ripening. *Gerontology.* **2010**, *57* (6), 521–527.

Navarro, F. F.; Valero, A.; Martinez, C. H.; Gutierrez, P. A.; Gimeno, R. M. G.; Cosano, G. Z. Development of a Multi-classification Neural Network Model to Determine the Microbial Growth/No Growth Interface. *Int. J. Food Microbiol.* **2010**, *143* (3), 203–212.

Nisha, P.; Singhal, R. S.; Pandit, A. B. Kinetic Modelling of Colour Degradation in Tomato Puree (Lycopersiconesculentum L.). *Food Bioprocess Tech.* **2011**, *4* (5), 781–787.

Ohali, Y. A. Computer Vision Based Date Fruit Grading System: Design and Implementation. *J. King Saud Univ. – Comput. Inf. Sci.* **2011**, *23,* 29–36.

Oliveira, F.; Gallagher, M. J. S.; Mahajan, P. V.; Teixeira, J. A. Development of Shelf-Life Kinetic Model for Modified Atmosphere Packaging of Fresh Sliced Mushrooms. *J. Food Eng.* **2012**, *111,* 466–473.

Pathare, P. B.; Opara, U. L.; Al-Said, F. A. Colour Measurement and Analysis in Fresh and Processed Foods: A Review. *Food Bioprocess Tech.* **2013,** *6* (1), 36–60.

Pieglovanni, L.; Fava, P.; Siciliano, A. A Mathematical Model for the Prediction of Water Vapor Transmission Rate at Different Temperature and Relative Humidity. *Packag. Technol. Sci.* **1995,** *8,* 73–78.

Powell, S. M.; Ratkowsky, D. A.; Tamplin, M. L. Predictive Model for the Growth of Spoilage Bacteria on Modified Atmosphere Packaged Atlantic Salmon Produced in Australia. *Food Microbiol.* **2015,** *47,* 111–115.

Prabha, D. S.; Kumar, J. S. Assessment of Banana Fruit Maturity by Image Processing Technique. *J. Food Sci. Technol.* **2015,** *52* (2), 1316–1327.

Prasad, K. Advances in Non-Destructive Quality Measurement of Fruits and Vegetables. In *Postharvest Biology and Technology of Horticultural Crops: Principles and Practices for Quality Maintenance*; Mohammed W., Siddiqui., Eds.; Apple Academic Press and CRC Press, Taylor and Francis Group: Boca Raton, FL, 2015a; pp 51–87.

Prasad, K. Non-destructive Quality Analysis of Fruits. In *Postharvest Quality Assurance of Fruits: Practical Approaches for Developing Countries*; Ahmad, Mohammad Shamsher., Siddiqui., Mohammed Wasim., Eds.; Springer International Publishing: Switzerland, 2015b; pp 239–258.

Prasad, K.; Jale, R.; Singh, M.; Kumar, R.; Sharma, R. K. Non-Destructive Evaluation of Dimensional Properties and Physical Characterization of Carrisacarandas Fruits. *Int. J. Eng. Stud.* **2010,** *2* (3), 321–327.

Prasad, K.; Singh, Y.; Anil, A. Effects of Grinding Methods on the Characteristics of Pusa 1121 Rice Flour, *J. Trop. Agric. Food Sci.* **2012,** *40* (2), 193–201.

Quevedo, R.; Mendoza, F.; Aguilera, J. M.; Chanona, J.; Gutierrez-Lopez, G. Determination of Senescent Spotting in Banana (Musa cavendish) Using Fractal Texture Fourier Image. *J. Food Eng.* **2008,** *84,* 509–515.

Radzevicius, A.; Viskelis, P.; Viskelis, J.; Karkleliene, R.; Juskeviciene, D. Tomato Fruit Color Changes during Ripening on Vine. *Int. J. Biol. Biomol. Agric. Food Biotechnol. Eng.* **2014,** *8* (2), 112–114.

Riad, S.; Mania, V. Rainfall-Runoff Model Using an Artificial Neural Network Approach. *Math. Comput. Model.* **2004,** *40,* 839–846.

Rivera, N. V.; Blasco, J.; Perez, J. C.; Dominguez, G. C.; Flores, M. D. J. P.; Vazquez, I. A.; Cubero, S.; Rebollo, R. F. Computer Vision System Applied to Classification of "Manila" Mangoes During Ripening Process. *Food Bioprocess Technol.* **2014,** *7,* 1183–1194.

Sadegaonkar, V. D.; Wagh, K. H. Automatic Sorting Using Computer Vision & Image Processing For Improving Apple Quality. *Int. J. Innov. Res. Dev.* **2015,** *4* (1), 11–14.

Salguero, J. F.; Gomez, R.; Carmona, M. A. Water Activity in Selected High-Moisture Foods. *J. Food Composit. Anal.* **1993,** *6,* 364–369.

Sastry, S. K.; Buffington, D. E. Transpiration Rates of Stored Perishable Commodities: A Mathematical Model and Experiments on Tomatoes. *ASHRAE Trans.* **1982,** *88,* 159–184.

Sharma, M.; Jacob, J. K.; Subramanian, J.; Paliyath, G. Hexanal and 1-MCP Treatments for Enhancing the Shelf Life and Quality of Sweet Cherry (Prunusavium L.). *Postharvest Biol. Technol.* **2010,** *125,* 239–247.

Silva, P. R. S. D.; Tessaro, I. C.; Marczak, L. D. F. Modeling and Simulation of *Byssochlamysfulva* Growth on Papaya Pulp Subjected to Evaporative Cooling. *Chem. Eng. Sci.* **2014,** *114,* 134–143.

Singh, R. P. Scientific Principles of Shelf-life Evaluation. In *Shelf-life Evaluation of Food;* Blackie: London, 1994; pp 3–26.

Singh, S. P.; Bansal, S.; Ahuja, M.; Parnami, S.; Singh, H. Classification of Apples Using Neural Networks. *Int. J. Sci. Technol. Manag.* **2015,** *4* (1), 1599–1605.

Singh, Y.; Prasad, K. Moringaoleifera Leaf as Functional Food Powder: Characterization and Uses. *Int. J. Agric. Food Sci. Technol.* **2013,** *4* (4), 317–324.

Siripatrawan, U.; Jantawat, P. A Novel Method for Shelf Life Prediction of a Packaged Moisture Sensitive Snack Using Multilayer Perceptron Neural Network. *Expert Syst. Appl.* **2008,** *34.* 1562–1567.

Song, Y.; Vorsa, N.; Yam, K. L. Modeling Respiration-Transpiraion in a Modified Atmosphere Packaging System Containing Blueberry. *J. Food Eng.* **2002,** *53* (2), 103–109.

Susstrunk, S.; Buckley, R.; Swen, S. *Standard RGB Color Spaces,* Seventh Color Imaging Conference: Color Science, Systems, and Applications. *Soc. Imaging Sci. Technol.***1999,** *1,* 127–134.

Syal, S.; Mehta, T.; Priya, D. Design and Development of Intelligent System for Grading of Jatropha Fruit by Its Feature Value Extraction Using Fuzzy Logics. *Int. J. Adv. Res. Comput. Sci. Softw. Eng.* **2013,** *3* (7), 1077–1081.

Taoukis, P. S.; Labuza, T. P. *Novel Food Packaging Techniques;* Ahvenainen, R., Ed.; CRC Press*:* Boca Raton, FL, 2003.

Tchango, J. T.; Watier, D.; Eb, P.; Tailliez, R.; Njine, T.; Hornez, J. P. Modeling Growth for Predicting the Contamination Level of Guava Nectar by Candida Pelliculosa under Different Conditions of pH and Storage Temperature. *J. Ind. Microbiol. Biotechnol.* **1998,** *18,* 26–29.

Torrieri, E.; Perone, N.; Cavella, S.; Masi, P. Modelling the Respiration Rate of Minimally Processed Broccoli (*Brassica rapa var. sylvestris*) for Modified Atmosphere Package Design. *Int. J. Food Sci. Technol.* 2010, *45* (10), 2186–2193.

Toylan, H.; Kuscu, H. A. Real-Time Apple Grading System Using Multicolor Space. *Sci. World J.* **2014,** *2014,* 1–10.

Van Impe, J. F.; Nicolai, B. M.; Martens, T.; Baerdemaeker, J. D.; Vandewalle, J. Dynamic Mathematical Model to Predict Microbial Growth and Inactivation during Food Processing. *Appl. Environ. Microbiol.* **1992,** *58,* 2901–2909.

Waghmare, R. B.; Mahajan, P. V.; Annapure U. S. Modelling the Effect of Time and Temperature on Respiration Rate of Selected Fresh-cut Produce. *Postharvest Biol. Technol.* **2013,** *80,* 25–30.

Walter, E. H. M.; Kabuki, D. Y.; Esper, L. M. R.; Sant'Ana, A. S.; Kuaye, A. Y. Modelling the Growth of *Listeria monocytogenes* in Fresh Green Coconut (*CocosnuciferaL.*) Water. *Food Microbiol.* **2009,** *26,* 653–657.

Watada, A. E.; Ko, N. P.; Minott, D. A. Factors Affecting Quality of Fresh-cut Horticultural Products. *Postharvest Biol. Technol.* **1996,** *9,* 115–125.

Yam, K. L.; Papadakis, S. E. A Simple Digital Imaging Method for Measuring and Analyzing Colour of Food Surfaces. *J. Food Eng.* **2004,** *61* (1), 137–142.

Yousef, A. O. Computer Vision Based Date Fruit Grading System: Design and Implementation. *J. King Saud Univ. Comput. Inf. Sci.* **2011,** *23,* 29–36.

Zheng, C.; Sun, D. W.; Zheng, L. Recent Applications of Image Texture for Evaluation of Food Qualities—A Review. *Trends Food Sci. Technol.* **2006,** *17* (3), 113–128.

Zwietering, M. H.; Jongenberger, I.; Roumbouts, F. M.; van'tRiet, K. Modelling of Bacterial Growth Curve. *Appl. Environ. Microbiol.***1990,** *56* (6), 1875–1881.

CHAPTER 11

MATHEMATICAL MODELING FOR MICRO-PERFORATED FILMS OF FRUITS AND VEGETABLES USED IN PACKAGING

OLUWAFEMI J. CALEB, MARTIN GEYER, and
PRAMOD V. MAHAJAN*

*Department of Horticultural Engineering, Leibniz Institute for
Agricultural Engineering (ATB), Max-Ethy-Allee 100, Potsdam 14469,
Germany*

Corresponding author. E-mail: pmahajan@atb-potsdam.de

CONTENTS

ABSTRACT

Life processes of fresh fruits and vegetables continue after harvest due to on-going metabolic activities, including respiration and ripening which continue in cells or plant parts until senescence and death. Additionally, rapid quality deterioration and reduced shelf-life may also result from physiological disorders and presence of mechanical injuries, which represent major quality challenges for the marketing of fresh minimally processed produce. Overall, inadequate management of these quality challenges may result in reductions in availability, edibility, and freshness, which increases postharvest food losses and direct financial losses for the role players in the fresh produce industry. Micro-perforation is a useful technique to achieve safe modification of internal atmosphere of package for safe storage and quality retention of horticultural produce in comparison with conventional non-perforated MAP system.

11.1   INTRODUCTION

Rapid quality deterioration and reduced shelf-life are the major challenges facing the production and marketing of fresh fruit and vegetables (Hussein et al., 2015). Life processes of fresh fruits and vegetables continue after harvest due to on-going metabolic activities, including respiration and ripening which continue in cells or plant parts until senescence and death (Caleb et al., 2013). Additionally, rapid quality deterioration and reduced shelf-life may also result from physiological disorders and presence of mechanical injuries, which represent major quality challenges for the marketing of fresh minimally processed produce (Siddiqui et al., 2011). Overall, inadequate management of these quality challenges may result in reductions in availability, edibility, and freshness, which increases postharvest food losses and direct financial losses for the role players in the fresh produce industry (Mahajan et al., 2014).

Studies have shown that modified atmosphere packaging (MAP) offers the ability to delay quality degradation and extend shelf-life of fresh produce. MAP slows down respiratory activity, delay softening, ripening, and reduces the incidence of various physiological disorders and pathogenic infestations (Caleb et al., 2013). However, when product respiration does not correlate to the permeability properties of packaging polymeric film, accumulation concentration of carbon dioxide (CO_2) could result in a state of anaerobic respiration and ethanol accumulation in the packaged product. Most polymeric films commonly used in MAP of fruits and vege-tables are limited in permeability properties and this represents a critical limitation for highly respiring fresh produce such as strawberries, grapes, cherries, blueberries, sweet corn, spinach, citrus, asparagus, mushrooms, and broccoli. The levels of atmosphere attained using polymeric films are rarely sufficient to maintain quality and ensure extended shelf-life of produce during storage (Hussien et al., 2015). This results in anaerobiosis and development of undesirable off-odors under low O_2 and elevated CO_2 atmospheres are common occurrences that severely modify the volatiles profile of packaged produce (Caleb et al., 2013a). The use of perforations as a technique offers the possibility to overcome these limitations (Oliveira et al., 2012a, b). Perforation can be used to achieve higher transmission rates of gases and water vapor through commonly used polymeric films (Gonzalez et al., 2008).

Micro-perforation is a useful technique to achieve safe modification of internal atmosphere of package for safe storage and quality retention of horticultural produce in comparison with conventional non-perforated MAP system (Montanez et al., 2005). In addition, perforating a package serves other crucial functions such as reduction in cooling time and preventing condensation of water vapor inside the package. The technique involves the use of a single or multiple numbers of small holes pierced through the polymeric films (Hussien et al., 2015). Perforations vary in size from micro-perforations (50–200 µm diameter holes or tubes) to macro-perfo-rations with holes or tubes greater than 200 µm in diameter (Hussien et al., 2015). Various methods have been used in perforation of polymeric films, this include tube perforation, mechanical puncturing (with needle), and laser perforation (Gonzalez et al., 2008; Gonzalez-Buesa et al., 2012).

11.2   TYPES OF PERFORATION

11.2.1   LASER PERFORATION

Laser perforation uses heat energy to evaporate packaging film to produce small, clean holes that are sealed along the edges (Lazare & Tokarev, 2004; Allan-Wojtas et al., 2008). Laser perforation systems consist of three impor-tant components which include a medium that generates the laser light, a power source for energy discharge in excited form to the laser medium to emit laser beams, and an optical cavity that compresses the beam to stimu-late the emission of laser radiation (Mir, 2009). Perforation of polymeric films is achieved by standard CO_2 laser systems operated well for speeds of about 300' per minute. Technological advancement in power source from 20 to 2 kW and from split beam approach to beam compression with the use of polygon mirror, which ensures the consistency of the beam strength has further improve the laser systems (Mir, 2009). Although technological advancement has been gained in the development of efficient laser perforation systems for polymeric films, the cost of production has being a critical factor in commercialization.

11.2.2 MICRO-ELECTRIC DISCHARGE MACHINING (MICRO-EDM)

Micro-electric discharge machining (micro-EDM), also known as sparks eroding is a relatively new technique used for micro-perforation of polymeric films. With this method, plastic material is removed by melting and vaporization caused by a series of electrical discharges (sparks) provided by a generator to produce micro-perforations on film (Allan-Wojtas et al., 2008).

11.2.3 MECHANICAL PUNCTURING

This technique is also referred to as pin-perforation, which involves semi-automated mechanical perforation of polymeric film with sharp pointed cold or hot needles (Piergiovani et al., 2003). The use of mechanical puncturing with needles for packaging film has been extensively reported. This include packaged minimally processed pomegranate arils (Hussein et al., 2015), tomato (Li et al., 2010), capsicum (Pandey & Goswami, 2012), spinach, strawberries (Kartal et al., 2012), and broccoli (FernAndez-León et al., 2013). Mechanical puncturing with cold needle is a slow and time consuming method and usually produces large perforations ($\geq$ 1 mm in diameter). Cold needles punch rough and incomplete holes with the polymeric materials till attached as flaps that can cover the holes. While hot needles melt the film to form irregular holes and redeposit the melted polymeric material as a large rim around the edges (Allan-Wojtas et al., 2008). The successful application of mechanical needle perforation in packaging can be attributed to the polymeric film flexibility, cost effectiveness and the fact that it does not require complex technical details.

11.3 GAS EXCHANGE THROUGH MICRO-PERFORATIONS

Gas exchange through packaged fruits and vegetable is driven by a dynamic process, balanced by respiration of produce and gas permeation through the package (Mahajan et al., 2008). The exchange of gases between packaging film and surrounding atmosphere is driven by the partial pressure gradient across packaging film (Mullan & McDowell, 2011). The flow of gases through micro-perforations is a combination of both convection and diffusion mechanism (Gonzalez-Buesa et al., 2012). Gas exchange by diffusion starts with the sorption of molecules into barrier surface via diffusional molecular exchange, followed by desorption on the opposite surface (Hu et al., 2001; Rodriguez-Aguilera & Oliveira, 2009). In multi micro-perforated packages, gas exchange occurs through the perforations, while in packages with a low number of perforations, the gas flux is usually by a combination of transmissions through the polymer material and transmission through the perforations (Beaudry, 2008). Similarly, film thickness affects the effective gas exchange through the film, by slowing down the transmission/permeation rate (Fonseca et al., 2000).

Table 11.1 presents a summary of factors and variables that affects the effective exchange of gas through micro-perforation. The rate at which O_2 is consumed and CO_2 is produced by respiring fresh produce inside the package depends on the concentrations of O_2 and CO_2 at a given temperature (Mangaraj et al., 2012). Hence, the barrier properties of packaging films to O_2, CO_2, and water vapor plays a significant role in packaged fresh produce. Ratio of permeability for CO_2 and O_2 (perm selectivity (β)), generally vary from 2 to 8 for different polymeric films used in MAP (Gonzalez et al., 2008). Al-Ati and Hotchkiss (2003) suggested that the effect of perm selectivity can be estimated by calculating the equilibrium gas composition with different permeability coefficients. Furthermore, the respiration coefficient of packaged fresh fruits and vegetables can vary from 0.7 to 1.3 (Fonseca et al., 2002), and a relatively high concentration of CO_2 could be reached inside non-perforated packages (Gonzalez-Buesa et al., 2012). However, with micro-perforation, the scenario is different in that CO_2 diffuses 0.77 faster than O_2, thus resulting in more or less equal generation of gradient of gases with a perm selectivity value close to 1 (Ozdemir et al., 2005). Thus, with micro-perforation, elevated CO_2 concentrations in atmosphere can be achieved without the quantity of O_2 in the package dropping rapidly toward detrimental anaerobic conditions (Hussein et al., 2015).

TABLE 11.1 Factors and Related Variables Involved in the Design Process for a Micro-Perforation Packaging System.

Factors	Variables
Package-related	Film thickness
	Film surface area for gases exchange
	Volume of the package
	Number of micro-perforations
	Diameter of micro-perforation
	Film permeability
Storage conditions	Gas composition
	Temperature
	Atmospheric pressure
Product-related	Produce mass
	Produce density
	Respiration rate
	Desired equilibrium gases composition
	Transpiration rate
	Postharvest treatments

11.4 MATHEMATICAL MODELING OF MICRO-PERFORATED FILMS

Design of perforation mediated-MAP involves the application of mathematical models capable of predicting gas and water vapor permeability through the film, as a function of perforation combined with the adequate understanding of the produce's physiological responses (Fishman et al., 1996). Mathematical models are useful tools in the design and validation of effective packaging system (Montanez et al., 2005; Mahajan et al., 2007; Pandey & Goswami, 2012). Modeling gas exchange through micro-perforated films takes either of the two major assumptions: (1) that perforation is the major route of gas transport with the final gas transfer rate being the additive term of permeation through perforation and diffusion across the film (Mir & Beaudry, 2004; Montanez et al., 2010); or (2) that gas transfer only occur through the perforations, while assuming the film as impermeable (Montanez et al., 2010). Maxwell-Stefan equation has been used to determine diffusive flux of gases in cases where mass transport of gas through micro-perforated packaged produce is assumed to take place, based on first assumption (Chung et al., 2003; Rennie & Tavoularis, 2009; Gonzalez-Buesa et al., 2012).

Modeling of produce respiration and exchange of gas and water vapor or mass transfer through film and perforations is based on different physical laws. Graham's law of effusion, Fick's law and Stephan–Maxwell law of diffusion and/or a combination of more than one physical law have been used to predict permeation of gas and water vapor through non-perforated and perforated packaging systems (with micro-perforations) (Fishman et al., 1996; Kader & Watkins, 2000; Gonzalez et al., 2008). Knudsen diffusion and effusion and/or hydrodynamic flow laws have been used to adequately describe permeation of gas through perforated films (Zinderighi, 2001; Del-Valle et al., 2004; Gonzalez-Buesa et al., 2009), in the case of combined influence of gas diffusion and sorption. Table 11.2 presents a summary of mathematical models that have been developed to describe gas exchange of perforation-mediated packaging system.

Furthermore, most of the developed models assume the uniform production of micro-perforations that are round, within the required size range, and unobstructed (Allan-Wojtas et al., 2008). However, in many practical cases, there is always variability in the shape, size, and uniformity of the micro-perforations drilled on the film. Hence, the measured permeability of micro-perforated films often fails to agree with predicted values. Larsen and Liland (2013) reported that perforations made by the acupuncture needle of calculated area (mean value) of 6500 μm^2 in the Amcor P-plus PET/PE-film had the highest O_2TR and CO_2TR, almost threefold the values for the laser perforations in the PET/PE-film and BOPP-film. Similarly, Allan-Wojtas et al. (2008) found that microstructural characteristics such as shape and size of micro-perforations affected the O_2 and CO_2 transmission rates of polyethylene and polyester films. Therefore, it is important to take into consideration, the microstructural characteristics of perforations as important factors that may affect the modeling process. This will help to avoid discrepancy between experimental and predicted values.

11.5 DESIGN OF MICRO-PERFORATIONS

To successfully design micro-perforated films for fresh produce, consolidated knowledge of product respiration rate, gas permeability (O_2 and CO_2) of packaging material, optimum atmosphere (O_2 and CO_2) for the given product, along with other parameters such as package geometry and product weight would be required. The simplest concept of packaging design is to use the packaging film as the regulator of O_2 flow into the package and the flow of CO_2. Assuming that there is no gas stratification inside the package

TABLE 11.2 Mathematical Models for Predicting Exchange of Gases and Water Vapor Through Perforated Films.[a]

Basis law applied	Mathematical equation(s)	Number(s) of perforation (n)	Film thickness (l)	Reference
Stephan–Maxwell's law (modified)	$-\dfrac{P}{RT}\left(\dfrac{Y_{i,k+1}-Y_{i,k}}{\Delta x}\right)$ $= \varphi_{pi}\sum\limits_{\substack{j=1\\ j\neq i}}^{n}\left(\dfrac{Y_{j,k}+Y_{j,k+1}}{2D_{ij}}\right)$	0, 6, 992	30 μm	Lee et al. (2000)
Stephan–Maxwell's law	$\dfrac{P\delta C_i}{RT\delta l} = \varphi_{pi}\sum\limits_{\substack{j=1\\ j\neq i}}^{n}\dfrac{\varphi_{pi}C_j - \varphi_{pi}Y_i}{D_{ji}}$	1–5	0.00284, n ≤ 0.102 cm	Paul and Clarke (2002)
Fick's law	$WTR_z = -D\left[\dfrac{M_w A P_T}{RT p A lm}\right]$ $(pW_1 - pW_2)$	1, 3, 6, 12, 18, and 24 (holes per 38.5 cm²).	0.2, 0.5, 1.75 mm	Dirim et al. (2004)
Fick's law	$\dfrac{dO_2}{dt} = AkO_2\left(pO_2^{in} - pO_2^{out}\right)$	0.13 m² diffusion area	35 μm	Ozdemir et al. (2005)
Fick's law	$\dfrac{dV(t)}{dt} = n_p D_i\left(P_i^{out} - P_i^{in}\right)$	1 hole	0.012, 0.025 mm	Techavises & Hikida (2008)
Fick's law	$J_{fi} = -\dfrac{P}{RT}\dfrac{Q_iA\left(P_i^{in} - P_i^{out}\right)}{L}$	0–14	40 μm	Gonzalez-Buesa et al.(2009)
Knudsen's law	$J_{k,A} = D_{k,A}\dfrac{\partial c_A}{\partial X}$	3–6	0.2 mm	Del-Valle et al.(2004)

[a]Adapted from Hussein et al. (2015).

and that the total pressure is constant, the differential mass balance equations (Mahajan et al., 2007) that describe O_2 and CO_2 concentration changes in a package containing a respiring product are:

$$V_f \times \frac{d(y_{O_2})}{dt} = \frac{P_{O_2}}{e} \times A \times (y_{O_2}^{out} - y_{O_2}) - R_{O_2} \times M \tag{11.1}$$

$$V_f \times \frac{d(y_{CO_2})}{dt} = \frac{P_{CO_2}}{e} \times A \times (y_{CO_2}^{out} - y_{CO_2}) + R_{CO_2} \times M \tag{11.2}$$

where V_f is the headspace (free volume) in the package, e is the thickness of polymeric film, and M is the weight of the product (M); the subscripts O_2 and CO_2 refer to oxygen and carbon dioxide, respectively and the superscript "out" refers to external atmosphere.

At steady-state, the accumulation term in eqs 11.1 and 11.22 is zero, and these equations are reduced to:

$$y_{O_2}^{eq} = y_{O_2}^{out} + \frac{R_{O_2}^{eq} \times e \times M}{P_{O_2} \times A} \tag{11.3}$$

$$y_{CO_2}^{eq} = y_{CO_2}^{out} - \frac{R_{CO_2}^{eq} \times e \times M}{P_{CO_2} \times A} \tag{11.4}$$

Equations 11.1–11.4 are coupled with the models that describe the dependence of product respiration rate on gas composition, temperature, and models that describe the dependence of permeability of the packaging material on temperature. Thereby, it would be possible to calculate exactly the type of packaging material required, size and number of perforations that would result in the ideal packaging solution. A design protocol for the selection of packaging material using mass balance equation at steady state (eqs 11.3 and 11.4) was reported by Mahajan et al. (2007). In a plot of CO_2 versus O_2 concentration, the points that correspond to equilibrium will lie along a straight line which crosses the point (0.21, 0, and air composition) and has a slope equal to $-$ RQ/β, where RQ is respiration quotient RCO_2/RO_2. Therefore, to ensure that a given packaging system may be able to yield the required gas composition, its permeability ratio has to be such that the resulting straight line crosses the window of recommended gas atmosphere for the selected product. A polymeric film with high β results in an equilibrium atmosphere that is low in CO_2 and in O_2, while films with a low β (e.g., <2) tend to accumulate high levels of CO_2 without regard to absolute permeation rates. Mahajan et al. (2007)

further explained how to predict equilibrium gas composition and time required to achieve it. Finally, the mass balance eqs (11.1) and (11.2) were integrated with the mathematical models for respiration rate and for permeability, including micro- or macro-perforations, along with other parameters such as product weight, film area, film thickness, and package geometry. Considering the complexity of mathematical models and a large number of parameters involved in MAP design, an integrative mathematical tool called Pack-in-MAP® software was developed (Mahajan et al., 2007). The user-friendly software was developed in Matlab to solve mass balance equations (11.3 and 11.4) considering mathematical models for respiration rate and film permeability. The software calculates the respiration rate for that product at the given storage conditions and it recommends the best possible films and number and size of micro-perforations if required. Validation showed a good agreement between the software predictions and experimental data on fresh strawberries at 10°C (Sousa-Gallagher et al., 2013).

11.6   APPLICATION OF MICRO-PERFORATION IN PACKAGING SYSTEMS

Perforation-mediated packaging systems have been applied extensively on fresh fruit and vegetables as shown in Table 11.3. For instance, Almenar et al. (2007) found that micro-perforated films with one and three perforations provided adequate CO_2 and O_2 equilibrium concentrations. The authors observed that the use of polyethylene terephthalate (PET)/polypropylene film with one and three perforations (average diameter of 100 µm) heat-sealed on plastic cups (125 mL capacity) retained the quality of strawberries through the generation of adequate equilibrium concentrations of gases (4–13 kPa CO_2 and 5–18 kPaO_2). De Reuck et al. (2010) reported that the quality of litchi (cv. McLean's Red) was best maintained for up to 21 days in perforated packages with 10 perforations (0.6 mm diameter) at 2°C. Cliff et al. (2010) reported that the use of micro-perforation (2–100 µm diameter perforations per package of 14×14 cm^2) maintained physicochemical and sensorial quality of apple slices such as volatile compounds, soluble solids concentration, titratable acidity, color and relative juice loss for 21 days. Micro-perforated package was able to establish desired gas composition (14 kPa O_2 and 7 kPa CO_2) and lower in-package ethylene concentration.

However, further research and validation of micro-perforation technique is needed for product specific applications. The need for systematic approach in order to obtain a successful application of PM-MAP (Oliveira et

TABLE 11.3 Application of Perforation-mediated Packaging for Fresh Fruit and Vegetables over the Last 10 Years.

Produce	Perforation parameters		Gas composition (CO_2/O_2) (kPa)	Storage temp. (°C)	References
	No. of holes	Perforated area (A) / diameter (D)			
Litchi cv. Mauritius		0.00939% of 720 cm^2	6/17.0	2	Sivakumar and Korsten (2006)
Wild strawberries	1 and 2 holes	0.0785 m^2 perforated area	10/10	10	Almenar et al. (2007)
Loquat fruit		(20 × 30) cm^2 bag	16–18/2–4	2	Amoros et al. (2008)
Litchi cv. McLean's Red	10 holes	0.6 mm D	~5/~17	2	De Reuck et al. (2010)
Mandarin		~150 μm D	1.2/19.8	3	Del-Ville et al. (2009)
Fresh cut apple		2–100 μm D in 196 cm^2	7/14	5	Cliff et al. (2010)
Mango	80–100 holes	~50–70 μm D	17/9	12	Boonruang et al. (2011)
Fresh sliced mushroom	2 holes	0.33 mm D	11.5/3.6	10	Oliveira et al. (2012a)
Broccoli		625 cm^2 A	5/10	5	Fernandez-León et al. (2013)
Strawberries	7 and 9 holes	90 μm D	15/5	4	Kartal et al. (2012)
Cherry tomatoes	5 holes	200 μm D	4.0 ± 0.1 CO_2	20	Briassoulis et al. (2012)
Peaches	100 holes	200 μm D	3.3 ± 0.01 CO_2	20	
Pomegranate arils	0, 3, 6 and 9	160.1 cm^2	0.1–34/1.3–19.2	5	Hussein et al. (2015)

Adapted from Hussein et al. (2015).

al., 1998; Mahajan et al., 2007; Montanez et al., 2010). The three basic disciplines underpinning food packaging namely produce physiology, polymer engineering, and converting technology are required effective and resource-efficient, MAP design exists at the intersections of these three disciplines, creating innovative packaging solution that is driven by consumer demand and balanced with environmental sustainability.

11.7   FUTURE NEEDS

Micro-perforation techniques for optimizing packages have been proven as a potential tool to improve package efficiency and extend shelf-life of various horticultural produce. However, there are concerns to be addressed in order to gain consumer confidence. This includes the potential risk to permeation of moisture, volatile organic compounds, and microbial contamination through perforations, especially during wet or moist handling conditions (Del-Valle et al., 2004; Dirim et al., 2004). Other concerns are the loss of freshness due to the dynamic change in headspace volatile organic compounds (flavor/aroma) (Hussein et al. 2015). For instance, Del-Valle et al. (2004) reported a significant permeation of volatile organic compounds through porous packaging, which led to a loss of fresh aroma and rapid deterioration of organoleptic properties of packaged produce. Hence, future research should focus on investigating the role of micro-perforation on MA-packaged fresh produce on the flavor attribute and microbiological quality and safety. Although the development of nanocomposite- and bio-based films offer a new paradigm in food packaging, the need for systemic packaging approach is essential. The need for simplified but applicable mathematical models is also important toward better understanding of the significance of micro-perforations, film thickness, storage conditions, and other parameters discussed on packaging.

KEYWORDS

- modified atmosphere packaging
- micro-perforations
- laser perforation
- mechanical puncturing
- micro-perforated films

REFERENCES

Allan-Wojtas, P.; Forney, C. F.; Moyls, L.; Moreau, D. L. Structure and Gas Transmission Characteristics of Microperforations in Plastic Films. *Packag. Technol. Sci.* **2008,** *21,* 217–229.

Almenar, E.; Del-Valle, V.; Hernandez-Munoz, P.; Lagarón, J. M.; Catala, R.; Gavara, R. Equilibrium Modified Atmosphere Packaging of Wild Strawberries. *J. Sci. Food Agr.* **2007,** *87,* 1931–1939.

Amoros, A.; Pretel, M. T.; Zapata, P. J.; Botella, M. A.; Romojaro, F.; Serrano, M. Use of Modified Atmosphere Packaging with Microperforated Polypropylene Films to Maintain Postharvest Loquat Quality. *Food Sci. Technol. Int.* **2008,** *14,* 95–103.

Al-Ati, T.; Hotchkiss, J. H. The Role of Packaging Film Permselectivity in Modified Atmosphere Packaging. *J. Agr. Food Chem.* **2003,** *51,* 4133–4138.

Beaudry, R. MAP as a Basis for Active Packaging. In *Intelligent and Active Packaging for Fruits and Vegetable;* Wilson, C. L., Ed.; CRC Press, Taylor & Francis Group: Boca Raton, FL, 2008; pp 31–56.

Boonruang, B. K.; Chonhenchob, V.; Singh, S. P. Comparison of Various Packaging Films for Mango Export. *Packag. Technol. Sci.* **2011,** *25,* 107–118.

Briassoulis, D.; Giannoulis, A.; Mistriotis, A. In *Novel PLA EMAP System for Cherrytomatoes and Peaches Able to Regulate the Targeted in-package Atmosphere-Part II: Experimental and Numerical Validation,* Proceedings of the CIGR-AgEng 2012, International Conference of Agricultural Engineering, Valencia, 2012.

Caleb, O. J.; Opara, U. L.; Witthuhn, C. R. Modified Atmosphere Packaging of Pomegranate Fruit and Arils: A Review. *Food Bioprocess Tech.* **2012,** *5,* 15–30.

Caleb, O. J.; Mahajan, P. V.; Al-Said, F. A.; Opara, U. L. Modified Atmosphere Packaging Technology of Fresh and Fresh-cut Produce and the Microbial Consequences-A Review. *Food Bioprocess Tech.* **2013a,** *6,* 303–329.

Caleb, O. J.; Mahajan, P. V.; Al-Said, F. A.; Opara, U. L. Transpiration Rate and Quality of Pomegranate Arils as Affected by Storage Conditions. *CyTA J. Food.* **2013b,** *11,* 199–207.

Chung, D.; Papadakis, D. E.; Yam, K. L. Simple Models for Evaluating Effects of Small Leaks on the Gas Barrier Properties of Food Packages. *Packag. Technol. Sci.* **2003,** *16,* 77–86.

Cliff, M. A.; Toivonen, P. M. A.; Forney, C. F.; Liu, P.; Lu, C. Quality of Fresh-cut Apple Slices Stored in Solid and Micro-perforated Film Packages Having Contrasting O_2 Headspace Atmospheres. *Postharvest Biol. Technol.* **2010,** *58,* 254–261.

De Reuck, K.; Sivakumar, D.; Korsten, L. Effect of Passive and Active Modified Atmosphere Packaging on Quality Retention of Two Cultivars of Litchi (*Litchi Chinensis Sonn.*). *J. Food Quality.* **2010,** *33,* 337–351.

Del-valle, V.; Almenar, E.; Hern, P.; Gavara, R. Volatile Organic Compound Permeation through Porous Polymeric Films for Modified Atmosphere Packaging of Foods. *J. Sci. Food Agr.* **2004,** *942,* 937–942.

Del-Valle, V.; Hernández-Muñoza, P.; Catalá, R.; Gavara, R. Optimization of an Equilibrium Modified Atmosphere Packaging (EMAP) for Minimally Processed Mandarin Segments. *J. Food Eng.* **2009,** *91,* 474–481.

Dirim, S. N.; Ozden, H. O.; Bayındırlı, A.; Esin, A. Modification of Water Vapour Transfer Rate of Low-density Polyethylene Films for Food Packaging. *J. Food Eng.* **2004,** *63,* 9–13.

Fernandez-León, M. F.; Fernandez-León, A. M.; Lozano, M.; Ayuso, M. C.; Amodio, M. L.; Colelli, G.; González-Gómez, D. Retention of Quality and Functional Values of Broccoli 'Parthenon' Stored in Modified Atmosphere Packaging. *Food Control.* **2013,** *31,* 302–313.

Fishman, S.; Rodov, V.; Ben-Yehoshua, S. Mathematical Model for Perforation Effect on Oxygen and Water Vapour Dynamics in Modified Atmosphere Packaging. *J. Food Sci.* **1996,** *61,* 956–961.

Fonseca, S. C.; Oliveira, F. A. R.; Brecht, J. K. Modelling Respiration Rate of Fresh Fruits and Vegetables for Modified Atmosphere Packages: A Review. *J. Food Eng.* **2002,** *52,* 99–119.

Fonseca, S. C.; Oliveira, F. A. R.; Lino, I. B. M.; Brecht, J. K.; Chau, K. V. Modelling O_2 and CO_2 Exchange for Development of Perforation-mediated Modified Atmosphere Packaging. *J. Food Eng.* **2000,** *43,* 9–15.

Gonzalez, J.; Ferrer, A.; Oria, R.; Salvador, M. L. Determination of O_2 and CO_2 Transmission Rates through Microperforated Films for Modified Atmosphere Packaging of Fresh Fruits and Vegetables. *J. Food Eng.* **2008,** *86,* 194–20.

Gonzalez-Buesa, J.; Ferrer-Mairal, A.; Oria, R.; Salvador, M. L. A Mathematical Model for Packaging with Microperforated Films of Fresh-cut Fruits and Vegetables. *J. Food Eng.* **2009,** *95,* 158–165.

Gonzalez-Buesa, J.; Ferrer-Mairal, A.; Oria, R.; Salvador, M. L. Alternative Method for Determining O_2 and CO_2 Transmission Rates through Microperforated Films for Modified Atmosphere Packs. *Packag. Technol. Sci.* **2012,** *26,* 413–421.

Hussein, Z.; Caleb, O. J.; Jacobs, K.; Manley, M.; Opara, U. L. Effect of Perforation-mediated Modified Atmosphere Packaging and Storage Duration on Physicochemical Properties and Microbial Quality of Fresh Minimally Processed 'Acco' Pomegranate Arils. *LWT Food Sci. Technol.* **2015,** *64,* 911–918. DOI: 10.1016/j.lwt.2015.06.040

Kader, A. A.; Watkins, C. B. Modified Atmosphere Packaging - Toward 2000 and Beyond. *HortTechnology.* **2000,** *10* (3), 483–486.

Kartal, S.; Aday, M. S.; Caner, C. Use of Microperforated Films and Oxygen Scavengers to Maintain Storage Stability of Fresh Strawberries. *Postharvest Biol. Technol.* **2012,** *71,* 32–40.

Larsen, H.; Liland, K. H. Determination of O_2 and CO_2 Transmission Rate of Whole Packages and Single Perforations in Micro-perforated Packages for Fruit and Vegetables. *J. Food Eng.* **2013,** *119,* 271–276.

Lazare, S.; Tokarev, V. Recent Experimental and Theoretical Advances in Microdrilling of Polymers with Ultraviolet Laser Beams. *P. SPIE.* **2004,** 5662, 221–231.

Lee, D. S.; Kang, J. S.; Renault, P. Dynamics of Internal Atmosphere and Humidity in Perforated Packages of Peeled Garlic Cloves. *Int. J. Food Sci. Technol.* **2000,** *35,* 455–464.

Li, L.; Li, X. –H.; Ban, Z. –J. A Mathematical Model of the Modified Atmosphere Packaging (MAP) System for the Gas Transmission Rate of Fruit Produce. *Food Technol. Biotechnol.* **2010,** *48,* 71–78.

Mahajan, P. V.; Caleb, O. J.; Singh, Z.; Watkins, C. B.; Geyer, M. Postharvest Treatments of Fresh Produce. *Philos. Trans. Roy. Soc. A.* **2014,** *372,* 1–19.

Mahajan, P. V.; Oliveira, F. A. R.; Montanez, J. C.; Frias, J. Development of User-friendly Software for Design of Modified Atmosphere Packaging for Fresh and Fresh-cut Produce. *Innov. Food Sci. Emerg. Technol.* **2007,** *8,* 84–92.

Mahajan, P. V.; Rodrigues, F. A. S.; Leflaive, E. Analysis of Water Vapour Transmission Rate of Perforation-mediated Modified Atmosphere Packaging (PM-MAP). *Biosyst. Eng.* **2008,** *100,* 555–561.

Mangaraj, S.; Goswami, T. K.; Giri, S. K.; Tripathi, M. K. Permselective MA Packaging of Litchi (cv. Shahi) for Preserving Quality and Extension of Shelf-life. *Postharvest Biol. Technol.* **2012,** *71,* 1–12.

Mir, N. Film, Perforated. In *Packaging Technology;* Yam, K. L., Ed.; John Wiley & Sons: Hoboken, NJ, 2009; pp 486–488.

Mir, N.; Beaudry, R. M. Modified Atmosphere Packaging. In *The Commercial Storage of Fruits, Vegetables, and Florist and Nursery Stocks;* Gross, K. C., Wang, C. Y., Saltveit, M., Eds.; Agriculture Handbook Number 66: Washington DC, USDA, ARS, 2004.

Montanez, J. C.; Oliveira, F. A.; Frias, J.; Pinelo, Mahajan, P. V.; Cunha & Manso. In *Design of Perforation-mediated Modified Atmosphere Packaging for Shredded Carrots: Mathematical Modelling and Experimental Validation,* Proceedings of the Florida State Horticultural Society, CXVIII, 2005, 423–428.

Mullan, M.; McDowell, D. Modified Atmosphere Packaging. In *Food and Beverage Packaging Technology;* Coles, R., Kirwan, M., Eds.; Wiley-Blackwell Publisher: Oxford, UK, 2011; pp 157–293.

Oliveira F. A. R.; Fonseca S. C.; Oliveira J. C.; Brecht J. K.; Chau K. V. Development of Perforation-mediated Modified Atmosphere Packaging to Preserve Fresh Fruit and Vegetable Quality after Harvest. *Food Sci. Tech. Int.* **1998**, *4*, 339–352.

Oliveira, F.; Sousa-Gallaghera, M. J.; Mahajan, P. V.; Teixeira, J. A. Evaluation of MAP Engineering Design Parameters on Quality of Fresh-sliced Mushrooms. *J. Food Eng.* **2012a,** *108*, 507–514.

Oliveira, F.; Sousa-Gallaghera, M. J.; Mahajan, P. V.; Teixeira, J. A. Development of Shelf-Life Kinetic Model for Modified Atmosphere Packaging of Fresh Sliced Mushrooms. *J. Food Eng.* **2012b,** *111*, 466–473.

Ozdemir, I.; Monnet, F.; Gouble, B. Simple Determination of the O_2 and CO_2 Permeances of Microperforated Pouches for Modified Atmosphere Packaging of Respiring Foods. *Postharvest Biol. Technol.* **2005**, *36*, 209–213.

Paul, D. R.; Clarke, R. Modeling of Modified Atmosphere Packaging Based on Designs with a Membrane and Perforations. *J. Membr. Sci.* **2002**, *208* (1), 269–283.

Pandey, S. K.; Goswami T. K. An Engineering Approach to Design Perforated and Non-perforated Modified Atmospheric Packaging Unit for Capsicum. *J. Food Process. Technol.* **2012**, *3*, 187.

Piergiovani, L.; Limbo, S.; Riva, M.; Fava, P. Assessment of the Risk of Physical Contamination of Bread Packaged in Perforated Oriented Polypropylene Films: Measurements, Procedures, and Results. *Food Addit. Contam.* **2003**, *20*, 186–195.

Rennie, T. J.; Tavoularis, S. Perforation-mediated Modified Atmosphere Packaging Part I. Development of a Mathematical Model. *Postharvest Biol. Technol.* **2009**, *51*, 1–9.

Sousa-Gallagher, M. J.; Mahajan, P. V.; Mezdad, T. Engineering Packaging Design Accounting for Transpiration Rate: Model Development and Validation with Strawberries. *J. Food Eng.* **2013**, *119* (2), 370–376.

Siddiqui, M. W.; Chakraborty, I.; Ayal-Zavala, J. F.; Dhui, R. S. Advances in Minimal Processing of Fruits and Vegetables: A Review. *J. Sci. Ind. Res.* **2011**, *70*, 823–834.

Techavises, N.; Hikida, Y. Development of a Mathematical Model for Simulating Gas and Water Vapour Exchange in Modified Atmosphere Packaging with Macroscopic Perforations. *J. Food Eng.* **2008**, *85*, 94–104.

CHAPTER 12

CUSHIONING MATERIALS FOR FRUITS, VEGETABLES, AND FLOWERS

NEERU DUBEY and VIGYA MISHRA[*]

Amity International Centre for Post-Harvest Technology and Cold Chain Management, Amity University, Noida, Uttar Pradesh, India

[*]*Corresponding author. E-mail: vigyamishra.horticulture@gmail.com*

CONTENTS

ABSTRACT

In modern era of globalization, cushioning material has become an indispensable part of packaging especially for storage and transport, as it helps to captivate a proportion of kinetic energy which is generated whenever a package suffers impact or is dropped from a certain height. It serves to protect the produce from different types of damages, that is, vibration, impact, bruising, and abrasion. Cushioning materials are required at the three important steps of fresh produce management: keeping harvested produce into any rigid storage container, transportation from field to pack houses, and transportation of packed produce from pack house to destination markets. The classification of cushioning material is based on usage as space fillers, resilient and non-resilient cushioning materials, and origin, that is, natural materials or manufactured. Naturally derived cushioning materials in common use are paddy straw, leaves of various plants, coconut fiber, rice hulls, and agricultural waste while manufactured cushioning materials are molded into trays, crate liner and partitions, bubble film, foam net, etc. The research is needed to develop innovative cushioning materials with combined properties of both the natural and synthetic materials like improved biodegradability, recyclability, high resilience, low compression, low moisture absorption, and less susceptibility to microbes. Innovations in cushioning materials will help in exploiting new markets as the produce can be transported to long distances with minimal damage.

12.1   INTRODUCTION

Post-harvest losses in fruits and vegetables are a big challenge as global markets become a reality and the transport of produce takes place over long distances. Worldwide distribution of sensitive produce is faced with various levels of impacts from transport and handling. Despite a variety of packaging options available today, bruising damage is commonplace for perishable produce throughout the supply chain. Bruising mostly happens during handling and transportation. At the time of harvesting also, if proper cushioning is not provided, the fruits particularly tree fruits suffer damage due to free fall and rough handling. During transport, proper packaging of the produce is quite important as these commodities are very sensitive to mechanical damages which may occur due to jolting, compression, vibration, or even in transit. Tender fruits and vegetables and cut flowers are particularly susceptible to damage right from the farm up to when it reaches the market which results in loss of valuable produce and farm income. This leads to a demand for "cushioning materials" which serve an important role to protect the packed produce from mechanical damage. The proper filling of packages with cushioning material avoids the rubbing among the fruits thereby minimizing bruising damage.

It has long been recognized that perishable produce like fruits, vegetables, and flowers should be protectively packaged to prevent damage in transport. These commodities are normally packaged in large containers and when transported over long distances, the produce may rub against each other approximately thousands of times which results in substantial losses. Nowadays many fruits like oranges and grapefruit are being packaged in separate compartments in larger containers for damage-free handling during transport. By packaging in this manner fruits can be transported to great distances without damage.

When packaging horticultural commodities it is desirable to maintain the fruit out of contact with each other so that during transport they will not be abraded and seriously damaged by virtue of repeated rubbing against each other. Typical packaging operations are currently being employed to pack perishable horticulture produce at random in a container after which the container is sealed and transported to the market. Whatever the distance traveled from orchard to consumer, the commodity suffers damage to some extent and in some instances, the loss rate varies from 10 to 30%. This is obviously very disadvantageous from economic and nutritional perspective.

It is obvious that any method capable of economically, efficiently, and automatically packaging perishable commodities while maintaining them completely separate from each other would be very desirable than loose packaging, for example, fruits and vegetables are being manually packaged in trays with cling films. Such kind of arrangements is helpful in preventing the substantial damages that inevitably occur when packages are shaken and jarred during transport. The chapter entails functions and properties of cushioning materials, classification of cushioning materials and their usage for fruits, vegetables, and flowers.

12.2   DEFINITION

Cushioning material may be defined as "anything that provides support against mechanical damages during, harvest, handling, storage and transport". Cushioning materials help to captivate a proportion of kinetic energy which generates whenever a package suffers impact or is dropped from a certain height. The fundamental role of a cushioning material is to reduce the forces created during sudden contact of one surface with another, which prevents compression or deforming damages and minimizes damaging impact forces.

12.3   FUNCTIONS OF CUSHIONING MATERIALS

An ideal cushioning material for perishable commodities should be able to dissipate the heat of respiration of the produce, free from infections, and physiologically inactive. It should also be non-hygroscopic and should also not promote corrosion. Environment compatibility and cost-effectiveness are also a desirable characteristic for an ideal cushioning material. Most importantly, it should provide mechanical protection against shock and rubbing and must preserve the commodities inherent properties, for instance, the taste in case of food, and should not contain substances and materials which could eventually interfere with the packed commodities. The first consideration in selecting cushioning materials is the shock protection requirement. It should be used to dissipate the shock which will be transmitted through the blocking of the commodities. Apart from protection of delicate goods against abrasion, surface of the commodities may be damaged by blocking, strapping, or container surfaces which are also protected by cushioning.

The major functions of cushioning material for perishable produce are:

- Protection from mechanical damage.
- Protection from damage due to vibration and compression.
- Protection of commodity from rubbing against each other.
- Protection from infection transfer.
- Protection of moisture vapor barriers at point of contact.
- Filling of void space in the container.

The most important property of cushioning material is recovery which ensures that the package contents continue to be protected even when repeatedly subjected to similar stresses. They should not contain any aggressive constituents which change the pH level and contributes to corrosion.

12.4   PHYSICAL AND CHEMICAL PROPERTIES OF CUSHIONING MATERIALS

The chemical and physical properties of cushioning materials are many and they display both desirable and undesirable characteristics. These characteristics vary in importance for different categories and types of applications. A characteristic desirable for one application may be harmful for another. The important characteristics which should be considered while selecting a cushioning material are listed below.

12.4.1 RESILIENCE

Resilience is the material's ability to undergo deformation on application of a load and the ability to recover rapidly and almost completely on removing the load. Resilience refers to a material's ability to recover from a compressional load, while rebound means the degree to which an impacting body bounces off. Polyurethane foam and polyethylene foam show a higher amount of resilience compared to expanded polystyrene and wood excelsior while least resilience is exhibited by shredded paper.

12.4.2 RATE OF RECOVERY

The time taken by the cushioning material to return to its original shape after compression is known as rate of recovery. Some materials have a rapid rate of recovery, attributed to quick spring back action which may result in damage to the commodities. At the same time, it should not take too long a period to come to its original shape (Anonymous, IIEM).

12.4.3 COMPRESSION SET

Compression set is the permanent deformation of the material due to either the static load on the system or due to repeated transit compression. Cushioning materials having high compression set creates free moving space in the container (Anonymous, IIEM). Compression set is greatest in EPS and lowest in latex foam sponge rubber.

12.4.4 CUSHION FACTOR AND CREEP

Cushion factor is defined as the ratio of the maximum stress to the total energy absorbed by the unit volume of the material. Creep is defined as the gradual deformation of a cushioning material taking place over a period of time.

12.4.5 DAMPING

It is the periodic oscillations of a cushioning material before it comes to rest. A resilient cushioning material after being compressed and during its recovery should come to its original shape without any oscillations (Anonymous, IIEM).

12.4.6 DUSTING

Dusting results from the disintegration of the bonded fiber structure of the materials and these particles can enter into gaps and damage the produce.

12.4.7 CORROSION

The corrosive effect is undesirable in some cushioning materials and should be avoided. If this cannot be avoided, the item must be protected from such materials by a neutral liner. Cushioning materials with a high acidic or basic content must be enclosed within waterproof or water vapor proof barriers. The corrosive nature of the cushioning materials is determined by the Hydrogen/ion concentration-pH.

12.4.8 HYGROSCOPICITY

Hygroscopic cushioning materials will have less cushioning value at high moisture content than at lower moisture content. When these materials are used they must be protected against long exposure to high humidity by providing a sealed waterproof barrier. If not possible that the non-hygroscopic materials which respond less rapidly to moisture change, should be used. Source: Anonymous (IIEM).

12.4.9 MICRO-BIOLOGICAL PROPERTIES

Fungus resistance of some materials is low and allows the growth of mold, mildew, and other fungi. This is a significant characteristic in case of perishable commodities. Many materials can be treated to inhibit such growth.

12.4.10 PERFORMANCE AT LOW TEMPERATURES

As most perishable produce are transported at low temperature, low-temperature performance of certain materials makes them suitable for use in these commodities. Some materials at low temperature become soft and resilient and lose their cushioning ability.

12.4.11 DENSITY

Density is also an important property of a cushioning material which limits its usage due to its weight contributing to the total weight of a package and naturally increasing the cost of transport.

Properties of ideal cushioning materials for horticultural produce:

The important properties of cushioning material for perishable commodities are

- It should have flexibility.
- Should be able to dissipate the heat of respiration of the produce.
- Free from infection and should not act as a source for transfer of infection.
- Physiologically inactive.
- Non hygroscopic and should also not promote corrosion.
- Environment compatible and cost-effective.
- Provide mechanical protection against impact, compression, and rubbing.
- Packaging must preserve the commodities inherent properties, for instance the taste in case of food, and should not contain substances and materials which could eventually interfere with the packed commodities.

12.5 SELECTION CRITERIA FOR CUSHIONING MATERIAL

12.5.1 SENSITIVITY CLASSIFICATION

To design and dimension the cushioning material properly, it is important to know what stresses it can withstand without suffering damage. Since each produce differs in the levels of sensitivity, it is very difficult to provide a general classification of goods. The manufacturer will in each instance be able to provide precise details about the sensitivity of their produce.

The sensitivity classification of a produce is determined by the admissible g value. 1 g is the acceleration due to gravity (9.81 m/s^2), that is, the force which usually applies to an object on the earth (Saraswathy et al., 2008).

If an acceleration of 2 g is applied (e.g., during fast cornering), the weight of the object doubles. This is precisely what happens to an item for transport which is secured on the loading area of a truck or stowed in a sea container. However, in addition to acceleration, the duration of any impact must always also be taken into account. The longer the duration of an impact, the greater is the risk of damage.

12.5.2 STRESSES DURING TRANSPORT

The stresses which occur during transport are the second important criteria in the selection of a cushioning material. These stresses are highly variable in nature and it is extremely difficult to determine what they will be. The greatest stresses occur if the packaged items are thrown or dropped from a height particularly in horticultural crops. This is why the potential drop height of a package as a function of its weight is used as a measure of stress (Saraswathy et al., 2008).

The regulation of Deutsche Bahn (German railroad operator) and Deutsche Post (German postal authorities) define maximum drop heights for packages as given in Table 12.1.

TABLE 12.1   Maximum Drop Height for Packages.

Regulations	Weight of package	Maximum drop height
Deutsche Bahn	50 kg	52 cm
	75 kg	46 cm
	100 kg	40 cm
	150 kg	27 cm
	200 kg	15 cm
Deutsche Post	No weight limit	60–80 cm

Source: Deutsche Transport.

12.5.3 STATIC AREA LOAD

The cushioning material is exposed to both dynamic and static forces during transport and cargo handling, but only static stresses apply when the produce is stored. These stresses are known as the static area load acting upon a cushioning material, which is calculated from the weight of the package contents and its bearing area:

$$\text{Static area load} = \frac{\text{Weight of package contents}}{\text{Bearing area}}$$

The static area load is measured in kg/cm^2 and is important for the purpose of selecting a suitable cushioning material, as the material must not lose its recovery when at rest merely under the weight of the package contents (Saraswathy et al., 2008).

12.5.4   RECOVERY

Recovery of the cushioning material is a decisive indicator of the loading capacity of the cushioning material on repeated exposure to stresses. Quick-recovery materials will return to their original height immediately upon removal of a compressional load. Being highly resilient, they also return a fairly high percentage of the stored compressional energy in the process. Slow-recovery materials do not instantaneously recover their full thickness and, therefore, do not return stored energy. This low resilience makes them desirable for applications requiring low rebound and high-energy absorption, or damping.

12.5.5   SPECIFIC WEIGHT

Specific weight is stated in kg/m^3 and is a measure of the hardness of a cushioning material; the higher the specific weight, the harder is the cushioning material (Saraswathy et al., 2008).

12.5.6   RESONANCE BEHAVIOR

Much of the mechanical damage to packages or commodities in shipment can be attributed to shocks and vibrations encountered during transportation. These vibrations, which are generally random in nature, can cause the shipment, or critical elements within, to resonate. This has the potential for damage to or failure of the commodities due to the repetitive application of stresses. Therefore, it is imperative that the design of protective cushioning systems takes into account the resonant frequencies of critical elements within the commodities. Conventionally, this is achieved by the measurement of the linear frequency response function (FRF) or transmissibility of the commodities/cushion system. However, commonly used cushioning materials can exhibit strong non-linear behavior. This non-linear behavior is exacerbated when the cushioning system is placed under high static loads. Large static loads are often a result of attempts to minimize the environmental impact (as well as the associated economic benefits) of packaging materials by reducing the amount of cushioning material used. Non-linear behavior can have significant implications for the design and optimization of protective packaging systems, especially when trying to evaluate the transmissibility of the system (Parker, 2008).

12.5.7 STRESS RANGE OF THE CUSHIONING MATERIAL

Each cushioning material has a specific stress range within which it exhibits optimum effectiveness. Cushioning curves, which are the plot of maximum impact deceleration against static area load, are used to select suitable cushioning materials. These cushioning curves may be used to determine the cushioning thickness which will provide sufficient shock absorption. Cushioning curves are plotted for a specific drop height. These curves indicate, for example, that a 5 cm thickness of plastic foam cushioning is required to reduce impact forces to the admissible level of at most 30 g. The area required to provide cushioning beneath a packaged item may then easily be calculated (Alders, 1995).

The other criteria's to be used are:

- Resistance of the commodities against damage by shock.
- Weight on bearing surface of the item to be considered. When too much weight is concentrated on a cushioning material it compresses to such a degree that it does not absorb the impact or shock energy. Cushion, therefore, must be sufficient to allow for compression under the weight of the article.
- Shape of the surface to be cushioned.
- Shock absorbing capacity of the cushioning material, including the effect of moisture thereon.
- Susceptibility of the material to corrosion due to moisture absorbed by hygroscopic properties of cushioning material.

When the cushioning material has been selected, the next step is to determine as to how it should be used. Before the correct thickness of cushioning material can be specified for a given application, it is necessary to know certain things about the articles being packaged, such as its dimensions, its weight, and some measure of its fragility. The symbol "G" is used to denote the fragility factor or the pull of gravity and is defined as the forces imposed upon an item, using the static weight of the item as the basic force to measure the degree of shock or vibration sustained in impact. The article that can withstand a maximum force equal to 50 times its own weight has a fragility factor of 50 and one that can withstand a force 100 times its own weight has fragility factor of 100 and so on.

12.6 TYPES OF MECHANICAL DAMAGES IN FRUITS AND VEGETABLES

Damages suffered by fruits/vegetables or flowers are numerous and are often broadly grouped into impact, abrasion, compression, and vibration damage, based on the type of force acting on the fruit (Sitkei, 1986). Impact damage is characterized by a quick application of force, which may occur when fruit is dropped (on ground or container or any stationary fruit), when an object drops onto fruit, or when fruit rolls into a barrier while compression damage involves a static applied force, such as the weight of stacked fruit above another fruit. Abrasive damage results from the rubbing of fruit against each other, or against some other surface like package wall. Similarly, vibration damage occurs when fruits are subjected to vibration forces, for example, during transport. This type of stimulus can result in, or exacerbate impact, abrasion, and compression injuries. Other types of damage that do not fall into these categories include cuts, tears, scratches, and punctures. These are severe forms of damage, generally caused by poor equipment or handling, and may contribute significantly to postharvest losses (Table 12.2).

12.6.1 IMPACT DAMAGE

Impact is a frequent occurrence during the harvesting, postharvest handling, packing, transportation, and distribution of fruit, and has been identified as the most important cause of mechanical injury in fruits (Altisent, 1991). Impact can occur between two fruits or against other surfaces (hard surface of storage container or package) during postharvest operations. Many impacts

TABLE 12.2 Containers and Type of Mechanical Damages During Transport.

Container	Type of damage	Examples
Paper and woven sacks	Impact damage during loading and unloading	Splitting of seams and material causing leaking and spillage loss.
Fiberboard boxes		Splitting of seams, opening of flaps causing loss of containment function. Distortion of shape reducing stacking ability.
Wooden cases		Fracture of joints, loss of containment function.
Cans and drums		Denting, rim damage. Splitting of seams and closures causing loss of containment and spoilage of contents.
Plastic bottles		Splitting or shattering causes loss of contents.
Fiberboard boxes	Compression damage through high stacking	Distortion of shape, seam splitting causing loss of containment and splitting of inner cartons, bags, and foil wrappings.
Plastic bottles		Distortion, collapse, and sometimes splitting, causing loss of contents.
Woven sacks	Vibration	Dispersing of contents.
CFB cases		Boxes become compressed and lose their cushioning qualities. CFB boxes are prone to impact damage.
Sacks—woven and paper	Snagging, tearing, and hook damage	Loss of containment function—spillage (more severe with paper sacks).
Tins		Punctured, loss of contents.

use to occur during the transfer of fruit, for example, from tree into harvest buckets, into containers for bulk transport and during placement onto and movement along the packing line. Impact may also occur after fruits are packed, due to dropping of boxes or rolling and sliding of cartons during transport. Impacts against tree branches during harvest would also be a potential cause of damage. Impact damage would be a particular risk if fruit were tossed downwards, with a high starting velocity. Other possible causes of impact during harvest may include the emptying of picking bags into crates. Mechanical harvest creates additional opportunities for impact, between fruit and against branches and catching surfaces (Sitkei, 1986). Fresh market fruits are very rarely harvested mechanically, so this is currently not an issue of concern. However, rising labor costs make this an increasingly attractive alternative, which may be of future importance. The most important type of impact in practice is of fruit against a stationary, rigid surface (Sitkei, 1986). This type of impact could occur when fruits are dropped into a storage bin or when they roll into the sidewall of a packing line. Cushioning can reduce the damage caused by impact, but injuries can still result if the forces are great enough. The impact of one fruit against another stationary fruit can also be an important cause of damage. This can occur during harvesting and transfer, when fruits are falling onto other stationary fruits. It can also occur on the packing line if fruits are unable to shift their center of gravity (e.g., if wedged into a corner). The impact damage of fruits can also occur after packing, due to dropping of boxes during handling, and the movement of fruit within boxes during transport. In the retail sector, further mechanical damage can occur due to rough handling of the fruit by staff or customers. Impact damage is thus a concern throughout the postharvest chain, from harvest to marketing.

12.6.3 ABRASION DAMAGE

Abrasion is generally a less important form of post-harvest mechanical damage than impact or compression, but is well recognized as a pre-harvest injury, often termed wind rub (Bagshaw et al., 1995). Abrasion can occur with the movement of two fruit surfaces against each other, a common type of pre-harvest injury due to windy conditions. Similarly, rubbing of the fruit against branches or leaves can also cause pre-harvest abrasion. Postharvest abrasion between fruits may occur due to vibration during transport (Hilton, 1994), for example, if one fruit is stationary, while another moves slightly, causing a rub between the two. Abrasion against other surfaces, such as picking bags, storage bins, or packing line components is also possible, with rough or irregular surfaces most likely to cause damage. Although postharvest damage due to abrasion is generally less of a concern than other types of mechanical damage, it can cause significant and economically important damage in some crops, such as pears (Mellenthin & Wang, 1974).

12.6.4   VIBRATION DAMAGE

Vibration damage occurs when fruit move around or rotate within a package due to a vibration during the transport. Damage can be caused by fruit striking against other fruit or packaging, giving an impact injury, or by rubbing of the fruit against another surface, resulting in abrasion. Vibration injury may cause only one of these types of damage, or all the three. For example, abrasion of the skin, with a smaller amount of compression damage and little impact injury happened due to vibration during transport of kiwifruit (Lallu et al., 1999). Vibration injury most commonly occurs during vibration developed during transport due to the interaction of the road and vehicle. Fruit will vibrate when the frequency of vibration reaches a certain level. If the resonance frequency of the fruit column is the same as the excitation frequency of the vehicle or road, the acceleration of the fruit can increase considerably due to resonance, and severe damage can result. In stacked or palletized cartons, the vibration can be directed up through the stack, increasing in magnitude at higher levels (Sitkei, 1986). Because of this, displaced cartons and vibration injury are most common at the top of stacks. Vibration injury within a box of fruit is also restricted to the top layers, as these fruits are most capable of movement. Trucks are mostly used for the transportation of fruits and vegetables which create vibrations that would be encountered during the transfer of stacked crates from orchard to packing shed, and during transportation to market.

12.6.5   COMPRESSION DAMAGE

Compression forces result from static weight or pressure acting on a product. The long duration and static loads involved in compression often result in a creep phenomenon. Creep occurs where the load on the fruit is constant, but the tissue continues to deform over time. Thus for compression injuries, both the duration and the magnitude of the load are important in determining the extent of injury. The static pressure of fruit, caused by the weight of fruit above, is the major cause of compression damage (Sitkei, 1986). This may occur in bulk bins, particularly if the depth of stacked fruit is high. It has been suggested that fruits like litchi are susceptible to this type of compression damage, therefore, field container depths of up to 50 cm are recommended to avoid damages. (Batten & Loebel, 1984; Greer, 1990) recommends a depth of 50 cm. Compression damage caused by stacked fruit is localized to the bottom layers of the stack, as these receive the greatest load from above. Packing fruit into cartons reduces the compression forces acting on stacked fruit, with the carton absorbing most of the load. However, box failure can result in fruit carrying the load of the box above. Box failure generally occurs due to the placement of excessive weight on the box or a decrease in the strength of the box. Compression can also occur during packing due to lidding pressure. When pressure is applied to one or more fruit in a container, the force is transmitted to the adjacent fruit. Closed lids can transmit pressure down through several layers of fruit (Schoorl, 1974). Overfilling of boxes makes this type of compression injury more likely.

12.6.6 DAMAGES DURING LOADING/UNLOADING

The main damage during loading and unloading operations is Impact injury which is caused by harsh dropping of packages during loading and unloading. Even a drop of few centimeters will cause an injury to the perishable produce. The effect of impact forces usually results in bruises, permanent damage, and lower perceived quality. Bruise sensitivity has also been reported to increase with storage time (Brown, 1993). The major cause of these losses is mechanical damage (bruising) due to impact. This impact could result from either vibration or sudden drop of the produce from certain heights. Impact sensitivity of fruits and vegetables was defined as having components namely bruise threshold and bruise resistance (Bajema & Hyde, 1998). Bruising in fruits and vegetables occurs when the produce rubs against each other, packaging containers, parts of processing equipment, and the tree (Altisent, 1991).

12.6.7 OTHER CAUSES OF DAMAGE

Severe mechanical injury may disrupt the integrity of the fruit skin, causing a cut, scratch, or puncture. A cut occurs when a sharp edge penetrates the product, without significant crushing of the tissue, as would be the case with a sharp secateurs or knife blade. Scratches are generally accompanied by tissue crushing. Punctures are caused by pointed objects, such as nails/pins used inside the package/containers or stems or thorns which penetrate the fruit surface and damage the tissue of the fruit. These severe types of damages may result from poor handling or equipment. In some fruits hand destalking (pulling on the fruit, rather than twisting) may also cause damage. It was found that mechanical destalking can cause substantial vibration damage in litchi (Bryton, 2004).

12.7 CLASSIFICATION OF CUSHIONING MATERIALS

Cushioning materials are available mainly as granular materials, bulk fibers and strips, matted fiber-textures, cellular structures and molded, or formed structures (Fig. 12.1). They are mainly made of cellulose materials or foamed plastics. Cushioning materials can broadly be classified into two ways: (1) Based on uses and (2) Based on origin. On the basis of uses three basic categories of cushioning materials are space fillers, resilient cushioning materials and non-resilient and rigid cushioning materials. On the basis of origin cushioning materials can be classified into—natural and synthetic types.

FIGURE 12.1   Cushioning of fresh produce for sale in local market.

12.7.1 CLASSIFICATION BASED ON USES

Today, effective and inexpensive packaging materials are designed to absorb the shock induced by in-transit handling, to protect exterior surfaces from abrasion and to fill voids in packing. The first step in designing a protective packaging solution is the selection of a cushioning material that is appropriate for the application. For perishables, the cushioning material to be used should be able to prevent vibration, impact, and compression damages. Prevention of movement of the item within the pack during transport and handling is essential component. To achieve this, specially designed inserts in the pack are used. These can be made of thermoformed plastic or of cardboard. Many options are possible depending upon fragility of produces to be packed, shocks, and vibrations upon which produce is subjected, value of produce packed.

12.7.2 SPACE FILLERS

These are mainly used to fill the voids in the packages as dunnage to prevent reorientation of the items and sometimes to absorb liquid commodities spilled from the broken unit containers.

12.7.3 RESILIENT CUSHIONING MATERIALS

These are used to protect packed commodities from damage due to repeated shocks. Materials falling in this group have compressibility and must be able to return to their original condition after each shock. They should be able to absorb shock energy without exerting too much force on the item. The force-displacement curve of the resilient cushioning materials falls into three groups as linear, tangent, and anomalous type.

12.7.4 NON RESILIENT AND RIGID CUSHIONING MATERIALS

These are used for protecting packaged items from a single severe shock as is experienced in an air dropping. These materials absorb shock by the collapse of their structure and cannot return to their original shape after the shock. The force-displacement curve of these materials falls in the category of anomalous type. The force is constant up to the collapse of their structure and increases very fast. They have low compression set for normal loads and have little or no resilience. However, they are resilient if they are used below their crushing force. They are normally used to cushion items intended for air dropping.

12.7.5 CLASSIFICATION BASED ON ORIGIN

On the basis of origin the cushioning materials can be classified as natural and synthetic.

12.7.6 NATURAL CUSHIONING MATERIALS

The natural cushioning materials are derived from plants and widely used in packaging fresh commodities. Natural cushioning materials are mostly used for fresh horticultural produce. The examples of natural cushioning material are paper shreds and agricultural waste commodities like straw, hulls, leaves, etc. In a study, a net made of dry banana string, an agricultural waste wrapping for apples, was shown to save the fruit from damage. This study mentioned problems of fungi attack due to the wrapping on the skin of the fruit (Jarimopas et al., 2004).

12.7.7 PAPER SHREDS

Various types of cushioning materials are used along with distribution packages to reduce damage and loss due to mechanical injury during handling and transportation. Sheets of newspaper or shredded newsprint are commonly used as a lining material in bamboo and plastic crates or CFB boxes (FAO, 2011). Shredded papers or tissues are the most economical and eco-friendly packaging solution for packaging perishable produce safely to transport them over long distances (Fig. 12.2). In Mango, paper cuttings are also used as a cushioning material inside the box and the quantity will vary depending upon the size and the capacity of the boxes (Singh et al., 2016; Bhardwaj & Sen, 2003).

FIGURE 12.2   Cushioning of fruits with shredded newspaper.

12.7.8 AGRICULTURAL WASTE

Agricultural waste, such as paddy straw, leaves, coconut fiber, cotton, and rice hulls, are sustainable and compostable. Rice hulls, coconut fiber, and wood straw have been used to prevent damages to papayas and mangoes during the distribution (Castro et al., 2012; Castro et al., 2014). Banana leaves and Teak leaves have been used as cushioning materials for guava fruits during storage (Chandra & Kumar, 2012). These are examples of the sustainability movement. Some agricultural waste commodities used as a cushioning material for fruits and vegetables are discussed below:

12.7.9 PADDY STRAW

Paddy straw is a natural cushioning material and widely used for packaging of fruits and vegetables for near or in country markets. Cushioning by paddy straw is done by keeping the cushioning materials between rows of fruits and layers of fruits inside the boxes. The function of paddy straw is to fix the fruits and vegetables inside the packages and prevent them from damaging, when there is a vibration and impact. It also serves as a cushion during unloading the produce on the ground (Fig. 12.3).

FIGURE 12.3 Paddy straw used as cushioning material for watermelons.

CFB boxes with paddy straw as cushioning material has been found to be the best packaging material for packing and the transportation of kokum fruits (Raorane, 2003). Khasi mandarin was packed with paddy straw and waste paper as cushioning material in bamboo basket and three-ply cardboard boxes and it was found that 2% $CaCl_2$ solution dip along with packing in cardboard boxes with paddy straw as cushioning material recorded more shelf-life (33.7 days) with good fruit quality (ICAR, 2009).

12.7.10 COCONUT FIBER

Among all the natural fibers available for cushioning material, coconut fiber is becoming one of the most used in the development of environment friendly produces, probably due to its characteristic of being an agricultural waste. In general, coconut fiber presents good properties in reinforcing composite, acoustic, and thermal insulation materials. Coir or coconut fiber is a natural fiber extracted from the husk of coconut and used as a cushioning material for perishable produce (Fig. 12.4). Coir is the fibrous material found between the hard, internal shell and the outer coat of a coconut. Coconut fiber and wood straw are being used to prevent damages to papayas and mangoes during the distribution (Castro et al., 2014).

FIGURE 12.4 Coconut fiber cushions.

12.7.11   LEAVES

Leaves of various plants (particularly Cassia, Polyalthia, Litchi, and Mango) are used as cushioning materials. Chandra and Kumar (2012) used banana leaves and teak leaves as wrapping materials for guava while neem (*Azadirachta indica*) leaves, rice straw, and bamboo leaves as cushioning materials for guava fruits during storage. In a study leaves of *Azadirachta*, melia, mentha, walnut, banana, basooti, and camphor were also used as cushioning materials in apple cv. Starking Delicious packages and it was found that fruits cushioned with camphor leaves were superior over non-cushioned fruits in retaining most of the quality characteristics (Chauhan & Babu, 2011; Chauhan et al., 2012). This effect of camphor leaves was attributed to volatiles generated by these botanicals which are used to destroy incipient infection on fruits, which may cause rotting during storage (Saxena et al., 1981).

12.7.12   RICE HULLS

Rice hulls are natural and environment-friendly waste commodities. They are light, biodegradable, difficult to burn and less likely to allow moisture to propagate. Two types of rice hulls are used-loose rice hulls and bagged rice hulls (Fig. 12.5). The problem with loose rice hulls is that when they are subjected to moisture they become denser thus less effective in shock absorption and are harder to handle due to its small grain and lightweight. In addition, they tend to stick to the commodities inside the bag. Bagged rice hulls increase the shock absorption capacity, are easier to handle, and provides double protection. In a study, rice hulls were used as a cushioning material inside a plastic tote and its impact absorption property was compared to a 3/16 inch bubble wrap and a 0.129 inch anti-vibration rubber pad. It was reported that 1 inch thick rice hulls sealed in plastic bag reduced impact acceleration by 41% as compared to 39% of bubble wrap and 42% of anti-vibration pad with the same thickness, respectively, which was due to trapped air inside the sealed plastic bag. In addition, rice hull was reported to be a good thermal insulating material and useful in protecting some temperature-sensitive commodities during the distribution by placing bagged rice hulls in all sides of a tote or box (Malasri et al., 2014). Garcia (1982) used rice hull ash of different particle sizes and moisture content as cushioning for tomatoes.

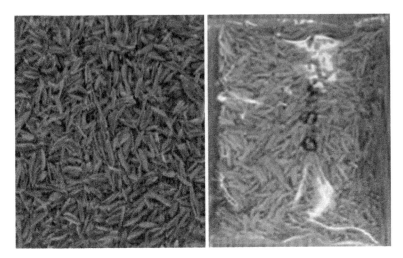

FIGURE 12.5   Loose and bagged rice hulls to be used as cushioning material.

Advantages of naturally derived cushioning materials:

- Low cost, environment friendly.
- Good damping characteristics.
- Good shock absorption capacity.
- Easily available, naturally derived.
- Some of them have anti-fungal and anti-bacterial effects.

Disadvantages of naturally derived cushioning materials are-

- When they become wet they lose their cushioning property.
- Provide good medium for the growth of fungi and bacteria.
- Resilience is poor.
- Poor fungus resistance.
- High compression set.

12.8   MANUFACTURED CUSHIONING MATERIALS

12.8.1   TRAY OR CELL PACKING

Pulp trays and cells are used for packing and transportation of apples, pears, tomatoes, melons, stone fruit, and many other fruits (Fig. 12.6). Each tray is designed specifically to offer the maximum cushioning and protection to the produce and is available in a wide range of colors and sizes. These are recyclable, efficient, and sustainable. Produce is placed in an individual compartment of a tray or cell stacked in a transport container. Tray or cell packing offers protection of produce from stacking pressure as well as from impact with each other. However, with its loose packing density and head-space in each compartment, produce could be damaged due to vibration during transportation. Trays are generally made from molded pulp paper or plastic. Cells are usually formed from corrugated board. Produce should be of uniform size in order to fit properly in molded tray compartments. The tray pack pattern has become increasingly common for produce packing as it is a simple and convenient system, with an appealing appearance.

FIGURE 12.6   Molded paper trays for cushioning of apples.

12.8.2   CRATE LINERS

Crate liners are used as cushioning materials in plastic crates during transport and storage and provide cushioning to individual commodities and liners. These are made of three-ply corrugated fiberboard to reduce the bruising and other duration damage of fruits and vegetables (Fig. 12.7) (Saran et al., 2013). Raut (1999) reported that CFB boxes with partition and tissue paper as cushioning material were the best packing material for packing and transportation of sapota fruits. Jarimopas et al. (2007) compared foam net and corrugated board as cushioning material for apple and reported that both plastic and paper-based protective wraps can be effective in providing protection against bruising from impacts. The best protection was achieved with the single face corrugated board with flutes on the outside. Fully mature sapota (*var. Cricket Ball*) packed in CFB boxes of size 400 × 300 × 150 mm, five ply rate with paper pieces as cushioning material was superior for road transportation as these even after transported to a distance of 400 km by road, ripened in 5 days after transportation and showed less spoilage, low PLW, and more firmness as compared to those packed in CFB boxes of different sizes, in traditional bamboo basket and polythene bags. Similarly fully mature aonla (*Var. Krishna*) packed in CFB boxes of non-telescopic type of size 400 × 300 × 150 mm with partition in between was found to be superior for road transportation and transported to a distance of 525 km by road and stored at room temperature for 10 days after transportation had less spoilage and more firmness compared to those packed in traditional bamboo basket and gunny bags (Anonymous, 2010).

FIGURE 12.7   Corrugated fiberboard liners for providing cushioning to fruits and vegetables.

12.8.3   FOAM NET

Several types of polymeric foams are used for cushioning. The most common are: expanded polystyrene, polypropylene, polyethylene, and polyurethane. This makes the fruits attractive and provides the necessary cushioning and separation of fruits from each other. These are soft, light in weight, and provide sufficient ventilation. Peleg (1985) described good interior packaging as the one which treats a fruit as separate units, avoids fruit-to-fruit contact, and absorbs the impact energy. At present, foam nets function well as one of the commercial packaging solutions (Fig. 12.8). Foam nets are also being used in "Pattern packing or Place packing". It is a kind of packing in which produce is generally placed by hand in a pattern, in the package (Fig. 12.8). This kind of packaging generally provides higher density packing as compared to random packing. Pattern packs are less vulnerable to produce damage, are more uniform and appealing in appearance. Pattern packing is generally used for relatively expensive, premium quality produce for high-end, or export markets (FAO, 2011).

FIGURE 12.8   Fruits with foam net cushioning and in pattern packing (with crate liners).

12.8.4 BUBBLE FILMS

Bubble films consist of two plastic films; one of which is completely flat and the other has small, round indentations. These two films have been heat sealed together, containing the necessary air. Bubble films are mainly used inside packaging containers to form the lining of the container and acts as a heat deflector and insulator. Bubble cushioning material provides outstanding protection from damage caused by shock, vibration or abrasion, and is easy to use. Bubble films are easy to handle, non-hygroscopic, versatile, and insensitive to extreme climatic conditions and have ideal shock absorption characteristics. The major disadvantage is that these are susceptible to pointed and sharp objects which make it subject to puncture. Plastic bubble film is used as cushioning in between the banana hands and to line the inside faces of the box (Fig. 12.9) (Source: APEDA).

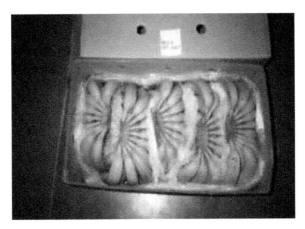

FIGURE 12.9   Bananas packed with bubble film cushioning.

12.8.5 SYNTHETIC FOAMS

Foam is a simple cushioning or blocking and bracing process for a variety of items in varying shapes and sizes. The foam is dispensed into carton lined with high-strength film. The film is folded over and the commodities are placed on the rising foam. A second sheet of film is placed over the commodities and more foam is dispensed. The benefit is that the customer receives the commodities undamaged. Styrofoam is a commonly used cushioning material for fruits and possesses a high mechanical strength in terms of compression, tension, bending, and shear. It has a smooth surface with high water and vapor resistant properties. It is also resistant to most acids and salts, and can be used repeatedly for an extended period of time. Wasala et al. (2015) used styrofoam sheets (8 mm in thickness) between the bottom and the top-most banana bunch layers on a truck and transported 120 km for 5 h during the daytime at an average daytime temperature and RH of $29 \pm 4°C$ and $68 \pm 5\%$, respectively. The truck was a medium-sized open truck with a leaf spring suspension, which is commonly used for transporting construction materials. In comparison to fresh banana leaves the styrofoam sheets reduced the physical damage to the bananas, with the quality parameters maintained at the preferred level (Fig. 12.10). One inch polyurethane pad placed on top of bulk transported peaches dampened in-transit vibration and reduced injury by 15% (O'Brien & Guillou,1969). Thick soft foam padding resulted in minimal color change and significantly reduced cracking in lychee (Philippa, 2004). Fresh lychee skin gave the best control of cracking. Styrofoam crates are also used for vegetable in local markets of Jordan (Fintrac, 2012).

(a)   (b)

FIGURE 12.10   Quality of banana as affected by cushioning materials: (a) banana leaves (control) and (b) styrofoam layer.

12.8.6   THERMOFORMED PLASTIC TRAYS

These are trays made from thermoplastics which can be softened by heating and hardened by cooling. These are generally made of polystyrene, polypropylene (PP), and polyvinyl chloride (PVC). This offers rigid packaging as well as a great cushioning effect and immobilizes the produce within the pack (Fig. 12.12). These trays are clean, neat in appearance, and light in weight. They give a cushioning effect to the commodities packed inside. The trays can be easily molded in any size and shape. The materials used can be easily cleaned, re-used, and is also recyclable.

FIGURE 12.11   EPS (styrofoam) crates being used in vegetable market.

FIGURE 12.12   EPS trays for dates.

Advantages of synthetic cushioning materials are

- Highly resilient.
- Good shock absorption capacity.
- Some are resistant to moisture.
- Can be molded according to produce.

- Resistant to microbial infection.
- Retain their cushioning ability in unfavorable circumstances also.

Disadvantages of synthetic cushioning materials are

- High cost is a constraint.
- Not easily available at farm level.
- Take high time to degrade.
- Non environment friendly.

12.8.7 MOISTURE ABSORBER

Moisture absorbers are internationally used in containers and protects valuable produce against fungi and corrosion; used in exports of fruits and vegetables, spices, groundnuts, and tea. Singh et al. (2016) used desiccant mixture of bentonite 0.55 g + sorbitol 0.25 g + $CaCl_2$ 0.20 g/g as moisture absorber to increase the shelf-life of white button mushrooms and reported that mushrooms packed in polypropylene pouches with 12 g of absorber showed best results with respect to weight loss and whiteness (Fig. 12.13).

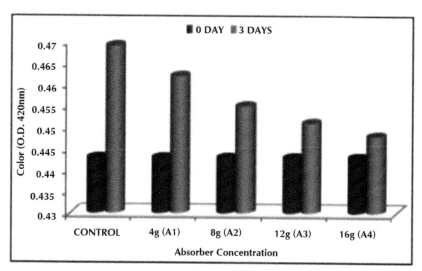

A = Absorber or desiccant mixture of bentonite 0.55 g + sorbitol 0.25 g + $CaCl_2$ 0.20 gg-1

FIGURE 12.13 Effect of different concentration of moisture absorber on degree of whiteness (color) of white button mushroom stored at ambient condition (22 + 2°C) (Singh et al., 2016).

12.9   OPERATION SPECIFIC USE OF CUSHIONING MATERIAL FOR FRUITS AND VEGETABLES

Cushioning materials are used in many stages during postharvest handling operations. However, in order to maintain postharvest quality of produce it becomes necessary at three main stages: (1) at the time of keeping harvested produce into any rigid storage container. All crates have hard surfaces and while keeping produce inside, there is a chance of dropping off from little height, causing impact bruising popularly called touching marks; (2) during transportation from field to pack houses. In general, plastic crates are used for transportation from field to pack house and the distance may vary from very short to too long. Based on the road conditions, there would be impact and vibration bruising; this type of bruising may not be visible immediately, but after few days, browning/pitting or blackening symptoms may develop resulting in rotting of produce. Cushioning materials if used in plastic crates reduce these bruising and touching marks drastically; (3) during transportation of packed produce from pack house to destination markets. Loading, unloading, and transportation jerks are also responsible for bruising. Therefore, it is recommended to use cushioning material in packages to preserve postharvest quality of fresh produce (Ahmad & Siddiqui, 2015).

Generally, cushioning materials are used during harvesting, handling, and transportation of produce.

12.9.1   USE OF CUSHIONING MATERIAL DURING HARVESTING AND HANDLING

Handling operations make up most of the steps in harvesting, grading, and packing of harvested produce. These operations may result in mechanical injury to fruit caused by impact from falls. Fruit blemishes and injury are usually noticed during harvesting. Quite often the fruits fall to the ground becoming subject to field infestation and mechanical damages (FAO, 1989). These damages can be prevented by use of cushioning materials. Surface marking is associated with abrasion and vibration. Abrasion may occur when fruit rubs against some surface like the picking bag or bucket, the field lug or bin, or even another fruit. Abrasion in the field can be reduced or eliminated by good supervision of harvest crews.

The use of cushioning material is important when fruits are harvested by shaking the trees and it may fall from a great height. During harvesting, there is possibility of produce suffering mechanical damage as a result of free fall to the ground (Fig. 12.14). Cushioning material used to cover fruit-catching frames should have two important requirements: (1) It should absorb enough impact force to prevent the bruising of fruit, and (2) retain its cushioning ability long enough to justify the expense because rapid deterioration of the cushioning material through weathering can quickly make the material useless.

(a)   (b)

FIGURE 12.14   Cushioning of fruits: (a) use of thick tarpaulin sheets for mango harvesting and (b) leaves and straw in storage shed for collecting peach and plum.

To prevent this, farmer uses net harvester and gently lower the produce on the ground or they collect the fruit on a tarpaulin sheet or a strong cloth. Leaf-shaped mango harvester consists of a GI frame attached to telescopic aluminum handle, a V-blade of high carbon steel, a divider to guide the fruits and a nylon netted receptacle to collect fruits. It functions through the actions of holding, pulling, and shearing.

Handling activities at the farmer's field or storage go down is also carried out and the dried leaves, thick sheets are used as cushioning material. When fruits or vegetables are difficult to catch, for example, mangoes or avocados, a cushioning material is placed beneath the tree to prevent damage to the fruit when dropping from high trees (FAO, 1998). When lettuce is field packed, several wrapper leaves are left on the head to help cushion the produce during transport (Lisa Kitinoja & Kader, 2001).

12.9.2 CUSHIONING MATERIAL DURING TRANSPORT/ LOADING/UNLOADING

Impact and compression damages are widely noticed during transport. Non-refrigerated transport also adds to the losses suffered by the commodity. The most common cushioning material used are Paddy straws (Fig. 12.14), jute bags and leaves of plants are used as the cushioning materials on trucks and storage area in the marketplace to prevent damage during transport, during loading, and unloading operations.

Dry grass, paddy straw, leaves, sawdust, paper shreds, etc., are some examples of cushioning materials which are used for fruits and vegetables. Sheets of newspaper or shredded newsprint are commonly used as a lining material in bamboo, plastic crates, and in corrugated fiberboard boxes. Leaves are also used as a lining material inside bulk containers. Plastic foam netting is commonly used for wrapping individual fruits before filling in corrugated boxes, partitions, and pads made of corrugated board are used to accommodate fruits in layer. Honeycomb material used for packing fruits totally separated from adjacent fruit can be highly advantageous.

FIGURE 12.15 Cushioning of watermelon fruits in market yard and vehicles.

The other materials used during transport are foam nets, bubble films, EPS strays, styrofoam crates, and other synthetic materials.

12.10 CUSHIONING MATERIALS FOR CUT FLOWERS

Cushioning materials are equally important for cut flowers as they are very sensitive to mechanical damages. There are many shapes of packing containers for cut flowers, but most are long and flat and a full telescoping design (top completely overlaps the bottom) (Fig. 12.16). This design restricts the depth of the flowers in the box, which may in turn reduce physical damage of the flowers. In addition, flower heads can be placed at both ends of the container for better use of space. With this kind of flower placement, whole layers of newspaper have often been used to prevent the layers of flowers from injuring each other. The use of small pieces of newspaper to protect only the flower heads, however, is a better practice, since it allows for more efficient cooling of flowers after packing. It is critically important that containers be packed in such a way that transport damage is minimized. Some growers anchor the commodities by using enough flowers and foliage in the box so that the package, after banding, holds itself firmly. Materials used for sleeving include paper (waxed or unwaxed), cotton (in case of orchids), corrugated card (smooth side toward the flowers) (Fig. 12.15) and polyethylene (perforated, unperforated and blister). Sleeves can be performed (although variable bunch size can be a problem), or they can be formed around each bunch using tape, heat- sealing (polyethylene), or staples.

FIGURE 12.16 Cushioning of different cut flowers.

Gladioli, snapdragons, and some other species are often packed in vertical hampers to prevent geotropic curvature which reduces their acceptability. Specialty flowers, such as anthurium, orchid, ginger, and bird-of-paradise are packed in various ways to minimize friction damage during transport. Frequently, the flower heads are individually protected by paper or polyethylene sleeves. Cushioning materials such as shredded paper, paper wool, and wood wool may be distributed between the packed flowers to further reduce damage (Reid, 2016). Halevy et al. (1978) reported that cut flowers of Bird of Paradise (*Strelitzia*) stems are bunched and packed singly. Some shippers also place a waxed paper sleeve on each inflorescence to protect them during transportation. Because of the weight of stems, dry or moist shredded paper is packed around the stems to reduce bruising. Similarly, cut flowers of Anthurium are packed in cardboard cartons and lined with polyethylene sheets and layers of newspaper to provide cushioning. Ten dozen or more anthuriums may be packed in a $102 \times 43 \times 29$ cm^3 carton with spathe face down and their stems interwoven and moistened, shredded newspaper is inserted to provide a cushion and maintain humidity. Some growers also cover each inflorescence with a waxed tissue or polyethylene envelops to reduce mechanical damage due to the spadix oppressing the spathe (Akamine, 1976).

Small heliconias are bunched in 5 or 10 s and may be sleeved in plastic film or netting to prevent bruising. Heavy heliconias are usually packed with moist shredded newspaper or between layers of newspaper. Bunches of Small heliconias are kept over the large bunches to provide additional cushioning (Criley & Paul, 1993). Some cushioning materials for cut flowers are given in Table 12.3.

12.11   FUTURE THRUST

Although cushioning materials are very important component for packaging of perishable horticulture produce, a little work has been done in this area. Besides, most of the work on cushioning material has been done in temperate crops therefore, it is suggested to be developed for tropical and subtropical crops too. Further studies are required to determine the appropriate cushioning material for every fruit, vegetable, and flower so that these produce can be safely transported to long distances with minimum bruising and impact damage. The research is needed to develop innovative cushioning materials with combined properties of both the natural and synthetic materials like improved biodegradability, recyclability, high resilience, low

TABLE 12.3 Some Cut Flowers and Their Cushioning Materials.

Cut flower	Botanical name	Cushioning material	References
Alstroemeria	*Alstroemeria*	CFB sleeve	Reid (2016)
Anthurium	*Anthurium andreanum*	Shredded newspaper and other shredded paper	Pritchard et al. (1991)
Bird of paradise	*Strelitzia* sp.	Waxed paper bags or moist shredded paper	Halevy et al. (1978)
Red ginger	*Alpinia purpurata*	Moistened shredded newspaper	Criley and Paul (1993)
Heliconia	*Heliconiapsitta corum* and *H. angusta*	Moistened shredded newspaper	Criley and Paul (1993)
Orchids	Orchid	Shredded paper shredded wax paper	Sugapriya (2009)
Roses	*Rosa* sp.	Plastic + waxed paper + soft corrugated card sleeves	Reid (2016)
Tulip	*Tulipages neriana*	Waxed paper	Reid (2016)

compression, low moisture absorption, and less susceptibility to microbes (natural cushioning material). Innovations in cushioning materials will help in exploiting new markets as the produce can be transported to long distances with minimal damage.

12.12   CONCLUSION

Fresh horticultural produce like fruits, vegetables, and flowers are very delicate in nature and, therefore, are easily damaged by mechanical factors like rubbing in between the commodities, vibrations, and compressions during harvesting, handling, and transportation. At this time proper packaging of these commodities plays an important role in protecting the commodities from damages. Cushioning materials are very important component of packaging especially for fresh fruits, vegetables, and flowers. These serve to protect the produce from mechanical damages and bruising which occurs due to rubbing of produce against each other/package wall by filling of void spaces in the package. These may be natural (paddy straw, rice hulls, coconut fibers, mango leaves, etc.), or synthetic (styrofoam, bubble sheets, molded paper trays, etc.). Use of agro waste materials like camphor leaves provides antimicrobial effect in addition to cushioning.

KEYWORDS

- cushioning material
- mechanical damage
- compression
- impact
- resilience
- water resistant

REFERENCES

Ahmad, M. S.; Siddiqui, M. W., Eds.; Factors Affecting Postharvest Quality of Fresh Fruits. In *Postharvest Quality Assurance of Fruits: Practical Approaches for Developing Countries;* Springer International Publishing: Switzerland, 2015; p 256.

Akamine, E. Post-harvest Handling of Tropical Ornamental Cut Crops in Hawaii. *Hort. Sci.* **1976,** *11* (2), 125–127.

Alders, A. W. C. *Reefer Transport & Technology;* Rotterdam Marine Chartering B. V. Agents: Krimpena/dYssel, Netherlands, 1995.

Altisent, M. R. In *Damage Mechanisms in the Handling of Fruits: Progress in Agricultural Physics and Engineering;* John, M., Ed.; Commonwealth Agricultural Bureaux (CAB) International: Willingford, UK, 1991; pp 231–255.

Anonymous. IIHR Annual Report 2009–10. Indian Institute of Horticultural Research: Bangalore, Karnataka, 2010.

Bagshaw, J. S.; Underhill, S. J. R.; Fitzell, R. D. Lychee Disorders. In *Tropical Fruits: Post-harvest Diseases of Horticultural Produce;* Coates, L., Cooke, T., Persley, D, Beattie, B., Wade, N., Ridgway, R., Eds.; Qld Department of Primary Industries: Brisbane, Australia, 1995; Vol. 2, pp 43–45.

Bajema, R. H.; Hyde, G. M. Instrumented Pendulum for Impact Characterization of Whole Fruit and Vegetable Specimen. *Trans. ASAE.* **1998,** *41* (5), 1399–1405.

Batten, D. J.; Loebel, M. R. Lychee Harvesting and Post-harvest Handling. Agfact H6.4.1, Department of Agriculture: NSW, Australia, 1984.

Bhardwaj, R. L.; Sen, N. Z. Zero Energy Cool Chamber Storage of Mandarin (*Citrus reticulata* cv. Nagpur Santra). *J. Food Sci. Technol.* **2003,** *40,* 669–672.

Brown, G. K.; Schulte, N. L.; Timm, E. J.; Armstrong, P. R.; Marshall, D. E. Reduce Apple Bruise Damage. *Tree Fruit Postharvest J.* **1993,** *4* (3), 6–10.

Castro, C. D. P. C.; Faria, J. A. F.; Dantas, T. B. H. Testing the Use of Coconut Fiber as a Cushioning Material for Transport Packaging. *Mater. Sci. Appl.* **2012,** *3* (3), 151–156.

Castro, C.; Faria, J.; Dantas, T. Evaluating the Performance of Coconut Fiber and Wood Straw as Cushioning Materials to Reduce Injuries of Papaya and Mango during Transportation. *Int. J. Adv. Packag. Technol.* **2014,** *2* (1), 84–95.

Chandra, D.; Kumar, R. Qualitative Effect of Wrapping and Cushioning Materials on Guava Fruits during Storage. *Hort. Flora Res. Spect.* **2012,** *1* (4), 318–322.

Chauhan, S. K.; Thakur, K. S.; Jawa, N. K.; Thakur, K. P. Botanical Formulation and Extracts Based on Plant Leaves and Flower, a Substitute for Toxic Chemical and Waxes for Shelf Life Extension and Quality Retention of Apple cv Starking Delicious in India. *J. Hortic. For.* **2012,** *4* (12), 190–100.

Chauhan, S. K.; Babu, D. R. Use of Botanicals: A New Prospective for Enhancing Fruit Quality Over Chemicals in an Era of Global Climate Change. *Asian J. Environ. Sci.* **2011,** *6* (1), 17–28.

Deutscher Transport-Versicherungs-Verbande, V.: Die Ware in der Transportversicherung, Hamburg, 1990–1994.

Food and Agricultural Organization. *Prevention of Post-harvest Food Losses Fruits, Vegetables and Root Crops a Training Manual.* FAO Training Series. No. 17/2FAO, Code: 17, AGRIS: J11 ISBN 92-5-102766-8, 1998.

FAO. *Packaging in Fresh Produce Supply Chains in Southeast Asia;* RAP Publication 2011/20, Food and Agriculture Organization of the United Nations Regional Office for Asia and the Pacific Bangkok, 2011.

Garcia, J. L. Storage of Tomato (*Lycopersiconesculentum*) in Rice Hulls Ash of Different Particle Sizes and Moisture Content. M.Sc. Thesis, University of Philippines, LosBanos, Laguna, 1982.

Greer, G. N. *Growing Lychee in South Queensland;* Department of Primary Industries: Brisbane, Australia, 1990.

Halevy, A. H.; Kofranel, A.M.; Besemer, S.T. Postharvest Handling Methods for Bird-of-paradise Flowers (*Strelitziareginae* Ait.). *J. Amer. Soc. Hort. Sci.* **1978,** *103* (2), 165–169.

Hilton, D. J. In *Impact and Vibration Damage to Fruit during Handling and Transportation,* ACIAR Proceedings (Postharvest Handling of Tropical Fruits), 1994; Vol. 50, pp 116–126.

ICAR Annual Report. *Report on Updation of Technical Specifications of Packaging for Fresh Fruits and Vegetables for Export.* Agricultural & Processed Food Commodities Export Development Authority: New Delhi, 2008–2009.

Jarimopas, B.; Mahayosanan, T.; Srianek, N. Study of Capability of Net Made of Banana String for Apple Protection against Impact. *Eng. J. Kasetsart.* **2004,** *17* (51), 9–16.

Jarimopas, B.; Singh, S. P.; Sayasoonthorn, S.; Jagjit Singh. Comparison of Package Cushioning Materials to Protect Post-harvest Impact Damage to Apples. *Packag. Technol. Sci.* **2007,** *20,* 315–324.

Lallu, N.; Rose, K.; Wiklund, C.; Burdon, J. Vibration Induced Physical Damage in Packed Hayward Kiwifruit. *Acta Hortic.* **1999,** *498,* 307–312.

Lisa Kitinoja; Adel Kader, A. *Small-Scale Postharvest Handling Practices: A Manual for Horticultural Crops,* 4th ed.; University of California: Davis, CA, 2001.

Malasri, S.; Stevens, R.; Othmani, A.; Harvey, M.; Griffith, I.; Guerrero, D.; Johnson, M.; Kist, M.; Nguyen, C.; Polania, S.; Qureshi, A.; Sanchez-Luna, Y. Rice Hulls as a Cushioning Material. Cloud Publications. *Int. J. Adv. Packag. Technol.* **2014,** *2* (1), 112–118. Article ID Tech-305 ISSN 2349–6665.

Mellenthin, W. M.; Wang, C. Y. Friction Discoloration of 'd'Anjou' Pears in Relation to Fruit Size, Maturity, Storage and Polyphenoloxidase Activities. *Hort. Sci.* **1974,** *9,* 592–593.

O'Brien, M.; Guillou, R. An In-transit Vibration Simulator for Fruit-handling Studies. *Trans. ASAE.* **1969,** *12* (1), 94–97.

Peleg, K. *Produce Handling, Packaging and Distribution;* AVI Publishing Company Inc.: Westport, CT, 1985.

Pritchard, M. K.; Hew, C. S.; Wang, H. Low-temperature Storage Effects on Sugar Content, Respiration and Quality of Anthurium Flowers. *J. Hort. Sci.* **1991,** *66,* 209–214.

Philippa, B. Optimising the Postharvest Management of Lychee (Litchi Chinensis Sonn): A Study of Mechanical Injury and Desiccation. Ph.D. Thesis, Faculty of Agriculture, Department of Crop Sciences, The University of Sydney, Sydney, 2004.

Raorane; GeetaPundalik. Studies on Growth, Flowering, Fruiting and Some Aspect of Harvest Handling of Kokum (*Garcinia indica* choisy). MSc Thesis, Dr. BalasahebSawant Konkan KrishiVidyapeeth, Dapoli, District. Ratnagiri, 2003.

Raut, V. U. Studies on Maturity Indices, Harvesting, Integrated Post Harvest Handling and Processing of *Sapota w.* Kalipatti. Ph.D. Thesis, Dr. B.S. Konkan KrishiVidyapeeth, Dapoli, M.S., India, 1999.

Reid, M. S. Cut Flowers and Greens. In *Commercial Storage of Fruits, Vegetables and Florist and Nursery Stocks;* Gross, K. C., Wang, C. Y., Saltveit, M., Eds.; United States Department of Agriculture: Washington, DC, 2016; pp 659–708.

Saraswathy, S.; Preethi, T. L.; Balasubramanyan, S.; Suresh, J.; Revathy, N.; Natarajan, S. *Post Harvest Management of Horticultural Crops;* Agrobios: India, 2008; xxxii, p 544. ISBN: 817754322.

Saxena, R.; Liquido, N. J.; Justo, H. B. *Natural Pesticides from the Neem Tree* (*Azadirachta-indica*)*;* Proceeding of International Neem Conference RattachEngem, Germany, 1981.

Schoorl, D. Packaging of Fruits and Mechanical Damage to Fruit. Thesis, University of Queensland, 1974.

Singh, N.; Mishra, V.; Vaidya, D. Optimization of Moisture Absorber with Packaging Material to Increase the Shelf-life and Quality of White Button Mushroom (*Agaricusbisporus*) at Ambient Condition. *Int. J. Multidisciplinary Res. Dev.* **2016,** *3* (11), 42–46.

Sitkei, G. *Mechanics of Agricultural Materials-Developments in Agricultural Engineering;* Elsevier: Amsterdam, 1986; Vol. 8, p 487.

Sugapriya, S. Evaluation of Dendrobium Orchids under Greenhouse Condition. Greenhouse Condition. MSc Thesis, University of Agricultural Sciences, Dharwad, 2009.

Wasala, W. M. C. B.; Dharmasena, D. A. N.; Dissanayake, C. A. K.; Tilakarathne, B. M. K. S. Feasibility Study on Styrofoam Layer Cushioning for Banana Bulk Transport in a Local Distribution System. *J. Biosyst. Eng.* **2015,** *40* (4), 409–416.

CHAPTER 13

NANO-ENABLED PACKING OF FOOD PRODUCTS

VIGYA MISHRA[1], NEERU DUBEY[1*], SIMPLE KUMAR[1], and NEHA SINGH[2]

[1]Amity International Centre for Post Harvest Technology and Cold Chain Management, Amity University, Noida, Uttar Pradesh, India

[2]Warner School of Food and Dairy Technology, SHIATS, Allahabad, Uttar Pradesh, India

*Corresponding author. E-mail: needub@gmail.com

CONTENTS

ABSTRACT

Nanotechnology research and application in the food industry is getting focused on food packaging. With the increasing global customer base, food retailing is transforming. However, with the move toward globalization, food packaging requires longer shelf-life, along with monitoring food safety and quality based upon international standards. Application of nano-based packaging material is helpful in achieving the same. Nano-based packaging modifies the physical properties of packaging material thereby providing enhanced gas and water vapor barrier, improving mechanical and heat-resistance, controlling microbial growth, delaying oxidation, detection of food spoilage, creating nano-biodegradable packaging materials, and providing brand protection. Nanotechnology has a prospective revolution in the food industry. It is also being used in innovative development of biosensors for detection of pathogens and chemical contaminants. Besides, this new technology has also raised serious concerns about toxicological aspects of nanoparticles in food, with emphasis on the risk assessment and safety issues. Also, it reflects the urgent need for regulatory framework capable of managing any risks associated with implementation of nanoparticles in food technology. This chapter focuses on tremendous benefits of nanotechnology in food industry in terms of food packaging, safety, and quality control as well as its safety concerns related to its migration in food system.

13.1   INTRODUCTION

The food and beverage packaging industry is growing day-by-day with innovations in packaging materials. It includes biodegradable packaging, edible packaging, active packaging, etc. Packaging industry has been a recent focus for nanomaterial applications. Even though it is a newly emerging technology, various market surveys have predicted it to increase continuously. It estimated that up to 400 companies around the world are developing possible applications of nanotechnology in food and food packaging (Neethirajan & Jayas, 2011). According to a new report published by Persistence Market Research, the global nano-enabled packaging market for food and beverages industry was worth USD 6.5 billion in 2013 and is expected to grow at a CAGR of 12.7% during 2014–2020 to reach an estimated value of USD 15.0 billion (CNBC, 2014). The European Institute for Health and Consumer Protection has also revealed that the use of nanomaterials in the food packaging market is expected to reach USD20 billion by 2020. Nanomaterials are increasingly being used in the food packaging industry due to the wide range of advanced functional properties they incorporate in packaging materials. It is also being expected that advances in the world of nanotechnology will continue to fuel the growth of the global nano-enabled packaging for food and beverages market in the coming future. The boosting demand for fresh packaged food since past few years has been a primary factor fueling the demand for nano-enabled packaging. Besides, introduction of innovative applications to the packaging industry is also expected to boost the nano-enabled packaging market for food and beverages. Manufacturers and retailers are showing interest in innovative shelf-stable packaging. This packaging is intelligent, safer, and traceable, which is expected to compel consumers to prefer it over other available packages in the market.

By application, the global nano-enabled packaging market for food and beverages is segmented into fruits and vegetables, beverages, prepared foods, meat products, bakery, and others. Increasing demand for freshness especially in case of fruits and vegetables is anticipated to boost the market for nano-enabled packaging.

Although nano-enabled packaging technology is gaining upright grip in the food and beverages industry, the revenue share of the nano-enabled packaging market for food and beverages is merely 3–4% of the total revenues of the food and beverages industry. Many companies offer nano-enabled packaging solutions but the buyer's cluster is limited, which prompts at the untapped scope for market growth.

13.2   SCOPE OF NANOTECHNOLOGY IN THE FOOD INDUSTRY

Nanotechnology is now conquering the food industry and establishing great potential. It has several uses in the food industry right from food production to processing and up to its packaging. A number of food companies are developing and incorporating nanomaterials in foods and food packaging which is making a difference not only in the taste of food, but also providing food safety and health benefits. Applications of nanotechnology in the food industry include: (1) encapsulation and delivery of substances in targeted sites; (2) increased flavor; (3) coating of antibacterial nanoparticles in packaging materials; (4) detection of contamination due to microbes; (5) improved food storage; (6) tracking; (7) tracing; and (8) brand protection (Chellaram et al., 2012).

Nanomaterials in foods can be naturally produced as well as intentionally added by man-made materials. Naturally produced nanomaterials could be of animals and plant origin, for instance, casein micelles in milk (Aguilera, 2014) and pectin nanostructures in fruit (Zhang et al., 2008). Intentionally added nanomaterials may come from two sources; naturally occurring or engineered material sources, which are not generally present in a food substance (Magnuson et al., 2011). A number of engineered nanomaterials are being used in the food industry because of their expected benefits. For example, nanoscale grains of salt have been developed which is helpful in reducing salt consumption by increasing its surface area and therefore even a small amount of it can give human taste buds the same original savory taste (Rasouli & Zhang 2006). Nano-supplements are being incorporated in the food system by encapsulation techniques for nutritional and drug delivery systems effectively (Chellaram et al., 2012).

Use of nanomaterials in food packaging is already a reality. Nowadays, nanomaterials are being used for several kinds of food packaging, since they represent a new alternative additive source for improving polymeric properties of packaging materials (Fig. 13.1).

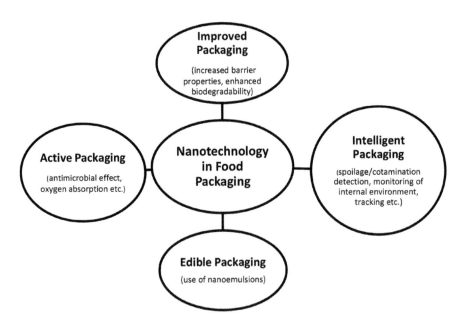

FIGURE 13.1   Applications of nanotechnology in food packaging.

Current major applications of nanomaterials in food packaging include:

- Enhancement of barrier properties through the incorporation of nano-fillers (e.g., nanoclay).
- "Active" food packaging with controlled release of active substances such as antibacterial to improve shelf-life of food (e.g., nanosilver).
- Improvement of physical characteristics to make the packaging more tensile, durable, or thermally stable (e.g., nano-titanium dioxide (TiO_2) and titanium nitride (TiN)).

Potential future applications include the use of "smart" packaging (in the form of nano-sensors, labels, etc.), as well as polymer composites incorporating nano-encapsulated substances allowing consumers to modify food depending on their own nutritional needs or tastes.

13.3   TYPES OF NANOMATERIALS USED FOR FOOD PACKAGING

Packaging of fresh foods like fruits and vegetables requires protection, barrier properties against gas/moisture/vapor, tampering resistance, and improved strength (especially in case of bio-based packaging) and many other special physical, chemical, or biological characteristics. Nano-food packaging materials are being helpful in extending the shelf-life of food by providing above mentioned properties, improving food safety, alerting consumers about contamination, or spoilage of food and even by releasing preservatives in the packages.

Currently, a variety of nanomaterials have been introduced to food packaging as functional additives. These include nanocomposites, nanolaminates, nanofillers, nanofibers, nanotubes, etc. Due to differences in physical and chemical characteristics, each nanomaterial provides distinct properties to the host material. For example, silver nanoparticles (AgNPs) are engineered mostly for antimicrobial and sterilization purposes. Other types of nanomaterials include nanoclays, layered silicates (naturally occurring fine-grained minerals), metal oxide nanoparticles such as zinc oxide and TiO_2, and nano-TiN, (EFSA, 2012). Nano-TiN is generally synthesized by heating TiO_2 particles in a nitrogen-containing gas at high temperature (Dong et al., 2011).

A brief account of some important types of nanomaterials used for packaging has been given below:

Nanocomposites: A nanocomposite is a multiphase material derived from the combination of two or more components, including a matrix (continuous phase) and a discontinuous nano-dimensional phase with at least one nano-sized dimension (i.e., with less than 100 nm). Polymer nanocomposites promise a new crop of stronger, more heat resistant, and high barrier materials (Luduena et al., 2007). For packaging, nanocomposites are prepared with improved properties like enhanced gas barrier and improved biodegradability. Mechanical, thermal, and biological properties of nanocomposites often vary markedly depending upon their component materials (Table 13.1). Polymer-based nanocomposites have been reported to achieve the same or better barrier properties than their conventional composite counterparts (Avella et al., 2005, 2007; Mihindukulasuriya & Lim, 2014). These nanocomposites are reinforced with small quantities (typically up to 5% by weight) of nanoparticles, which have very high aspect ratios (L/h > 300) (Chaudhry et al., 2008; FAO/WHO, 2009). They are incorporated in addition to the traditional fillers and additives.

TABLE 13.1   Nanocomposites and Their Properties.

Type	Nanofillers	Properties due to nanofillers	References
Bio-nanocomposites			
Polylactic acid (PLA)	Clays and calcium carbonate	–	Gacitua et al. (2005)
Gelatin	Silver nanoparticles	–	Rafician et al. (2014)
Zein	MMT nanofiber mats	–	Park et al. (2013)
Corn starch	nanodiamond	–	Park et al. (2004)
Starch	Cashew tree gum	–	Pinto et al. (2014)
Starch	clay	–	Tang et al. (2008)

Nanocomposites have also been researched for their use in so-called "active" food packaging, which refers to the controlled release of active substances from the food packaging materials. An example of nanocomposites used for "active" packaging is polymer composites with antimicrobial nanomaterials, for example, silver, zinc oxide, and magnesium oxide. Another most well-known example of a nanocomposite currently used in "active" food packaging is nanosilver coated plastics which confers antimicrobial properties to improve food and beverage shelf-life (de Azeredo et al., 2011; Fortunati et al., 2013; Valipoor et al., 2013).

Apart from conferring barrier properties to extend the shelf-life of food other nanocomposites confer various physical characteristics to make the packaging more tensile, durable, or thermally stable (Beltran et al., 2014). Examples of other nanocomposites include UV-absorber nanocomposites containing alumina and TiN which is used to improve strength of packaging materials, nano-calcium carbonate-polymer composites, nano-chitosan-polymer composites, biodegradable nanoclay composites of starch and polylactic acid, biodegradable cellulose nano-whiskers, and other gas-barrier coatings (e.g., nano-silica) (Sanchez-Garcia et al., 2010; Smolander & Chaudhry, 2010). Some materials are already commercially available and are being used by beverage companies in certain countries.

Nanofillers: These are the materials used as fillers in making nanocomposites or used as a coating with synthetic and biopolymers. Nanofillers can be classified into organic and inorganic. Organic nanofillers include nano-clay, proteins, polysaccharides, etc. while inorganic nanofillers are metals, metal oxides, etc. (Table 13.2). Other examples of nanofillers potentially used in food packaging include alumina (e.g., wheel-shaped alumina plate-lets used as fillers for plastic materials), nano-precipitated calcium carbonate (to improve mechanical properties, heat resistance, and printing quality of polyethylene (PE)), polyhedral oligomeric silsesquioxane (POSS) nanoclay (to improve barrier properties), zinc oxide calcium alginate nanofilms (used as a food preservative), and silica/polymer hybrids (to improve oxygen-diffusion barriers for plastics) (Smolander & Chaudhry, 2010; Bajpai et al., 2012). Mode of action of some important nanofillers majorly being used in nano packaging of foods have been discussed below:

1. *Nanoclay:* A well-known example of one of the first nanocomposites to be explored for use in food packaging. Clay (i.e., bentonite) is a naturally occurring substance with platelets whose thicknesses are in the nanoscale size range. Bentonite has a long history of permitted use as a food additive at levels up to 5% w/w in Europe. The dispersal of sheet-structured nanoclay into the polymer matrix creates the enhanced barrier properties of homogeneous polymer due to the increase of tortuous pathway against penetrating molecules (Adame & Beall, 2009). A classic property of nanoclay is to improve the mechanical and barrier properties of plastic/biodegradable packaging. It is being incorporated with nylons, polyolefins, copolymers, epoxy resins, polyurethane, PE, terephthalate, etc.

   The most widely studied type of clay fillers is montmorillonite (MMT), a hydrated alumina-silicate layered clay consisting of an

TABLE 13.2 Different Nanofillers for Food Packaging Applications.

Classification	Type of nanofiller	Examples, if any	Properties	References
Organic	Nanoclay	MMT	Antimicrobial properties, water barrier properties	Pinto et al. (2014)
		Cloisite Na+	Optimized surface properties and food contact applications	Shin et al. (2014)
		Cloisite 30B	Optimized surface properties and food contact applications	Kanmani and Rhim (2014); Shin et al. (2014)
		Cloisite 20A	Optimized surface properties and food contact applications	Shin et al. (2014)
	Natural biopolymers	Chitosan	antimicrobial properties	Martelli et al. (2013)
		Cellulose	Gas barrier	Shakeri and Radmanesh (2014)
		Cellulose nanofibrils	High gas barrier material	Rafieian et al. (2014)
	Natural antimicrobial agents	Nisin	Antimicrobial	Imran et al. (2012)
Inorganic	metal	Silver	Antimicrobial properties towards wide range of microorganisms; stable and low volatile at high temperature	Incoronata et al. (2011); Kanmani and Rhim (2014); Youssef et al. (2014)
		Copper	Antimicrobial properties	Conte et al. (2013)
		Gold	Antimicrobial properties	Youssef et al. (2014)
	Metal oxides	ZnO	Deodorizing and antibacterial properties	Kanmani and Rhim (2014); Nafchi et al. (2013); Rouhi et al. (2013)
		MgO	Antibacterial	Sanuja et al. (2014)
		Ag_2O	Antibacterial	Tripathi et al. (2011)
		SiO_2	provides excellent barrier properties	Bharadwaj et al. (2002)

edge-shared octahedral sheet of aluminum hydroxide between two silica tetrahedral layers (Weiss et al., 2006). The imbalance of the surface negative charges is compensated by exchangeable cations (typically Na^+ and Ca^{2+}).

2. *Layered Silicate:* The layered silicates commonly used in nanocomposites consist of two-dimensional layers, which are 1 nm thick and several microns long depending on the particular silicate. Its presence in polymer formulations increases the tortuosity of the diffusive path for a penetrant molecule, providing excellent barrier properties (Bharadwaj et al., 2002).

3. *Silver Nanoparticles (AgNPs):* AgNPs have a larger surface area per mass than micro-scale silver particles or bulk silver material, the potential to release silver ions is also greater than that of bulk silver (Marambio-Jones & Hoek, 2010).

4. *Nano-ZnO and Nano-TiO$_2$:* The photocatalytic reaction of nano-ZnO and nano-TiO2 attributes to generation of reactive oxygen species (ROS), resulting in the oxidation of cytoplasm of bacterial cells and leading to cell death. It was reported that ZnO is relatively more efficient and attractive over silver due to the less toxicity and cost effectiveness (Duncan, 2011; Silvestre et al., 2011).

5. *Nano-TiN:* It is widely used for mechanical strength and processing aid particularly for polyethylene terephthalate (PET) (Chaudhry & Castle, 2011) (Table 13.3).

TABLE 13.3 Morphology and Dimensions of Nanofillers.

Type of nanofiller	Morphology/structure	Density
Montmorillonite (MMT)	Platelets	2.6 gcm^{-3}
Layered double hydroxide (LDH)	Platelets with distinct hexagonal shape	1.5 gcm^{-3}
Carbon nanotubes (CNT)	tubular, seamless cylinder of graphene sheet	1.3−1.5/1.5/1.8 −2.0 gcm
Cellulose nanowhiskers (CNW)	Rod like	~1.6 gcm^{-3}
Microfibrillated cellulose (MFC)	Highly fibrous network consisting of bundles of nanofibrils	~1.6 gcm^{-3}
Bacterial cellulose (BS)	Ribbon-shaped nanofibers forming a highly fibrous network	~1.6 g cm^{-3}
Chitin whiskers (CHW	Slender rods	~1.5 gcm^{-3}
Starch nanocrystals (SNC)	Platelet-like nanoparticles; often forming few μm size aggregates	~1.55 g cm^{-s}

13.4   APPLICATION OF NANO-BASED FOOD PACKAGING

It can generally be divided into three main categories (Silvestre et al., 2011; Duncan, 2011):

1.   Improved Packaging: For intelligent packaging, physical properties of the packaging material can be enhanced by using certain nanomaterials like nanoclay. Improvement in barrier and mechanical properties can also be achieved by blending of nanomaterials with polymer. Nanocomposites for food packaging are prepared using MMT minerals as nanoclays with nylon, PE, polyvinyl chloride (PVC) and starch. For improved packaging, nanolaminates/nanocomposites are formed with improved barrier properties, temperature and humidity resistance, biodegradability, etc.

Barrier Protection: Fresh produce especially fruits and vegetables are living commodities which continue respiration even after harvest. This physiological activity may reduce the shelf-life of produce by causing senescence and decay. The shelf-life of fresh fruits and vegetables can be enhanced by packing/storing them in an inert and low oxygen atmosphere which results in reduced the rate of respiration, reduced microbial growth, and retarded enzymatic spoilage by changing the gaseous environment surrounding the food product. It is very difficult to achieve especially in case of biodegradable packaging where biopolymers are used. Biopolymers are one of the favorable alternatives to be exploited and developed into eco-friendly food packaging materials due to their rapid degradability (Tang et al., 2012). But, major bottlenecks in the use of biopolymers as food packaging materials are poorer mechanical, thermal, and barrier properties as compared to the conventional non-biodegradable materials made from petroleum which led to the use of the nanocomposite concept (Di Maio et al., 2013; Kanmani & Rhim, 2014; Reddy & Rhim, 2014; Sanuja et al., 2014). Biopolymers can be incorporated with nanofillers to form bio nanocomposites with improved physical properties (Table 13.4). Bio-nanocomposite is a multiphase material comprising two or more constituents which are continuous phase or matrix particularly biopolymer and discontinuous nanodimensional phase or nanofillers. Nanofillers are incorporated in the polymer matrix of the substances due to their large surface area which favors the filler matrix interactions and its performance. Also, the nano-reinforcement acts as small, barriers for gases

TABLE 13.4 Physical and Biological Properties Incorporated by Different Nanoparticles in Packaging Material.

Nanoparticles	Properties	References
Layered silicates (enhance polymer performance)	• Layered silicates commonly used in nanoparticles consist of two-dimensional layers.	Bharadwaj et al. (2002); Cabedo et al. (2004)
	• Available low cost and simple processability.	
	• Providing excellent barrier.	
Cellulose-based two types-microfibrils and whiskers	• Obtained from cellulose providing excellent barrier.	Azizi Samin et al. (2005)
carbon nanotubes (CNTs)	• Consist of one atom thick single wall nanotube.	Kim, et al. (2008)
	• Having extraordinary high aspect ratios and elastic modulus.	Zhou et al. (2004)
Silica (SiO2)	• Improves mechanical and barrier properties.	Wu et al. (2002)
	• Addition of SiO_2 into propylene (PP matrix) improves tensile strength, modulus, and elongation.	
Starch nanocrystals (SNC)	• Show platelet morphology with thickness 6–8 nm.	Kristo and Biliaderis (2007)
	• Addition of SNC improved tensile strength and modulus but decreased their elongation.	
Chitin/chitosan nanoparticles	• Chitin whiskers to soy protein isolate (SPI) thermoplastics improved tensile properties and water resistance.	Lu et al. (2004); De Moura et al. (2009)
	• Addition of nanoparticles significantly improved mechanical and barrier properties of the films.	
AgNP-based antimicrobial nanocomposites	• Possess strong toxicity to a wide range of microorganisms.	Duncan (2011)
Nanoclays	Creates the enhanced barrier properties of homogeneous polymer due to the increase of tortuous pathway against penetrating molecules.	Duncan (2011)

TABLE 13.4 *(Continued)*

Nanoparticles	Properties	References
Nano-ZnO and nano-TiO$_2$	• The photocatalytic reaction of nano-ZnO and nano-TiO$_2$ attributes to generation of ROS, resulting in the oxidation of cytoplasm of bacterial cells and leading to cell death. • ZnO is considered as relatively more efficient and attractive over silver due to the less toxicity and cost effectiveness.	Duncan (2011); Silvestre et al. (2011)

by complicating the path of the material. The incorporation of nano-fillers such as silicate, clay, and TiO_2 to biopolymers may improve not only the mechanical and barrier properties of the biopolymer but also offers other benefits in food packaging as antimicrobial agent, biosensor, and oxygen scavenger (de Azeredo, 2009). The development of bio-nanocomposite materials for food packaging is important not only to reduce environmental problems but also to improve the functions of the food packaging materials.

If we take an example of synthetic polymers, polypropylene (PP) (a type of polyolefin) films are often used in the food industry owing to their better transparency, brilliance, low specific weight and chemical inertness in comparison to other polymers. However, PP (like other polyolefins and other polymers) is also characterized by low barrier properties (i.e., an inherent permeability to gases and other small molecules), which results in poor protection of packaged foods. To improve barrier deficiencies of PP and other polymers, a second component such as a polymer blend or multilayer, filler, etc. is added to it (Duncan, 2011; Han et al., 2011; Manikantan & Varadharaju, 2011). Polyamide-based nanoclay have been developed largely and commercialized under the trade names Durethan, Imperm, Aegis, and noted for their durability and protection (Chellaram et al., 2012).

Biodegradability: Pollution is the most burning issue of present era due to its major role in global warming and climate change. Nowadays the production and use of non-biodegradable materials or plastics as food packaging materials have significantly increased. These types of materials are usually derived from petroleum products, hence they cause waste disposal problem (Avella et al., 2005) as well as changes in the soil nature and accumulation of toxic gases in atmosphere leading to global warming. This led to the invention of biodegradable packaging materials. But the problem with these materials was the lack of mechanical strength and high permeability to water and gases. Nanoparticles of proteins, carbohydrates, and lipids obtained from animal and plant materials, metal oxides, and carbon nanotubes (CNT) can be effectively used to incorporate biodegradability in the synthetic polymers. In addition collagen, zein, and cellulose from corn are being synthesized into nanofibers and being added along with nanoclays to use it for comfort packaging.

Bio-degradable bio-nanocomposites prepared from natural biopolymers such as starch and protein exhibit advantages as a food packaging material by providing enhanced organoleptic characteristics such as appearance, odor, and flavor. Plantic Technologies Ltd, Altona, Australia has manufactured and is selling biodegradable and fully compostable bioplastics packaging. This is constructed from organic cornstarch using nanotechnology (Neethirajan & Jayas, 2011). The unique advantages of natural biopolymer packaging include their ability to handle particulate foods, act as carriers for functionally active substances, and provide nutritional supplements (Rhim & Ng, 2007). Nanomaterials offer an opportunity to enhance the mechanical and thermal properties of packaging to improve the protection of foods from undesirable mechanical, thermal, chemical, or microbiological effects.

2. **Active Packaging:** Active packaging may be defined as "packaging in which subsidiary constituents have been deliberately included in or on either the packaging material or the package headspace to enhance the performance of the package system" (Robertson, 2005). Nanomaterials are also being used to interact directly with the food or the environment of package to allow better protection of the product. For example, AgNPs and silver coatings can provide antimicrobial properties, with other materials being used as UV protectors (Table 13.5). The bio-nanocomposite can be used as an active food packaging whereby the food packaging can interact with food in some ways by releasing beneficial compounds such as antimicrobial agent, antioxidant agent, or by eliminating some unfavorable elements such as oxygen or water vapor. The bio-nanocomposite can also be a smart food packaging whereby it can perceive property of the packaged food such as microbial contamination or expiry date and uses some mechanism to register and convey information about the quality or safety of the food (de Azeredo et al., 2013).

TABLE 13.5   Nanofillers for Active Packaging of Foods.

Nanofiller	Detection properties	References
Carbon nanotube	Micro-organisms, toxins, and spoilage of food	Han et al. (2011)
TiO_2	Oxygen scavenger	Xiao et al. (2004)
ZnO	UV barrier and thermal stability	Kammani et al. (2011)

Nanolaminates: Nanotechnology provides food scientists with a number of ways to create novel laminate films suitable for use in the food and dairy industry. A nanolaminate consists of two or more layers of materials with nanometer dimensions that are physically or chemically bonded to each other. A variety of different adsorbing substances could be used to create the different layers, including natural polyelectrolytes (proteins and polysaccharides), charged lipids (phospholipids and surfactants), and colloidal particles (micelles, vesicles, and droplets). It would be possible to incorporate active functional agents such as antimicrobials, anti-browning agents, antioxidants, enzymes, flavors, and colors into the films. These functional agents would increase the shelf-life and quality of coated foods.

3. **Intelligent/Smart Packaging:** Intelligent packaging has been designed to sense some specific biochemical or microbial changes in the food, for example detecting specific kind of pathogens developing in the food or specific gases produced from spoiling food. Intelligent packaging interacts with the environment, and indicates on a condition of packaged food like freshness and presence of gases. Recently, the innovative application of some nanotechnology in design of smart or intelligent packaging to enhance communication aspect of package has been reported (Mihindukulasuriya & Lim, 2014). This smart packaging might increase efficiency of information transfer during distribution. The response generated due to changes related with internal or external environmental factor will be recorded through specific sensor. Nanostructures can be applied as reactive particles in packaging materials. Reactive nanofillers incorporated in polymers can act as a sensor by showing the ability to respond to environmental changes like temperature, humidity, O_2 and CO_2 levels, degradation, and contamination of food products (Table 13.6). These responses are monitored especially during storage of food in order to ensure the freshness of the food. Sometimes, sealing defects in food packaging are left undetected, at this time the sensor is important to indicate undesirable changes in the food product and to aware consumers if the food has gone spoiled or unpalatable. Some "smart" packaging has also been developed to be used as a tracking device for food safety, to avoid counterfeit and provide brand protection.

Nanosensors: Nanosensors incorporated into the food packaging materials may help to track any physical, chemical, or even biological modification during food processing phase and throughout the supply chain, for example, to monitor temperature or humidity over time and then to provide relevant information of these conditions. Nanosensors in plastic packaging can detect gases given off by food when it spoils and the packaging itself changes color to alert the consumer. These films are packed with "silicate nanoparticles" to reduce the flow of oxygen into the package and the leaking of moisture out of the package to stay food fresh. Smart packaging with specialized nanosensors and nano-devices has been designed to detect toxins, food pathogens, and chemicals.

Application of polymer nanocomposites, antimicrobial packaging, and nanocoated films is more advanced and some nano packaging products are already on the market. The new versions of the analyzer provide an option for automatic determination of the fruit ethylene. The current sensor using electrocatalysis and nanotechnology is a new and promising technology for affordable detection of ethylene (for apple, avocado, pear, and kiwi), which will enable research in areas where ethylene could not be measured before due to lack of portable, sensitive, and near real-time measurement equipment.

Time–temperature indicators or integrators (TTIs) are an example of nanosensors. These are designed to monitor, record, and translate whether a certain food product is safe to be consumed or not, in terms of its temperature history. It is very important for the foods which need to be stored under some specific storage temperatures for example, milk based products. For instance, if a product is supposed to be frozen, a TTI can indicate whether it had been inadequately exposed to higher temperatures and the time of exposure. The communication is usually manifested by a color development (related to a temperature dependent migration of a dye through a porous material) or a color change (using a temperature dependent chemical reaction or physical change) Timestrip® has developed a system (iStrip) for chilled foods, based on gold nanoparticles, which is red at temperatures above freezing. Accidental freezing leads to irreversible agglomeration of the gold nanoparticles resulting in loss of the red color. Leak indicators based on oxygen and carbon dioxide can be used to monitor the quality of food. These indicators have a change of color as a result of a chemical or enzymatic reaction by the presence or absence of these gases (Bratovcic et al., 2015).

Similarly, non-toxic and irreversible O_2 sensors to assure O_2 absence in oxygen-free food packaging systems have also been developed, for example, packaging under vacuum or nitrogen. Lee et al. (2005) developed an UV-activated colorimetric O_2 indicator which uses TiO_2 nanoparticles to photosensitize the reduction of methylene blue (MB) by triethanolamine in a polymer encapsulation medium using UVA light. Upon UV irradiation, the sensor bleaches and remains colorless until it is exposed to oxygen, when its original blue color is restored. The rate of color recovery is proportional to the level of oxygen exposure (Bratovcic et al., 2015).

Several biosensors have been designed for detection of most common food pathogens, for example, *Listeria* monocytogenes, *Escherichia coli,* and *Salmonella* sp. as well as mycotoxins in food (Duran & Marcato, 2013). Application of nanoparticles for detection of foodborne pathogens and their toxins has been reported. Furthermore, aflatoxins produced by *Aspergillus flavus* and *A. parasiticus* that contaminate food products could be detected by the use of magnetic nanoglodimmuno sensor.

13.4.1 EDIBLE NANO-BASED PACKAGING

Generally, edible films and coating can be described as thin-continuous layer of edible component expended as coating or a film among the food material to facilitate a barrier to mass transfer. They primarily diverge in the mode of formation and application to foods. Edible coating is done with the help of liquid film forming solution onto the surface of the food products. They are created by casting or by traditional plastic extrusion techniques.

Films made from edible comprises water-soluble polysaccharides (hydrocolloids) and lipids. The most commonly used polysaccharides are cellulose derived such as alginates, pectins, starches, chitosan, and other polysaccharides. Polysaccharide gives variety of properties like rigidity, firmness, thickening quality, viscosity, adhesive, and gel-forming capability. Different types of lipid complexes like animal and vegetable fats have been employed for synthesizing the edible films and coatings. Mainly used lipid molecules are waxes, acylglycerols, and fatty acids. Lipid films provide a barrier against moisture and add gloss touch to the confectionary products. Edible coating and films are the convenient means for adding additives and supplementary ingredients to amend the color, texture, and additionally prevents the microbial growth. For the material in which nano-coating is required, for that the nanoparticles are used on the packaging material cover or can be incorporated as a nanolaminate layer. Reports comparing multilayer films with nano-coating show advantages such as reduced material usage and stress-free film conversion method.

13.5   COMMERCIALIZATION OF NANO-ENABLED FOOD PACKAGING IN MARKET

Companies are already producing nano-based packaging materials that are extending the life of food and drinks and improving food safety (Table 13.6). While the nanofood industry struggles with public health concerns over safety, the food packaging industry is moving full-speed ahead with nanotechnology products. The examples may be- active or "smart" packaging that promises to improve food safety and quality and enhances shelf-life of the product or intelligent packaging that would be able to alert the consumer if any spoilage/contamination occurs inside the package or the food responds to any change in internal environment of the package and self-repairs holes and tears. Nanocor is one of the major global suppliers of nanoclays specifically designed for plastic nanocomposites and has commercialized nanoclay-based resins and packaging products developed with the MMT minerals for foods. An account of nano packaging materials present in the market is given below:

- Durethan® KU2-2601, a composite of polyamide 6 (nylon 6) and nanoclay was developed for the barrier film and coating in packaging. Durethan® can be applied in various areas of packaging, from ordinary foodstuff to the medical field since the clay nanoparticles are dispersed throughout in a polymer matrix, providing excellent properties (Duran & Marcato, 2013; Cushen et al., 2012).
- Aegis™ OXCE is nylon 6-nanoclay composite and is being used for high-oxygen barrier packaging for beer and flavored alcoholic beverages.
- Imperm®, nylon nanocomposites were developed by using MXD6 (nylon polymer from Mitsubishi Gas Chemical Company Inc.) and is applicable in a barrier layer for thermoformed containers or a multi-layer PET bottle which can be used for beer, liquor, or small carbonated soft-drink beverages (Amico, 2004; Nanocor, 2008). It is currently commercialized by Miller Brewing (USA) in their plastic beer bottles; particularly in Miller Lite, Miller Genuine Draft, and Ice House brands (Chaudhry, et al., 2008).
- "Plantic® R1" Tray developed by Plantic Technologies Ltd. Is a thermo formed starch-based nanoclay composite. The tray has been used in Cadbury® Dairy Milk™, and Milk Tray™ chocolates and Marks and Spencer Swisschocolates (Plantic Technologies Limited, 2007).

TABLE 13.6 An Overview of Commercially Available Nano-enabled Food Packaging.

Nanocomposite type	Commercial name or trademark of the package	Improved functionality obtained from product	Commercial application
Nanoclay-nylon 6 nanocomposite	Durethan® KU2-2601	Gas and moisture barrier, strength, toughness and abrasion, and chemical resistance	Foodstuffs
Nanoclay-nylon 6 composite	Aegis™ OXCE developed by Honeywell polymer	Oxygen scavenger used in high-oxygen barrier packaging	Beer and flavored alcoholic beverage
Nanoclays nylon	Imperm® being commercialized by Miller Brewing (USA)	Oxygen and water barrier, but higher haze and lower clarity	Beer and alcoholic beverages
nanoclay composite starch	Plantic® thermoformed trays	Biodegradable	Used in Cadbury® Dairy Milk™ and Milk Tray™ chocolates and Marks and Spencer Swiss chocolates

TABLE 13.6 *(Continued)*

Nanocomposite type	Commercial name or trademark of the package	Improved functionality obtained from product	Commercial application
Nanosilver Plastic	Window® nanosilver food containers,	Antimicrobial	
Nanostarch Nanosilver	EcoSphere Biolatex® Fresh box nanosilver box	Natural and biodegradable excellent Antimicrobial properties against various bacteria and fungus	McDonald's Hamburger clamshells in US Storage and packing of fresh fruits and vegetables
Nanosilver	Baby dream silver-nano-baby bottle	Antibacterial and deodorizing	Used as baby milk bottle

TABLE 13.6 *(Continued)*

Nanocomposite type	Commercial name or trademark of the package	Improved functionality obtained from product	Commercial application
Nanosilver	Anson nano freshness keeping film, storage bag and container	Antibacterial and freshness keeping	Home storage of foods
Nanosilver	Fresher Longer™ storage bags and containers	Antibacterial and freshness keeping	Boxes for storage of fresh fruits and vegetables and boxes for other type of foods

- Baby Dream Co. Ltd., an infant product company in South Korea, has developed a baby mug and milk bottle with AgNPs. The company claims that this container maintains 99.9% germ suppression, deodorization, and freshness even without sterilization (Momin et al., 2012).
- Baoxianhe nanosilver storage box has been launched in China by Quan Zhou HuZheng Nano Technology Co. Ltd.
- Anson Nano-Biotechnology (Zhuhai) Co. Ltd. began nano packaging marketing with Anson Nano Antimicrobial Storage Series; plastic bags and film for fresh food.
- In Taiwan, Song Sing Nano Technology Co. Ltd. is using nano-ZnO particles for anti-mold plastic wrap. It claims to sterilize completely under indoor light conditions.
- EcoSynthetix, a global bio-based material company, has developed a new technology, called EcoSphere Biolatex® which involves producing a natural base binder from annually renewable resources such as corn or potato starch by modifying the starch molecule. This nanomaterial is being used for McDonald's hamburger clamshells in the United States, replacing traditional adhesive.

13.6 MIGRATION OF NANOMATERIAL IN FOOD

Despite the increased marketing efforts in the nanotechnology sector, research into nanotechnology of food and food-related products has only just started to develop. The main reasons for the late incorporation of nanotechnology into the food sector are issues associated with the possible labeling of the food products and consumer-health aspects.

Safety assessment of nanomaterials used in food packaging first requires an understanding of potential exposure via migration into food. If there is no exposure, it follows there is no risk of adverse effects in consumers. Migration of nanomaterial constituents or the nanoparticles themselves from polymer nanocomposites into food or food simulants has been evaluated by many researchers using standard migration tests. These tests are based on European standardized methods used to evaluate migration from food packaging and are carried out using different food simulant solutions, characterized by varying levels of water solubility and acidity. The methods are used for common packaging materials and have not been validated for nanomaterials.

Over the last few years, several studies have reported the migration of nanoparticles into foodstuffs but an adequate toxicological data is not yet available and safety assessments are still in progress (Table 13.7). Most of these studies are focused on nanosilver as there are concerns by the public and government about its safety and health effects. Migration of silver from commercial nanosilver/nano-TiO$_2$ containers into a wide range of food samples have been investigated by Metak and Ajaal (2013). Food samples selected included fresh apples, white sliced bread, fresh carrot, pre-packed soft cheese, modified atmosphere packaging milk powder, and fresh orange juice. Samples were stored in nanosilver/nano-TiO$_2$ or control containers at 40°C for 7 or 10 days. After the exposure period, food samples were processed and analyzed for silver and titanium content by ICP-MS. SEM-EDX and X-ray diffraction analysis of the containers confirmed the presence of silver and TiO$_2$ nanoparticles 20–70 nm in size, with some aggregated (100 nm), incorporated into the polymer (but not coated onto the surface). Overall, insignificant levels of Ag and Ti were measured in the food samples after 7 and 10 days exposure to the containers. The highest level of silver was measured in orange juice (5.7 ± 0.02 μg/L vs. 0.16 ± 0.01 μg/L in controls), with others ranging from approximately 2–5.2 μg/L. The concentration in orange juice is approximately equivalent to 0.0057 mg/kg food. The highest level of Ti was also found in orange juice (2.5 ± 0.03 μg/L).

TABLE 13.7 Migration Studies on Nano-based Food Packaging Materials.

Type of polymer	Nanomaterial used	Food/simulant used for testing	If NPs detected? (yes/no)	References
PET bottles	Nanoclay	Simulant (3% acetic acid, 10% ethanol, isooctane)	No	Chaudhary et al. (2008)
PLA films	Nanoclay	Simulant (95% ethanol)	No	Schmidt et al. (2009)
PP container	Nanosilver (AgNPs)	Food (apple, sauce, and pizza) Simulant (3% acetic acid)	No	Chaudhary et al. (2008)
LDPE film	(AgNPs)	Simulant (3% acetic acid, 10% and 95% ethanol)	No	Bott et al. (2014)
HDPE film	(AgNPs)	Simulant (distilled water, 3% acetic acid, 10%, and 95% ethanol)	Yes	Artiaga et al. (2015)
PE films (made with Agion™ Ag ion fills)	(AgNPs)	Simulant (3% acetic acid)	Yes	Cushen et al. (2014)

Some reports have indicated that nanosilver may harm human cells by modifying the function of mitochondria, increasing membrane permeability, and generating ROS. Song et al. (2011) studied the migration of nanosilver from PE packaging with food simulants by inductively coupled plasma mass spectrometry (ICP-MS) determination and the results showed that the amount of migration of silver slightly increased with time and temperature in 3% (w/v) acetic acid prior reaching a steady state. However, in 95% (v/v) ethanol, the amount of nanosilver migration depended on time, while temperature did not show any significant effect. Within this test limitation, they suggested that further migration studies of nanoparticles need to be performed with real food samples, as AgNPs are not likely to be used in highly acidic foods.

13.7 SAFETY CONCERNS RELATED TO NANOPACKAGING

The concern about effect of nanomaterials on health and environment welfare came into existence as the new properties of nanomaterials was started being exploited (Helland et al., 2006). The reports have been published which informs about harmful effect on humans on exposure of nanomaterial (Borm et al., 2006). However, the concern related to health and environmental risk by nanomaterials is in their primary screening stage, where side effects are not well-defined. Due to this, usage of natural biodegrading material and recycling of nanomaterial are prohibited. It is also uncertain that many nanoparticles have the ability to cease the development of beneficial bacteria that are present in the environment and in the digestive tracts like probiotics (Kuzma et al., 2006). The greater chemical reactivity and bioavailability of nanomaterials may result in greater toxicity of nanoparticles compared with the same unit of mass of larger particles of the same chemical composition, as the materials that are 300 nm or less in diameter can be taken up by individual cells and nanomaterials which measure less than 70 nm can even be taken up by a cell's nuclei, where they can cause major damage (Chen & von Mikecz, 2005; Geiser et al., 2005).

The British Standards Institution, International and European Committee for Standardization, and OECD have undertaken all the ethical, social, and demand issues related to nanosafety including the vocabulary related to nanotechnology, classification, validated measurement and characterization procedures. The major concern is whether the nanomaterial used in packaging and targeted delivery system has an adverse effect on the protection of food. The successive query is related to the environmental issue which is associated with recycling and change in the recycling systems because of the usage of nanomaterial. No effective instrument is yet developed to test packaging material or track nanomaterial within the food. There is scarcity of knowledge about the influence of nanomaterial in waste disposal. Additionally, there is no attention given to nanomaterial that is present in the packaging system during recycling, incinerating for energy revival, and liberation in landfills (Tiede et al., 2009). The government federations like European Union (EFSA; The European Food Safety Authority) and the United States of America (US-FDA; The Food and Drug Administration) gave the report illustrating one advantage of nanomaterial for packaging which is limited to the material made of plastics. The EU and United States are the model of regulatory bodies which pass the agreement by assessing the safety norms enquiring about constituents required for plastic food packaging. The assessment of safety is entirely based on the chemistry and toxicity details which are provided by the aspirant according to the data requirements passed by government agencies.

13.8   FUTURE PERSPECTIVE AND CONCLUSION

The importance of nanotechnology is increasing for the food industry. New protocols are being explored which assure valid results and applications in the domain of food packaging system. Developing countries are also benefited as they can establish new market for novel nanomaterials and manufacturing processes as nanotechnology has great potential. It is projected that the next generation of intellectual packaging will allow an improved monitoring of the flow, protection, and excellence of food materials. With addition of suitable nano-particles, it is easy to synthesize packaging material of intense tensile strength, resistance, and thermal performance. The incorporated nanosensors in the packaging system warn the consumer on spoilage. The ability of the field of nanotechnology and its benefits are needed to be explored in the concern of industry and human health. Similarly, definite policies have to be made by the food organizations to launch proper and safe commercialization of nano-based food applications. The edible nanolaminates got the access in the fresh fruits and vegetables, confectionary products, and formulation, due to its ability to provide barrier from water, lipids, gases, odors, and off-taste. Besides, many nanomaterials provide extensive properties to the packaging material like antimicrobial activity, oxygen reducing activity, immobilization of enzyme, and warning of degradation of the contaminated food. The two main concern of importance involves issue related to food safety and quality that may affect the consumer and second issue is related to environment safety. The international and regional industries have now reduced the gap by delivering and enclosing the rule of nanotechnology usage in foods, food additives, and packaging material.

Nevertheless, it remains unknown what levels of nano-exposure we are currently facing, what levels of exposure could harm human health or the environment, if there is any safe level of nano-exposure, and whether or not nanomaterials will bioaccumulate along the food chain. Therefore, until we have a more comprehensive understanding of the biological behavior of nanomaterials, the regulation and limitation of nanomaterial applications in food contact materials must be clarified.

13.9 FUTURE THRUSTS

Potential future applications of nanotechnology relevant to the area of food packaging include:

- Use of CNT or carbon nanodots for conductive and reinforcement applications in nanocomposites (Chaudhry et al., 2008; Das Purkay-astha et al., 2014,).
- Nanocellulose conferring barrier properties for use as an economical biodegradable food packaging material (Li et al., 2015).
- Development of Materials for "smart" food packaging, for example, nanosensors, biosensors, and labels to signal the condition of food by detecting food pathogens and triggering a color change in the packaging to alert the consumer about contamination or spoilage.
- Development of polymer composites incorporating nano encapsulated substances (e.g., enzymes, catalysts, oils, adhesives, polymers, inorganic nanoparticles, biological cells, flavor and color enhancers, vitamins, etc.), which would allow consumers to modify the food depending on their own nutritional needs or tastes, or act as an alternative antimicrobial application for improving shelf-life of food (Duncan, 2011; Liang et al., 2012).
- None of the commercial packaging material has yet been developed and tested for fresh fruits and vegetables. Therefore, there is a strong need to develop and commercialize nano-based smart packaging to sense the ripening-related changes, pathogen infection, and edible quality of fresh fruits and vegetables.

- Migration testing methods specifically for nano-based packaging materials need to develop.
- Framing laws for making nano-labeling mandatory is very essential which needs to be done in the future.

13.10 CONCLUSION

Nanotechnology offers tremendous opportunities for innovative developments in food packaging that can benefit both consumers and industry. Nanomaterials like nanosilver particles have tremendous potential towards preservation and shelf-life extension of foods. Its application shows considerable advantages in improving the properties of packaging materials in the form of improved, intelligent, and active packaging. But we are still in the early stages of research and require continued research and testing to better understand the advantages and disadvantages of nanotechnology for its use in packaging materials. The food and agricultural industry have been investing billions of dollars into nanotechnology research as well as there are a number of nanofood products already available in the market. But as there is no law, making nano-labeling mandatory everywhere in the world, it is impossible to tell how many commercial food products are actually containing nano ingredients. Therefore, rules and regulations need to be framed for making nano-labeling mandatory and conducting more investigations on the migration of nanomaterials in food to make it safer and acceptable among consumers.

KEYWORDS

- **nanotechnology**
- **nano-based packaging**
- **nanocomposites**
- **nanofillers**
- **edible packaging**

REFERENCES

Adame, D.; Beall, G. W. Direct Measurement of the Constrained Polymer Region in Polyamide/Clay Nanocomposites and the Implications for Gas Diffusion. *Appl. Clay Sci.* **2009**, *42*, 545–552.

Aguilera, J. M. Where is the Nano in Our Foods? *J. Agr. Food Chem.* **2014**, *62*, 9953–56.

Amico, E. D. Chasing Nanocomposites. *Chem. Week.* **2004**, *10*, 6–9.

Artiaga, G.; Ramos, K.; Ramos, L.; Cámara, C.; Gómez-Gómez, M. Migration and Characterisation of Nanosilver from Food Containers by AF4-ICP-MS. *Food Chem.* **2015**, *166*, 76–85.

Avella, M.; De Vlieger, J. J.; Errico, M. E.; Fischer, S.; Vacca, P.; Volpe, M. G. Biodegradable Starch/Clay Nanocomposite Films for Food Packaging Applications. *Food Chem.* **2005**, *93*, 467–474.

Avella, M.; Bruno, G.; Errico, M. E.; Gentile, G.; Piciocchi, N.; Sorrentino, A.; Volpe, M. G. Innovative Packaging for Minimally Processed Fruits. *Packag. Technol. Sci.* **2007**, *20*, 325–335.

Azizi Samir, M. A. S.; Alloin, F.; Dufresne, A. Review of Recent Research into Cellulosic Whiskers, Their Properties and Their Application in Nanocomposite Field. *Biomacromolecules.* **2005**, *6* (2), 612–626.

Bajpai, S.; Chand, N.; Chaurasia, V. Nano Zinc Oxide-loaded Calcium Alginate Films with Potential Antibacterial Properties. *Food Bioprocess Tech.* **2012**, *5*, 1871–1881.

Beltran, A.; Valente, A. J. M.; Jiménez, A.; Garrigós, M. A. C. Characterization of Poly(ε-caprolactone)-based Nanocomposites Containing Hydroxytyrosol for Active Food Packaging. *J. Agr. Food Chem.* **2014**, *62*, 2244–2252.

Bharadwaj, R. K.; Mehrab, A. R.; Hamilton, C.; Trujillo, C.; Murga, M.; Fan, R.; Chavira, A. Structure-property Relationships in Cross-linked Polyester–clay Nanocomposites. *Polymer.* **2002**, *43* (13), 3699–3705.

Blasco, C; Pico, Y. Determining Nanomaterials in Food. *Trend. Anal. Chem.* **2011**, *30* (1), 84–99.

Borm, P.; Klaessig, F.C.; Landry, T. D.; Moudgil, B.; PauluhnM, J.; Thomas, K.; Trottier, R.; Wood, S. Research Strategies for Safety Evaluation of Nanomaterials, Part V: Role of Dissolution in Biological Fate and Effects of Nanoscale Particles. *Toxicol. Sci.* **2006**, *90* (1), 23–32.

Bott, J.; Störmer, A.; Franz, R. A Model Study into the Migration Potential of Nanoparticles from Plastics Nanocomposites for Food Contact. *Food Packag. Shelf Life.* **2014**, *2*, 73–80.

Bratovcic, A.; Odobasic, A.; Catic, S.; Sestan, I. Application of Polymer Nanocomposite Materials in Food Packaging. *Croat. J. Food Sci. Technol.* **2015**, *7* (2), 86–94.

Cabedo, L.; Giménez, E.; Lagaron, J. M.; Gavara, R.; Saura, J. J. Development of EVOH–Kaolinite Nanocomposites. *Polymer.* **2004**, *45* (15), 5233–5238.

Conte, A.; Longano, D.; Costa, C.; Ditaranto, N.; Ancona, A.; Cioffi, N.; Scrocco, C.; Sabbatini, L.; Contò, F.; Del Nobile, M. A. A Novel Preservation Technique Applied to Fiordilatte Cheese. *Innov. Food Sci. Emerg. Technol.* **2013**, *19*, 158–165.

Chaudhry, Q.; Castle, L. Food Applications of Nanotechnologies: An Overview of Opportunities and Challenges for Developing Countries. *Trends Food Sci. Technol.* **2011**, *22* (11), 595–603.

Chaudhry, Q.; Scotter, M.; Blackburn, J.; Ross, B.; Boxall, A.; Castle, L.; Aitken, R.; Watkins, R. Applications and Implications of Nanotechnologies for the Food Sector. *Food Addit. Contam.* **2008**, *25* (3), 241–258.

Chellaram, C.; Murugaboopathi, G.; John, A. A.; Sivakumar, R.; Ganesan, S.; Krithika, S.; Priya, G.; Archana, H.; Chellaram, C. Impact of Marine Nanoparticles for Sustained Drug Delivery. *Ind. J. Innov. Dev.* **2012**, *1,* S8, 37–39.

Chen, M.; Von Mikecz, A. Formation of Nucleoplasmic Protein Aggregates Impairs Nuclear Function in Response to SiO2 Nanoparticles. *Exp. Cell. Res.* **2005**, *305* (1), 51–62.

Cushen, M.; Kerry, J.; Morris, M.; Cruz-Romero, M.; Cummins, E. Nanotechnologies in the Food Industry–Recent Developments, Risks and Regulation. *Trends Food Sci. Technol.* **2012**, *24* (1), 30–46.

Cushen, M.; Kerry, J.; Morris, M.; Cruz-Romero, M.; Cummins, E. Silver Migration from Nanosilver and a Commercially Available Zeolite Filler Polyethylene Composites to Food Simulants. *Food Addit. Contam. Part A.* **2014**, *31,* 1132–1140.

Das Purkayastha, M.; Manhar, A. K.; Das, V. K.; Borah, A.; Mandal, M.; Thakur, A. J.; Mahanta, C. L. Antioxidative, Hemocompatible, Fluorescent Carbon Nanodots from an "End-of-pipe" Agricultural Waste: Exploring Its New Horizon in the Food-Packaging Domain. *J. Agric. Food Chem.* **2014**, *62,* 4509–4520.

De Azeredo, H. M. C. Antimicrobial Nanostructures in Food Packaging. *Trends Food Sci. Technol.* **2013**, *30,* 56–69.

De Azeredo, H. M. C. D.; Nanocomposites for Food Packaging Applications. *Food Res. Int.* **2009**, *42,* 1240–1253.

De Azeredo, H. M. C. D. Mattoso, L. H. C.; McHugh, T. H. *Advances in Diverse Industrial: Nanocomposites in Food Packaging – A Review;* InTech: Croatia, 2011.

De Azeredo, H. M. C. Nanocomposites for Food Packaging Applications. *Food Res. Int.* **2009**, *42* (9), 1240–1253.

De Moura, M. R.; Aouada, F. A.; Avena-Bustillos, R. J.; McHugh, T. H.; Krochta, J. M.; Mattoso, L. H. C. Improved Barrier and Mechanical Properties of Novel Hydroxypropyl Methylcellulose Edible Films with Chitosan/Tripolyphosphatenanoparticlses. *J. Food Eng.* **2009**, *92,* 448–453.

Di Maio, L.; Scarfato, P.; Milana, M. R.; Feliciani, R.; Denaro, M.; Padula, G.; Incarnato, L. Bionanocompositepolylactic Acid/Organoclay Films: Functional Properties and Measurement of Total and Lactic Acid Specific Migration. *Packag. Technol. Sci.* **2013**, *27* (7), 535–547.

Dong, S.; Chen, X.; Gu, L.; Zhou, X.; Xu, H.; Wang, H.; Liu, Z.; Han, P.; Yao, J.; Wang, L.; Cui, G.; Chen, L. Facile Preparation of Mesoporous Titanium Nitride Microspheres for Electrochemical Energy Storage. *ACS Appl. Mater. Interfaces.* **2011**, *3* (1), 93–98.

Duncan, T. V. Applications of Nanotechnology in Food Packaging and Food Safety: Barrier Materials, Antimicrobials and Sensors. *J. Colloid Interface Sci.* **2011**, *363* (1), 1–24.

Duran, N.; Marcato, P. D. Nanobiotechnology Perspectives. Role of Nanotechnology in the Food Industry: A Review. *Int. J. Food Sci. Technol.* **2013**, *48* (6), 1127–1134.

EFSA. Scientific Opinion on the Safety Evaluation of the Substance, Titanium Nitride, Nanoparticles, for Use in Food Contact Materials. *EFSA J.* **2012**, *10* (3), 2641. http://www.efsa.europa.eu/en/efsajournal/doc/2641.pdf

FAO/WHO. *FAO/WHO Expert Meeting on the Application of Nanotechnologies in the Food and Agriculture Sectors: Potential Food Safety Implications;* Food and Agriculture Organization of the United Nations and the World Health Organization: Switzerland, 2009. http://www.worldvet.org/docs/FAO_WHO_Nano_Expert_Meeting_Report_Final.pdf

Fortunati, E.; Peltzer, M.; Armentano, I.; Jiménez, A.; Kenny, J. M. Combined Effects of Cellulose Nanocrystals and Silver Nanoparticles on the Barrier and Migration Properties of PLA Nano-Biocomposites. *J. Food Eng.* **2013**, *118* (1), 117–124.

Gacitua, W.; Ballerini, A.; Zhang, J. Polymer Nanocomposites: Synthetic and Natural Fillers a Review. *Maderas Cienc. Technol.* **2005,** *7* (3), 159–178.

Geiser, M.; Rothen-Rutishauser, B.; Kapp, N.; Schurch, S.; Kreyling, W.; Schulz, H.; Semmler, M.; ImHof, V.; Heyder, J.; Gehr, P. Ultrafine Particles Cross Cellular Membranes by Nonphagocytic Mechanisms in Lungs and in Cultured Cells. *Environ. Health Perspect.* **2005,** *113* (11), 1555–1560.

Han, D. Y.; Tian, Y.; Zhang, T.; Ren, G.; Yang, Z. Nano-Zinc Oxide Damages Spatial Cognition Capability via Over-enhanced Long-term Potentiation in Hippocampus of Wistar Rats. *Int. J. Nanomed.* **2011,** *6,* 1453–1461.

Helland, A.; Kastenholz, H.; Thidell, A.; Arnfalk, P.; Deppert, K. Nanoparticulate Materials and Regulatory Policy in Europe: An Analysis of Stakeholder Perspectives. *J. Nanopart. Res.* **2006,** *8* (5), 709–719.

Imran, M.; Revol-Junelles, A. M.; René, N.; Jamshidian, M.; Akhtar, M. J.; Arab-Tehrany, E.; Jacquot, M.; Desobry, S. Microstructure and Physico-chemical Evaluation of Nano-Emulsion-Based Antimicrobial Peptides Embedded in Bioactive Packaging Films. *Food Hydrocoll.* **2012,** *29,* 407–419.

Kanmani, P.; Rhim, J. W. Physicochemical Properties of Gelatin/Silver Nanoparticle Antimicrobial Composite Films. *Food Chem.* **2014,** *148,* 162–169.

Kim, J. Y.; Han, S.; Hong, S. Effect of Modified Carbon Nanotube on the Properties of Aromatic Polyester Nanocomposites. *Polymer.* **2008,** *49,* 3335–3345.

Kristo, E.; Biliaderis, C. G. Physical Properties of Starch Nanocrystalreinforced Pullulan Films. *Carbohyd. Polym.* **2007,** *68,* 146–158.

Kuzma, J.; VerHage, P. Nanotechnology in Agriculture and Food Production: Anticipated Applications. Washington, DC, The Project on Emerging Nanotechnologies, 2006. Available from: http://www.nanotechproject.org/process/assets/files/2706/94_pen4_agfood.pdf.

Lee, S. K.; Sheridan, M.; Mills, A. Novel UV-activated Colorimetric Oxygen Indicator. *Chem. Mater.* **2005,** *17* (10), 2744–2751.

Liang, R.; Xu, S.; Shoemaker, C. F.; Li, Y.; Zhong, F.; Huang, Q. Physical and Antimicrobial Properties of Peppermint Oil Nanoemulsions. *J. Agric. Food Chem.* **2012,** *60,* 7548–7555.

Ludueña, L. N.; Alvarez, V. A.; Vasquez, A. Processing and Microstructure of PCL/Clay Nanocomposites. *Mater. Sci. Eng. A.* **2007,** *460–461,* 121–129. http://dx.doi.org/10.1016/j.msea.2007.01.104

Magnuson, B. A.; Jonaitis, T. S.; Card, J. W. A Brief Review of the Occurrence, Use, and Safety of Food-Related Nanomaterials. *J. Food Sci.* **2011,** *76* (6), R126–R133.

Manikantan, M. R.; Varadharaju, N. Preparation and Properties of Polypropylene-based Nanocomposite Films for Food Packaging. *Packag. Technol. Sci.* **2011,** *24,* 191–209.

Marambio-Jones, C.; Hoek, E. M. V. A Review of the Antibacterial Effects of Silver Nanomaterials and Potential Implications for Human Health and the Environment. *J. Nanopart. Res.* **2010,** *12* (5), 1531–1551.

Martelli, M. R.; Barros, T. T.; de Moura, M. R.; Mattoso, L. H. C.; Assis, O. B. G. Effect of Chitosan Nanoparticles and Pectin Content on Mechanical Properties and Water Vapor Permeability of Banana Puree Films. *J. Food Sci.* **2013,** *78,* N98–N104.

Metak, A.; Ajaal, T. Investigation on Polymer Based Nano-Silver as Food Packaging Materials. *Int. J. Biol. Vet. Agric. Food Eng.* **2013,** *7,* 772–778.

Mihindukulasuriya, S.; Lim, L. T. Nanotechnology Development in Food Packaging: A Review. *Trends Food Sci. Technol.* **2014,** *40,* 149–167.

Momin, J. K.; Jayakumar, C. Prajapati, J. B. Potential of Nanotechnology in Functional Foods. *Emir. J. Food Agric.* **2012,** *25* (1), 10.

Nafchi, A. M.; Nassiri, R.; Sheibani, S.; Ariffin, F.; Karim, A. A. Preparation and Characterization of Bionanocomposite Films Filled with Nanorod-Rich Zinc Oxide. *Carbohyd. Polym.* **2013**, *96*, 233–239.

Nanocor. Nanocor Product Lines. [Online] 2008. http://www.nanocor.com/products.asp (accessed Jan 14, 2015).

Neethirajan, S.; Jayas, D. S. Nanotechnology for the Food and Bioprocessing Industries. *Food Bioprocess Technol.* **2011**, *4* (1), 39–47.

Oberdorster, G.; Maynard, A.; Donaldson, K.; Castranova, V.; Fitzpatrick, J.; Ausman, K.; Carter, J.; Park, J. H.; Park, S. M.; Kim, Y. H.; Oh, W.; Lee, G. W.; Karim, M. R.; Park, J. H.; Yeum, J. H. Effect of Montmorillonite on Wettability and Microstructure Properties of Zein/Montmorillonite Nanocomposite Nanofiber Mats. *J. Compos. Mater.* **2013**, *47*, 251–257.

Park, S.; Chung, M. G.; Yoo, B. Effect of Octenylsuccinylation on Rheological Properties of Corn Starch Pastes. *Starch.* **2004**, *56* (9), 399–406.

Pinto, A. M. B.; Santos, T. M.; Caceres, C. A.; Lima, J. R.; Ito, E. N.; Azeredo, H. M. C. Starch-Cashew Tree Gum Nanocomposite Films and Their Application for Coating Cashew Nuts. *LWT Food Sci. Technol.* **2014**, *62*, 549–554.

Plantic Technologies Limited. Cadbuer Schweppes and the Plantic R1 Tray. PlanticWebsite. 2007. http://www.plantic.com.au/Case%20Studies/Plantic_Cadbury_CS.pdf (accessed Jan 21, 2015).

Rafieian, F.; Shahedi, M.; Keramat, J.; Simonsen, J. Thermomechanical and Morphological Properties of Nanocomposite Films from Wheat Gluten Matrix and Cellulose Nanofibrils. *J. Food Sci.* **2014**, *79*, N100–N107.

Rasouli, F.; Zhang, W.; Philip Morris. Nanoscale Materials. U.S. Patent US20, 060, 286, 239 A1, 2006.

Reddy, J. P.; Rhim, J. W. Characterization of Bionanocomposite Films Prepared with Agar and Paper-Mulberry Pulp Nanocellulose. *Carbohyd. Polym.* **2014**, *110*, 480–488.

Rhim, J. W.; Ng, P. K. W. Natural Biopolymer-Based Nanocomposite Films for Packaging Applications. *Crit. Rev. Food Sci. Nutr.* **2007**, *47* (4), 411–433.

Robertson, J. M. C.; Robertson, P. K. J.; Lawton, L. A. A Comparison of the Effectiveness of TiO2 Photocatalysis and UVA Photolysis for the Destruction of Three Pathogenic Micro-Organisms. *J. Photochem. Photobiol. A Chem.* **2005**, *175* (1), 51–56.

Rouhi, J.; Mahmud, S.; Naderi, N.; Raymond Ooi, C. H.; Mahmood, M. R. Physical Properties of Fish Gelatin-based Bio-Nanocomposite Films Incorporated with ZnO Nanorods. *Nanoscale Res. Lett.* **2013**, *8*, 1–6.

Sanchez-Garcia, M. D.; Lopez-Rubio, A.; Lagaron, J. M. Natural Micro and Nanobiocomposites with Enhanced Barrier Properties and Novel Functionalities for Food Biopackaging Applications. *Trends Food Sci. Technol.* **2010**, *21*, 528–536.

Sanuja, S.; Agalya, A.; Umapathy, M. J.; Studies on Magnesium Oxide Reinforced Chitosan Bionanocomposite Incorporated with Clove Oil for Active Food Packaging Application. *Int. J. Polym. Mater. Polym. Biomater.* **2014**, *63*, 733–740.

Schmidt, B.; Petersen, J. H.; Bender Koch, C.; Plackett, D.; Johansen, N. R.; Katiyar, V.; Larsen, E. H. Combining Asymmetrical Flow Field-flow Fractionation with Light-scattering and Inductively Coupled Plasma Mass Spectrometric Detection for Characterization of Nanoclay Used in Biopolymer Nanocomposites. *Food Addit. Contam. Part A Chem Anal. Control Expo Risk Assess.* **2009**, *26* (12), 1619–1627.

Shakeri, A.; Radmanesh, S. Preparation of Cellulose Nanofibrils by High-Pressure Homogenizer and Zein Composite Films. *Adv. Mater. Res.* **2014**, *829*, 534–538.

Shin, S. H.; Kim, S. J.; Lee, S. H.; Park, K. M.; Han, J. Apple Peel and Carboxymethylcellu-
lose-Based Nanocomposite Films Containing Different Nanoclays. *J. Food Sci.* **2014,** *79,*
E342–E353.

Silvestre, C.; Duraccio, D.; Sossio, C. Food Packaging Based on Polymer Nanomaterials.
Prog. Polym. Sci. **2011,** *36,* 1766–1782.

Smolander, M.; Chaudhry, Q. Nanotechnologies in Food Packaging. In *Nanotechnologies in
Food;* The Royal Society of Chemistry: Cambridge, 2010; pp 86–101.

Song, H.; Li, B.; Lin, Q. B.; Wu, H. J.; Chen, Y. Migration of Silver from Nanosilver–Polyethylene Composite Packaging into Food Simulants. *Food Addit. Contam A.* **2011,** *28* (12),
1758–1762.

Tang, X.; Alavi, S.; Herald, T. J. Barrier and Mechanical Properties of Starch-Clay Nanocomposite Films. *Cereal Chem.* **2008,** *85,* 433–439.

Tiede, K.; Hassellöv, M.; Breitbarth, E.; Chaudhry, Q.; Boxall, A. B. Considerations for Environmental Fate and Ecotoxicity Testing to Support Environmental Risk Assessments for
Engineered Nanoparticles. *J. Chromatogr. A.* **2009,** *1216* (3), 503–509.

Tripathi, S.; Mehrotra, G. K.; Dutta, P. K. Chitosan-Silver Oxide Nanocomposite Film: Preparation and Antimicrobial Activity. *Bull. Mater. Sci.* **2011,** *34,* 29–35.

Valipoor Motlagh, N.; Hamed Mosavian, M. T.; Mortazavi, S. A. Effect of Polyethylene
Packaging Modified with Silver Particles on the Microbial, Sensory and Appearance of
Dried Barberry. *Packag. Technol. Sci.* **2013,** *26,* 39–49.

Weiss, J.; Takhistov, P.; McClements, D. J. Functional Materials in Food Nanotechnology. *J.
Food Sci.* **2006,** *71* (9), 215–223.

Wu, C. L.; Zhang, M. Q.; Rong, M. Z.; Friedrick, K. Tensile Performance Improvement
of Low Nanoparticles Filled-Polypropylene Composites. *Compos. Sci. Technol.* **2002,** *62,*
1327–1340.

Xiao-e, L.; Green, A. N. M.; Haque, S. A.; Mills, A.; Durrant, J. R. Light-Driven Oxygen
Scavenging by Titania/Polymer Nanocomposite Films. *J. Photochem. Photobiol. A Chem.*
2004, *162,* 253–259.

Youssef, A. M.; Abdel-Aziz, M. S.; El-Sayed, S. M. Chitosan Nanocomposite Films Based
on Ag-NP and Au-NP Biosynthesis by *Bacillus subtilis* as Packaging Materials. *Int. J. Biol.
Macromol.* **2014,** *69,* 185–191.

Zhang, L.; Chen, F.; An, H.; Yang, H.; Sun, X.; Guo, X.; Li, L. Physicochemical Properties,
Firmness, and Nanostructures of Sodium Carbonate-Soluble Pectin of 2 Chinese Cherry
Cultivars at 2 Ripening Stages. *J. Food Sci.* **2008,** *73* (6), N17–N22.

Zhou, X.; Shin, E.; Wang, K. W.; Bakis, C. E. Interfacial Damping Characteristics of Carbon
Nanotube-Based Composites. *Compos. Sci. Technol.* **2004,** *64* (15), 2425–2437.

INDEX

Printed and bound by CPI Group (UK) Ltd, Croydon, CR0 4YY

23/10/2024

01777705-0004